Advances in Geographical and Environmental Sciences

Series Editor

R. B. Singh, University of Delhi, Delhi, India

Advances in Geographical and Environmental Sciences synthesizes series diagnostigation and prognostication of earth environment, incorporating challenging interactive areas within ecological envelope of geosphere, biosphere, hydrosphere, atmosphere and cryosphere. It deals with land use land cover change (LUCC), urbanization, energy flux, land-ocean fluxes, climate, food security, ecohydrology, biodiversity, natural hazards and disasters, human health and their mutual interaction and feedback mechanism in order to contribute towards sustainable future. The geosciences methods range from traditional field techniques and conventional data collection, use of remote sensing and geographical information system, computer aided technique to advance geostatistical and dynamic modeling.

The series integrate past, present and future of geospheric attributes incorporating biophysical and human dimensions in spatio-temporal perspectives. The geosciences, encompassing land-ocean-atmosphere interaction is considered as a vital component in the context of environmental issues, especially in observation and prediction of air and water pollution, global warming and urban heat islands. It is important to communicate the advances in geosciences to increase resilience of society through capacity building for mitigating the impact of natural hazards and disasters. Sustainability of human society depends strongly on the earth environment, and thus the development of geosciences is critical for a better understanding of our living environment, and its sustainable development.

Geoscience also has the responsibility to not confine itself to addressing current problems but it is also developing a framework to address future issues. In order to build a 'Future Earth Model' for understanding and predicting the functioning of the whole climatic system, collaboration of experts in the traditional earth disciplines as well as in ecology, information technology, instrumentation and complex system is essential, through initiatives from human geoscientists. Thus human geosceince is emerging as key policy science for contributing towards sustainability/survivality science together with future earth initiative.

Advances in Geographical and Environmental Sciences series publishes books that contain novel approaches in tackling issues of human geoscience in its broadest sense — books in the series should focus on true progress in a particular area or region. The series includes monographs and edited volumes without any limitations in the page numbers.

More information about this series at http://www.springer.com/series/13113

Ram Kumar Mishra · R. B. Singh ·
Anupama Dubey
Editors

Sustainable Climate Action and Water Management

IGU UGI

Springer

Editors
Ram Kumar Mishra
City Campus
Institute of Public Enterprise
Osmania University Campus
Hyderabad, Telangana, India

R. B. Singh
Department of Geography
Delhi School of Economics
University of Delhi
New Delhi, Delhi, India

Anupama Dubey
City Campus
Institute of Public Enterprise
Osmania University Campus
Hyderabad, Telangana, India

ISSN 2198-3542 ISSN 2198-3550 (electronic)
Advances in Geographical and Environmental Sciences
ISBN 978-981-15-8236-3 ISBN 978-981-15-8237-0 (eBook)
https://doi.org/10.1007/978-981-15-8237-0

This Springer imprint is published by the registered company Springer Nature Singapore Pte Ltd.
The registered company address is: 152 Beach Road, #21-01/04 Gateway East, Singapore 189721, Singapore

Preface

Water and climatic actions are strongly correlated. Any change in the water cycle directly impacts climate change and vice versa. The unpredictable attribute of water availability is increasing every year. This has magnified the incidences of flooding, which deleteriously impacts the source of water points and sanitation facilities through contaminating it. Change in climate intimidates this erratic, fluctuating, and rising extreme weather pattern, which leads to greater ambiguity over a long term. Thus, it can be concluded that climate change is one among the key dynamic forces for water resource management. It is crucial to use water in a sustainable way to maintain our climatic cycle. Definition of sustainable development includes the demographic, economic, environmental, social, and technological forces. Water is the main driving force behind three major pillars of sustainable development known as environment, social, and economic developments. As per Sustainable Development Goal (SDGs) of the United Nations, the development of these pillars, social, economic, and environmental depends upon the obtainability and execution of water resources to accomplish the demand for water. Thus, the growing role of water resources in a rapidly fluctuating climate is very vital to attain the level of sustainable development.

Therefore, this volume has been designed to discuss various aspects of water resource, climate change, and sustainable development. By identifying the challenges and solutions of water resource and climate change in a specific region, the volume will help in exhibiting certain avenues to reach a better and sustainable future. The volume provides a unique medium to the practitioners and academicians to share their ideas in form of a paper by adopting different methodologies, approaches, and practices in the above-mentioned domains. This volume consolidates research work of experts and academicians, which will assist the Policy-Makers to access and further use the valuable information in the decision-making, about the changing contours of water management and climate change mitigation.

Hyderabad, India — Ram Kumar Mishra
New Delhi, India — R. B. Singh
Hyderabad, India — Anupama Dubey

Acknowledgements

This book arose from the work of all the paper contributors, who are working significantly in the area of climate change, water management, and sustainable development. We sincerely acknowledge all the professors, researchers, scientists, water and climate change specialists, managers, administrators, and directors without whose effort, this book could not have been written. We would like to thank the language editor at the Institute of Public Enterprise, India, for his valuable suggestions.

We would like to thank the staff at Springer for their help and support.

June 2020

Ram Kumar Mishra
R. B. Singh
Anupama Dubey

Contents

List of Figures

List of Tables

Part I
Water Resource Management: Issues and Challenges

Chapter 1
Role of Watershed Management in Inclusive Development of a Region: A Case Study of Velchal Watershed, Telangana, India

Anupama Dubey and Ram Kumar Mishra

Abstract The growing population at world level has worsened the increasing water crisis. NASA Satellite data revealed that 21 out of 37 large aquifers are severely water stressed. It is widely believed that an overall involvement of community, policy-makers at top level to beneficiaries at grass root level, may reduce the emerging problem of water scarcity and diminishing water quality. Availability of water is beneficial only if it is potable and potable water is beneficial if it is managed properly. It is pertinent to conserve and manage the water resource at sustainable level to reach the optimum level of inclusive development. As per Sustainable Development Goals (SDG) goal six (Clean Water and Sanitation), is directly related to the goal 1 (No Poverty), goal 2 (Zero Hunger), goal 3 (Good Health and Well-being), goal 4 (Quality Education), goal 5 (Gender Equality), goal 7 (Affordable and Clean Energy-Sustainable), goal 11: (Sustainable Cities and Communities Sustainable). Development of water resource at glocal level is now an interdisciplinary phenomenon and each goal of SDGs is interconnected. In nutshell, water resource development is unswervingly impacting the health, gender equity, education, livelihood and poverty alleviation scenarios. Government of India has developed and has been implementing many watershed management programmes since mid-1990 s to attain inclusive development in rural areas. But sometimes lack of maintenance and irregular monitoring, stakeholder's disinterest, unpredictable rainfall and lack of awareness become constraints in sustainable management of Watershed, impacting the inclusive development of a region. Therefore, this paper attempted to assess the impact of watershed management in inclusive development of a region by taking a case study of Velchal watershed in Rangareddy District of Telangana. This study is based on primary sources collected from various stakeholders of Velchal Watershed through Questionnaire Survey and Focused Group Discussions in the year 2017 supported by secondary data.

A. Dubey (✉) · R. K. Mishra
Institute of Public Enterprise, O.U. Campus, Hyderabad, India
e-mail: anupama@ipeindia.org

R. K. Mishra
e-mail: rkmishra@ipeindia.org

R. K. Mishra et al. (eds.), *Sustainable Climate Action and Water Management*,
Advances in Geographical and Environmental Sciences,
https://doi.org/10.1007/978-981-15-8237-0_1

Keywords Watershed · Impact assessment · Institutional · Physical · Economic and social development

1.1 Introduction

Water is the most crucial component of life. Mounting pressure on water has impacted the demand–supply ratio in primary, secondary and tertiary sectors. It is pertinent to improve and upgrade different processes of water management as per its topographic divisions. Geographical profile, land-use, demography, climatic conditions, technological developments determine the usages of water resource. Mounting pressure on increasing water usages can be resolved by proper management of water resource. But financial constraints are limiting the conservation and management of water resource in developing regions of the globe. Proper watershed management maintains the productive capacity of natural resource in watershed area by restricting the degrading process.

Earlier watershed management was only being considered as a geographical unit but gradual development of watershed schemes in India, it was realized that watershed management is a socio-political ecological entity, which is playing an important role in the inclusive development of a region. Although many issues have been covered by many scholars, such as Jain and Gandhi 2016; Bromely 1989; Farrington et al. 1999; Deshpande and Narayanamoorthy 1999; Kerr et al 2000; Vaidyanathan 1999, 2006; Reddy and Dev 2006; Biswas et al. 2005; Pascual et al 2009; Kshirsagar K G., and M P Madhusoodan, S. Chavan and R. Rathod 2003, but still research scope in the above is tremendous.

1.1.1 Study Area: Velchal *Watershed*

Velchal watershed project has been funded by NABARD and implemented through READ NGO at Velchal Village of Rangareddy District of Telangana. NABARD has completed the implementation of 203 watershed projects in the state of Telangana under Prime Minister's relief package. Ten districts have been covered under the project (09 districts in Telangana and Guntur in Andhra Pradesh). The total area covered under watershed is 1146.87 ha. The ridge area is found at 660 contour intervals adjacent to *Narsimhaswamigutta*. The main streams locally called *Pulilonka, Patlollalonka, Gudemvoduka and Chelmaya* flow into *Ramnathgudupally Cheruvu*. The watershed area is totally under sheet and gully erosion. Further, the slope is nearly 2–15% in normal cultivable areas. The soil is completely red gravel. The depth of the soil is 0.20 m to 0.35 m in cultivatable lands. The village is connected to Vikarabad and Sadashivapet road and it is 120 km away from Hyderabad. The Watershed area covers Velchal as the main village while *Morangapally*, *Ramanthgudpally*,

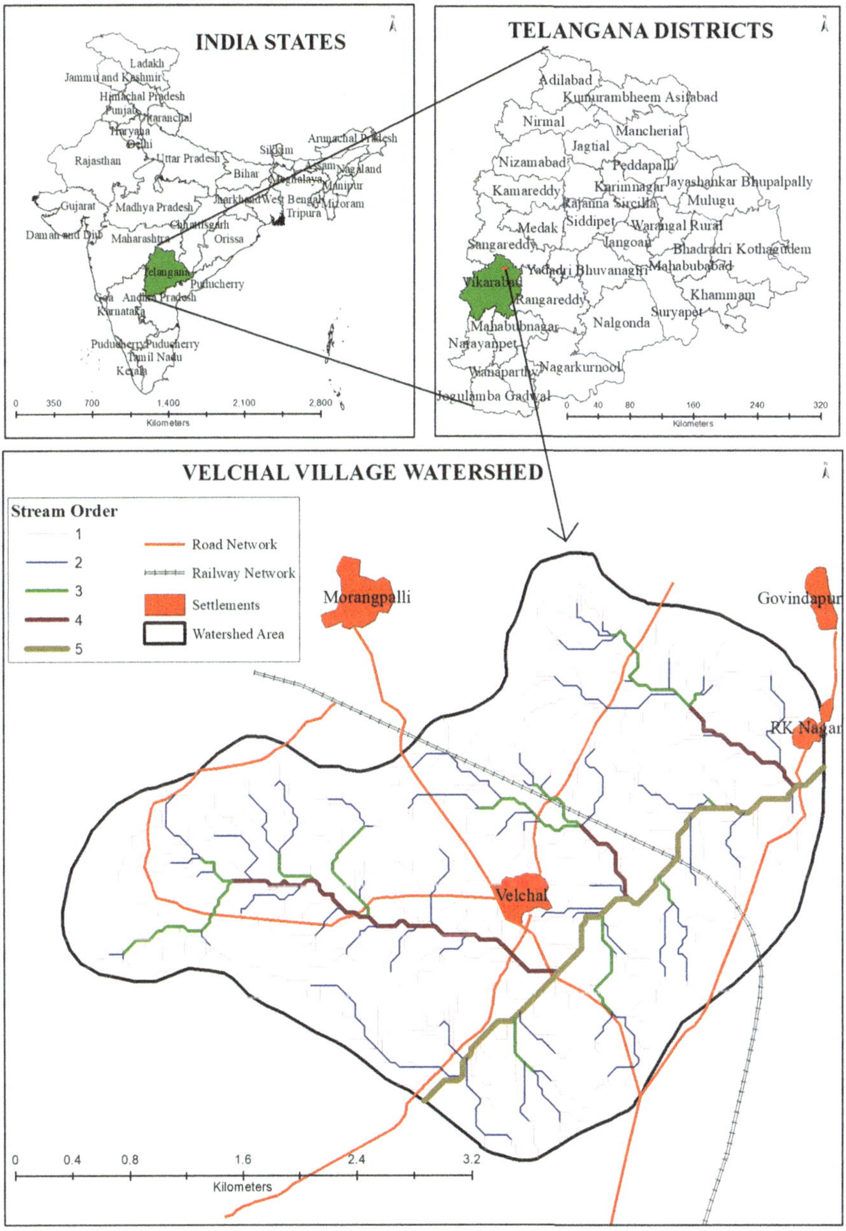

Fig. 1.1 Overview of Velchal Watershed. *Source* secondary Sources 2018

Mallareddyguda, Govindapur are habitations (Fig. 1.1). There are 447 households covered in this watershed.

1.2 Research Methodology

1.2.1 Sample Size

Approximately 25% of sample size has been taken covering 100 households in four sample villages. 25 households were selected on a random basis from each selected village.

1.2.2 Sampling Design

Purposive random sampling has been done to collect primary data. Questionnaire Survey and Focused Group Discussions were the major tools used to collect data on socio-economic and institutional parameters such as capital formation in the villages, poverty induced migration and gender aspects. Stakeholders included READ, farmers, women groups and villagers including village *Sarpanch.*

1.2.3 Data Source

The data source includes both primary and secondary data. To analyze the project indicators such as secondary data have been collected from READ NGO and other government departments because these were the only source of secondary data collection. A village was considered as strata to collect data on various parameters. Random sampling method was adopted to collect primary data as per the availability of Check Dams, Ponds, Talai, Earthen Field Bund, Contour Furrows, Stone Barrier, Diversion Channel and Vegetative Barriers. Participatory approaches were used to evaluate the direct and indirect impact of Watershed Development such as the change/improvement in environmental status and sustainable use of resources. Specific suggestions and recommendations provided for further considerations.

The observations were subcategorized as (a) Institutional, (b) Physical, (c) Economic, and (d) Social. Questionnaire Survey and Focused Group Discussions were the major tools to collect data.

Date of starting of the project was 01.12.2010 and scheduled date of completion of the project was 28.07.2014. This was further delayed by a few months and the final completion of the project was on 08.08.2015. The total area sanctioned under the project was 1146.87 and total area treated was 1026.87. During the field visit, it was observed that villagers were very satisfied with the planning process of Velchal

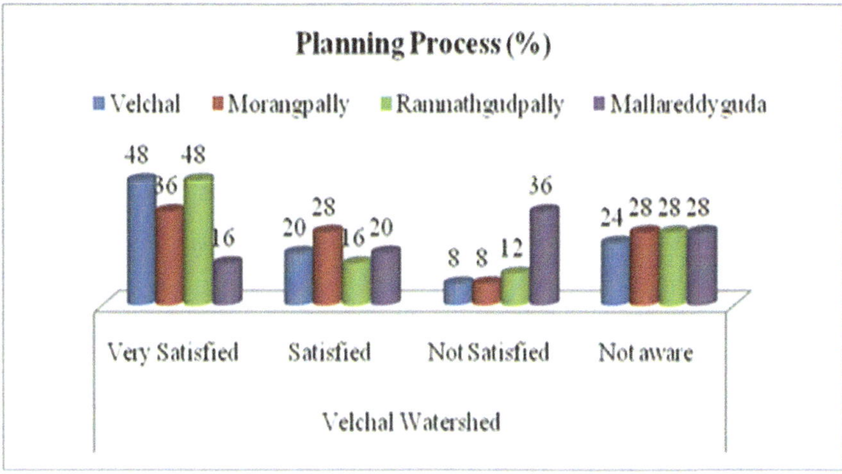

Fig. 1.2 Status of Planning Process at Velchal watershed. *Source* Primary Data, 2017

watershed. Most of the respondents informed that their living standard increased because of increased job availability at *Velchal* watershed. The supporting NGOs also contributed to the timely completion of the project. As per group discussions, people's participation was very high in Velchal watershed during all phases of development. The notable feature is that, after completion of the project they are still actively participating in maintenance and further development of Watershed region. Thus, in a nutshell, the said programme was finished in the given time frame and was found to be very effective (Fig. 1.2).

1.2.4 Role of Various Institutions and Their Effectiveness in Implementation and the Management Process

Data analysis indicates the role of various institutions was high at Morangpally and Ramanathgudpally village. 24% of the respondents at Velchal and Mallareddyguda expressed that role of institution is very high followed by 16% at Morangapally. But the most active village of Velchal watershed is Ramanathgudpally. On the other hand, it was really alarming that 24%–40% of the respondents were not aware of various institutions and functioning of watershed. In their opinion, awareness regarding various institutions is limited to kith and kin of watershed stakeholders (Fig. 1.3).

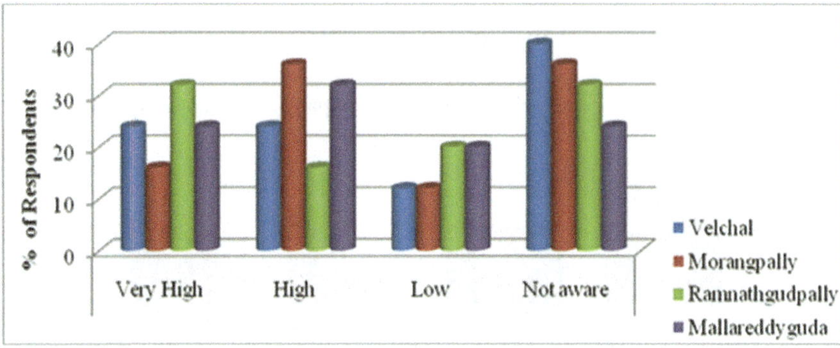

Fig. 1.3 Role of various Institutions. *Source* Primary Data, 2017

1.2.5 Role of Peoples' Institutions and Their Effectiveness

- As per the rules of watershed community, every family in the watershed should be benefited through the watershed programme. As per the policy, the amount is mostly given to the landless, single women, POP families, artisans etc.
- 40% of the respondents at Ramanathgudpally, 32% of respondents at Velchal and Mallareddyguda considered the role of institutions as high.
- 48% respondents at Morangapally and 28% of the respondents at Velchal and Mallareddyguda expressed that the role of peoples institutions is very low.
- At Moranpally, it was observed that committee members have not worked efficiently to spread awareness regarding people's institutions. Except in Velchal village, strength of people's institution is very weak (Fig. 1.4).

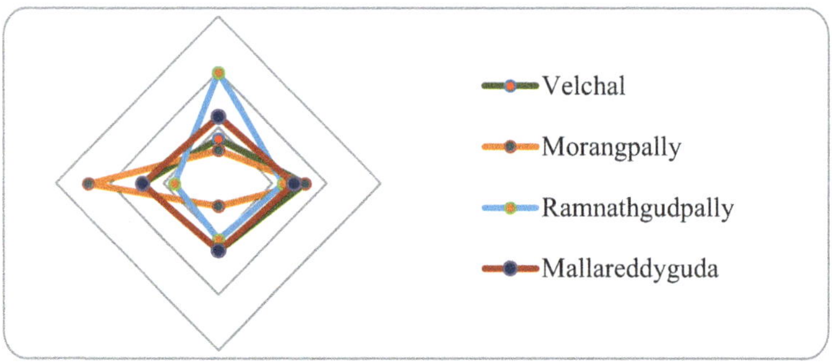

Fig. 1.4 Role of People's Institution. *Source* Primary Data, 2017

Table 1.1 Status of women empowerment at Velchal watershed

% of Respondents at Velchal watershed	Highly Empowered	Empowered	Not empowered	Not aware	Total
Velchal	12	16	40	32	100
Morangpally	20	8	48	24	100
Ramanathgudpally	20	12	28	40	100
Mallareddyguda	16	12	52	20	100

Source Primary Data, 2017

1.2.6 Empowerment of Women/Landless Labour

These are maintained under Self Help Groups by the Women of the village and loans are given only to one woman in a family. Until a loan is repaid completely, no other loan is sanctioned (Table 1.1).

The condition of women empowerment is not satisfactory at Velchal watershed. More than 40% of the respondents at Velchal and Morangpally told that there were not many improvements in women's condition after development of Velchal watershed. 52% of the respondents at Mallareddyguda agreed that empowerment of women is far from satisfaction. But still many female stakeholders are not participating actively in Watershed programme, which needs due attention of local administration. There is a need to conduct various awareness programmes by involving female participants.

1.2.7 Impact Evaluation Broadly in Three Areas Viz., Project, Socio-Economic and Institutional

Agriculture is the main profession of the majority of the villagers. Villagers are working under NREGS scheme under 100 days employment programme and worked under NABARD'S watershed development programme. The financial condition of the villagers remains very pathetic in the absence of employment (Fig. 1.5).

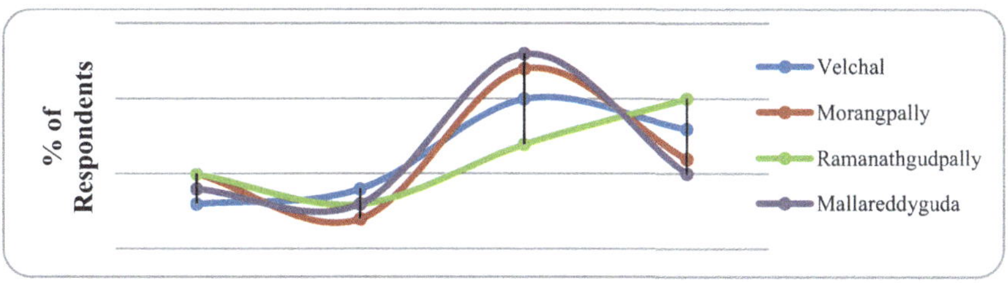

Fig. 1.5 Impact evaluation of Velchal watershed. *Source* Primary Data, 2017

Table 1.2 Change in number of wage days

Particular	Pre-watershed	Post-watershed
Wage days created by Watershed Project	–	32420
Agriculture wage days in a year (average)	84000	96000

Source Secondary data, 2016

Survey indicated that 20–40% of respondents were of positive opinion in the context of employment generation.

1.2.8 Changes in Employment Profile

The watershed programme has generated additional person days of 32420 days and also provided 12000 days wage opportunities for agricultural labour. Watershed interventions provided economic security to poor families, which enable them to earn in their own village. Farmers were getting 100–150 Rs/day on an average during pre-watershed, which increased to 250–350 Rs/day. This increase in wage rate has encouraged the small farmers/agri labourers to remain in their own native villages during non-monsoon period also (Table 1.2).

1.2.9 Social Changes at Velchal Watershed

1.2.9.1 Migration

Watershed development impacted migration in a very positive manner. A primary survey reveals that there is no increase in migration after watershed development. Not only this, all the villages at Velchal watershed are showing a decreasing trend. This percentage was less in other villages because of proximity to city areas. Ratio of in-migration is increasing as 68–80% of the migrants agreed that out-migration stopped tremendously during the last 5 years (Table 1.3).

The above table indicates that migration by the villagers from the watershed area reduced to a larger extent. Total migrants from the watershed area were about 42 and it got reduced to 12 and seasonal migrants reduced to 12 from 4 and 8 permanent migrants have left the village for better opportunities. The VWDC take pride as out migration stopped significantly during distress and drought times.

Table 1.3 Change in Migration

Change in migration	Velchal watershed			
	Velchal	Morangpally	Ramanathgudpally	Mallareddyguda
Increased	0	0	0	0
Decreased	36	20	32	24
Stopped	56	80	68	76
No change	8	4	4	4
Total	100	100	100	100

Source Primary Data, 2017

1.2.9.2 Livelihoods

The livelihood revolving fund released an amount of Rs. 1363400/-. The livelihood fund was given as grant to the community and loan to the beneficiary. So far an amount of Rs. 2704000 was disbursed to 242 members—POP, poor and single women in the first round, later loans were disbursed to small and marginal farmers. The loan amount was given as small amount ranging initially from Rs. 5000 to 10000 to take up small activity like purchasing small ruminants, petty business and artisans. Even the beneficiaries in the beginning were hesitant to take big amount, as they were not confident of their repaying capacity. Most of the members who had obtained loan have taken for the said purpose mentioned in the document. The small farmers have taken loan for agriculture inputs. Slowly when the community was confident of repaying, they have taken a bigger amount of Rs. 15000–20000 to take up activities like dairy development units, calf-rearing, carpentry machine, etc. It is also noticed that now members have upgraded their skills and have recently taken a bigger loan amount, like for example, firstly the dappu community has taken loans to purchase local musical instruments like local *drums* now upgraded to *band* along with uniform. Thus a cumulative effect was observed, the present amount released has increased twofold, the financial capital in the watershed which is showing an immediate result and is also stated by the community as ***the biggest asset*** left behind and will help in sustainability of the families (Tables 1.4 and 1.5).

Table 1.4 Status of Migration

S. No.	Type	Pre-watershed (number of HH)	Post-watershed (Number of HH)
1	Seasonal	12	4
2	Distress/drought	22	0
3	Permanent	8	8
4	Reverse migration		4

Source Rural Education and Agriculral Development (READ) Non Government Organisation,Vikarabad, India, 2010–2016

Table 1.5 Summary of Livelihood activities by wealth rank category

S. No.	Wealth rank	Post-watershed				Remarks
		Total Number of Households	Number of loans/units	Investments from Watershed programme (Lakhs)		
				Grant	Revolving	
1	POP	26	26	224000	220000	
2	Poor	110	110	540000	590000	
3	Middle	72	72	380000	350000	
4	Well off	24	24	169400	180600	
5	Nursery group	10	10	50000	0	
	Total	242	242	1363400	1340600	

Source Rural Education and Agriculatural Development (READ) Non Government Organisation, 2010–2016

Different activities such as Non-pest management, Multi-crop thresher, Seed cum fertilizer driller, Chaff cutter, Vegetable cultivation, Floriculture, Band group, Milk animals, Goat rearing, Back yard poultry, Plate making, Tea shop, Tailoring, Photographer, Vegetable seller, Bakery item seller, Concrete mixing grinder, Saloon, etc were supported under the livelihood and productivity enhancement measures of the watershed funds:

1.2.9.3 Social Norms

The village watershed committee has created three major norms (a) farmers in the village should utilize water with proper management as required for the crop, (b) the cattle are not allowed for free grazing and (c) each villager should discourage deforestation.

1.2.9.4 Sustainability Concerns

Coordination with other government departments and NGOs will be beneficial for inclusive development of that region. The watershed development programme has become a wide platform for the delivery of government schemes. The departments of agriculture, animal husbandry, horticulture, fishery, Nedcap are providing subsidy inputs to the community. Besides ICRISAT support, READ is giving improved seed, i.e., red gram and maize and farmers were taken to ICRISAT for exposure visit. Thus READ is closely working in convergence with the line departments and providing effective extension and inputs to the community.

Table 1.6 Status of quality of Life

Quality of life	Velchal Watershed			
	Velchal	Morangpally	Ramanathgudpally	Mallareddyguda
Improved	96	88	92	88
Not improved	0	4	4	4
No change	4	8	4	8
Total	100	100	100	100

Source Primary Data, 2017

Table 1.7 VWC status

Registered	Yes
Total number of members	11
Number of women members	4
Registration number and date (Registered under societies registration act)	AMC/DCO/RR/3076/2015, Dt 27.01.2015
If cooperative is registered, date and registration number	

Source Rural Education and Agricultural Development (READ) NGO, 2017

1.2.9.5 Quality of Life

Majority of the respondents realized that quality of life improved after watershed development. They are in a financially and socially stronger condition compared to pre-watershed development (Table 1.6).

Implementation of watershed programmes has changed the present and future growth of the whole catchment area in a very positive way. Respondents supported the above-mentioned view.

1.2.9.6 Present Status of VWC

Natural factors such as climatic conditions, rainfall status also affect the watershed. The present status of Velchal watershed was satisfactory because of good rainfall this year. Water accumulated in the check dams, canals and ponds at Velchal watershed would be used for agricultural practices (Table 1.7).

1.2.9.7 Activities Being Continued in Post-watershed Scenario

Post watershed the villagers increased the depth and width of the canal made under NABARD to increase the water level for storage when there are heavy rains. This

Table 1.8 Community Contribution (Regular Shramadaan)

Particulars	CBP	FIP
Total labour payment	Rs.505418/-	Rs.5041181/-
Total estimated value of Shramadan	Rs. 80867/-	Rs.912049/-
Total Shramadhan achieved	Rs. 80867/-	Rs.864591/-*
Shramadan as % of total labour cost	16%	16%

Source Rural Education and Agricultural Development NGO, Vikarabad, India, 2010–2016

made Velchal village to get more benefited with groundwater recharge with the recent heavy rains.

1.2.9.8 Status of Major Structures

The major structures like check dams and stone traps are wearing out due to low quality of construction. They need to be renovated immediately.

1.2.9.9 Status of Record Maintenance and Collection of Maintenance Fund

Records are maintained in ledgers in a proper way. An account in Grameena Bank in the village is also maintained for the maintenance fund.

1.2.9.10 Usage of Livelihood Fund, Recovery and Impact

The funds under livelihood are maintained by the committee members under the Watershed Programme. Many villagers took livelihood loans for setting up some income generation activities like a hotel, grocery shop, cutting shop, photography and videographer shop, musical band troop, tractor, buying cattle, goats, etc. The amounts were repaid by many villagers at the time of implementation of Watershed Project as they used to get some amount of money for the work completed by them. After the project was over, the villagers were having problems in earning during non-monsoonal period (Table 1.8).

1.2.9.11 Extent of Community Involvement

All the villagers are participating to do Shramadaan. They are still waiting for such projects.

1.2.9.12 Process Change

- There is active participation and good coordination by the villagers with the NGOs and they are equally benefited with timely recovery of bank loans.
- All the three VWC, SHGs and JLGs are effectively and actively participating in the village and in all the watershed activities.

1.2.9.13 Impact Evaluation of Natural Parameters

- The watershed is well developed and capable of utilizing the resources but less rainfall since the last few years is an obvious constraint in the management of watershed activities.
- It was also observed that the farmers changed the existing cropping pattern. Before watershed, farmers preferred only cotton and maize. Now they started cultivating paddy, red gram, Jawar along with vegetables and horticulture. But still uses of high crops are not very much prevalent in the watershed region.
- Rain Gauge instruments were installed at few places to measure rainfall. But still installation of more rain guages is in demand.
- Forest area is also covered under this watershed project and most of the land was converted to cultivation land.
- People in Velchal village are working on the NREGS to improve the works done under NABARD Watersheds Project like digging the depth of the Check Dam canals, Water Absorption Trenches, Digging of Wells.
- Secondary data reflect that the groundwater level has increased compared to pre-watershed development.
- Due to enhancement of land productivity and moisture maintenance capacity increased; it helped in increase of crop production and cropping intensity during watershed period.
- Agronomic interventions are low because of less rainfall and same is applicable in case of Market intervention. Use of high value crops is slightly less but there is immense scope for utilization of high crops during proper rainy seasons.
- There is a high vegetation coverage in watershed during good rains.
- Not much change in livestock status was observed due to water scarcity.
- All the villagers were united to do Shramadaan. They were still waiting for such new projects.
- The structural designs were very durable and could sustain heavy rainfall.

1.2.9.14 Impact Evaluation of Socioeconomic Indicators

- Number of man days increased and agricultural employment also increased.
- Open wells, Canals, Check dams, Trenches, Bundings, Non-Farm Bundings, etc., increased post-watershed development but their renovation is in high demand.
- Institutional credit flow increased from the livelihood programmes.

- Out migration not only stopped after watershed development but it also declined enormously since the last 4–5 years.
- Majority of the respondents realized that quality of life improved after watershed development. They are in a financially and socially stronger condition compared to pre-watershed development.

1.3 Suggestions

- The exact impact of the watersheds was observed during the survey as there was a heavy rainfall recently. The water storage at the Checkdams is seen; groundwater and bore wells also got recharged and increased. But the Checkdams that were constructed are getting damaged, they need to be renovated.
- The villagers expressed their happiness with the work done by them under NABARD with the support of READ NGOs and got benefited because of increased groundwater level. Therefore, this work could be documented as a successful case study to help further policy implementation and also for research purpose.
- All the villagers were actively participating in the Watershed Programmes with interest and got benefited with the amount they could get from the work done accordingly. The villagers are willing to work on such projects if given in the near future. Therefore, further coordination is required among policy implementers and grass root level stakeholders.
- All the villagers are involved in the JLGs and created five groups with five members in each group with two committee members in each group. This was a successful programme and many of them got benefited with the loans they used to get and with easy instalments in diminishing rate of interest.
- Some of the waste lands that were not in use due to less water have been brought into cultivation after the rains and groundwater recharge with the works done under NABARD. Many villagers are changing the cropping pattern for increasing the crop yield as the soil quality has improved. Social fencing, tree cutting and free grazing are decreased. Community involvement is increased. Durability and sustainability of water conservation techniques also increased after watershed development. Thus, there is a need for further investment either through public–private partnership or international agencies such as Asian Development Bank.
- Credit flows and hand loans decreased when LH loans are given. As the rate of interest was only 1 rupee and they used to have some work to do under NABARD, it was easy to repay the livelihood loans. Many people, who repaid the livelihood loans in time, are eagerly waiting to get more loans for their improvements. At the same time, they are also eagerly waiting for more works from NABARD such as watershed programmes.

- As per the observation, renovation of the major water conservation structures such as Check Dams, Canals, Stone Bunds, Canopy Coverage is required. Renovation will stop the extra leakages of surface water.
- Installation of RO water purifier is in high demand by the respondents to improve the quality of drinking water, as it was observed that villagers were facing problems because of deteriorating water quality.
- Issue of the LH loans for the 100% repaid beneficiaries so that the capital income can be utilized to increase.
- Every villager got benefited with this NABARD Watershed Project for around 5 years. Later, villagers worked under projects like NREGS but it was not so beneficial for them as they got benefited more from NABARD Project. Thus, it is also necessary to analysis per day wage and number of employment days of different schemes and views of beneficiaries.

Inclusive development of rural area in need of the day as villages of India are the first link supply chain in terms of resource and manpower. Still, close to 65% of the Indian population is depending upon agriculture for their livelihood. Growth of the rural economy is proportional to the gross national income of this country. It is pertinent to generate a permanent source of income at village level to sustain in future. Billions of informal workers in India migrated suddenly because of the impact of ongoing global pandemic (Covid-19) and most of them are jobless. It is really difficult to accommodate all of them in agriculture sector, which already is crowded. Thus, if Indian government renovates her existing agricultural infrastructure with implementation of more such schemes and gives high emphasis to the use and spread of indigenous techniques of water conservation and management, the severe problem of unemployment and overburdening of agriculture sector can be resolved to a larger extent.

References

Biswas AK et al (2005) Integrated water resources management in South and South-East Asia. Oxford University Press, New Delhi

Bromley DW (1989) Economic interests and institutions—The conceptual foundations of public policy. Basil Blackwell, New York

Deshpande RS, Ratnareddy V (1991) Watershed approach in fragile resource regions-an analytical study of Maharashtra, mimeograph series no. 33, Gokhale Institute of Politics and Economics: Pune

Farrington John CathrynTurton, James AJ (eds) (1999) Participatory watershed development: challenges for the twenty-first century. Oxford University Press, Delhi

Jain D, Gandhi VP (2016) Reforming watershed management institutions for inclusive and sustainable growth: role of institutional interaction and participative decision making. IIM Kozhikode Soc Manag Rev 5(1):22–40

Kerr J et al (2000) An evaluation of dry land watershed development projects in india, environment and production technology division. International Food Policy Research, Washington, D.C.

Khirsagar KG et al (2003) Evaluation Study of Wadgaonlakh Watershed in Osmanabad District of Maharashtra, Osmanabad

Pascual U et al (2009) Water agriculture and sustainable wellbeing. Oxford University Press, New Delhi
Reddy VR, MahendraDev S (2006) Managing water resources. Oxford University Press, New Delhi
Vaidyanathan A (1999) Water resource management. Oxford University Press, New Delhi
Vaidyanathan A (2006) India's water resources: contemporary issues on irrigation. Oxford University Press, New Delhi
Vaidyanathan A (2001) Watershed development: reflections on recent developments. In: Nair KN, Chattopadhyay, Srikumar (eds) Watershed Management for Sustainable Development Field experiences and Issues. Kerala Research Programme on Local Level Development, Trivandrum

Chapter 2
Rainwater Harvesting Systems as Disaster Risk Reduction (DRR) Tool for Drought and Flood Prone Areas of Sri Lanka

T. N. Ariyananda

Abstract Sri Lanka has been experiencing frequent droughts and floods, attributed to the effect of climate change. This adversely affects water resources resulting in groundwater depletion, water pollution and soil erosion, leading to siltation of water sources. The island receives an average rainfall of 2000 mm yearly. Out of the total water received by the island, around 60% escapes to the sea as run-off. Therefore, the simple solution in this case is to increase the retention of water that is received during the rainy periods to be used during the non-rainy periods (rainwater harvesting). Over the last 25 years, the traditional rainwater harvesting has been revived bringing much relief to households during times of drought, tsunami, Chronic Kidney Disease of unknown aetiology and resettlement for many people living in rural areas of Sri Lanka. Lanka rainwater harvesting (LRWHF) with the support of USAID launched a 3-year project to provide safe disaster-resilient drinking water through rainwater harvesting, to floods and drought-prone areas in three districts in the island. As a result of this project, safe drinking water was provided to nearly 17,900 people from 391 households, 48 schools and 10 rural hospitals through roof water harvesting units in the three districts. It has been reported that during the drought and floods in recent years, these households and institutes installed rainwater harvesting systems and were able to cope better.

Keywords Rainwater · Flood · Drought · Disaster · Climate

2.1 Introduction

In Sri Lanka, during the last few years, the frequency of climate-related disaster occurrence in the country has caused water stresses to people in almost all districts due to drought and floods. Heavy rainfall, floods and long droughts are frequent occurrences, which increased significantly over the past 10 years, leading to Sri

T. N. Ariyananda (✉)
Lanka Rain Water Harvesting Forum, Battaramulla, Sri Lanka
e-mail: tanuja@sltnet.lk

R. K. Mishra et al. (eds.), *Sustainable Climate Action and Water Management*, Advances in Geographical and Environmental Sciences,
https://doi.org/10.1007/978-981-15-8237-0_2

Lanka being positioned second in the recent ranking based on vulnerability to climate change (Global Climate Risk Index 2017).

Recent climate change prediction studies have indicated that Sri Lanka will experience high variability of rainfall (Jayasinghearachchi 2008). That means wet areas will get wetter and dry areas will be drier, and intense rain can be expected within a short period. Sri Lanka has an average annual rainfall of 2000 mm, which ranges from 900 mm in the dry zone and 5000 mm in the wet zone. The rainfall pattern in Sri Lanka is bimodal and depends on two monsoon seasons. The amounts of rainfall escaping to sea as run-off from wet zone are reported as 78.8% and dry zone is 51.1% (Sri Lanka Water Vision 2025). However, Sri Lanka experiences both spatial and temporal variation in rainfall causing water shortages during dry season and floods during rainy season. Therefore, a simple solution in this case is to increase the retention of water that is received during the rainy periods to be used during the non-rainy periods. Realizing the importance of rainwater harvesting as a solution to overcome water scarcity in the country government of Sri Lanka passed a national policy on rainwater harvesting in 2005 (Sri Lanka Rain Water Policy and Strategy 2005). Legislation that followed has made rainwater harvesting mandatory in a new building above a certain roof size in urban and municipality areas since 2009. Rainwater harvesting in the form of collection and storage in a tank or retention in the soil and ground will be a good option for communities facing water shortages during non-rain period.

Over the last 20 years, the traditional rainwater harvesting has been revived bringing much relief to households during times of drought, tsunami, Chronic Kidney Disease of unknown aetiology (CKDu) and resettlement of many people living in rural areas of Sri Lanka. This is evident from the experience of the project implemented by Lanka Rain Water Harvesting Forum in partnership with USAID, which constructed domestic rainwater harvesting systems and demonstration rainwater harvesting systems in schools and hospitals in the Northern Province and Uva Province.

2.2 The Project

A 3-year project to provide safe disaster-resilient drinking water to floods and drought-prone areas in the island is commenced in 2016. The project sought to provide more efficient disaster-resilient water supplies to vulnerable communities, strengthen the capacity of disaster-prone communities and institutionalization of Disaster Risk Reduction (DRR) policies and practices related to flood and drought management.

Location: The project was implemented in war-affected and drought and flood-prone area of North and Uva Province of Sri Lanka. Drought and flood effected areas selected are Kilinochchi, Moneragala and Badulla districts (Fig. 2.1). These districts are selected on the basis of high drought hazard area (UNDP Report 2012) having high poverty headcount index (Census and Statistics—Sri Lanka August

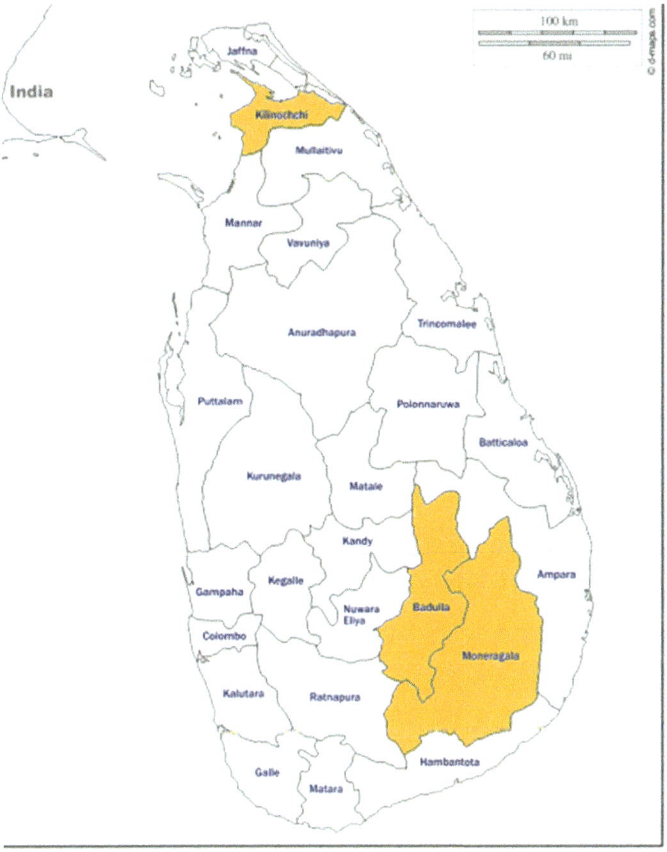

Fig. 2.1 Project locations

2015), percentage of people with access to unimproved/unsafe water, and war-affected community in the north. The percentage of people with piped water supply in northern province is 6.5% and Uva province is 16.2% (NWS&SB web site).

Selection: Project locations for implementation within districts were selected in consultation with District Secretary, National Water Supply & Drainage Board, Health Authorities, Disaster Mitigation Centre and other Government and non-governmental organizations. Accordingly, resettling areas in war affected areas, drought prone areas, areas with prevailing waterborne disease and areas without piped water supplies were selected.

Beneficiary households were selected from distance to nearest drinking water source, prevalence of waterborne disease, number of children, women-headed households (war widows), disability and willingness to contribute in kind in the form of unskilled labour.

2.2.1 Rainwater Harvesting Systems (RWHS)

Rainwater harvesting is defined as collection of run-off rainwater for domestic use, agriculture, soil conservation and environmental management. Rainwater collected from rooftop is stored for human use and run-off rainwater from land surfaces or rock catchments are collected for agriculture use or groundwater recharging.

It is estimated that an average roof area of 50 m^2 will provide a household with minimum 50 L of water per day with an 8000-L tank during the driest period from May to September in these districts (Mansure 1999). This will serve a household with 5 people a minimum of 10 L per day per person for drinking and cooking during the driest period. Roof run-off can also be collected to recharge the groundwater, thereby increasing the quality and quantity of water in the wells.

Domestic Rain Water Harvesting Systems were constructed in 391 households in the three districts in the North and Uva Province of Sri Lanka. Beneficiaries are all populations in the drought- and flood-prone areas of the districts, who do not have safe and reliable water supply systems. Households headed by women, with children and with disabled persons were given priority. The rainwater system consists of an 8000 L Ferro cement above ground tank, two lengths of gutters, a first flush devise and filter (Fig. 2.2). All householders who receive the rainwater harvesting system are trained on how to operate and maintain the system. Rainwater harvesting systems were constructed with unskilled labour and material contribution from the household.

Demonstration rainwater harvesting systems and groundwater recharging systems were installed in 49 schools in Killinochchi, Badulla and Moneragala districts. These act as demonstration centres for promotion of the concept as well as used for their daily water needs and during emergencies. School rainwater harvesting system consists of 30,000 L Ferro cement tank, first flush device, gutters from appropriate

Fig. 2.2 Domestic RWH system

roof and size, filter and tap/s to take water out (Fig. 2.3). If there is a shallow well nearby, the overflow from the rainwater tank is directed to the well for groundwater recharging. If there is no well nearby, then the roof water from the nearest building is directly sent to the well through a first flush and filter (Figs. 2.4 and 2.5).

In 10 Rural hospitals/Medical Clinic, where there is no regular water supply, RWH system of 10,000—16,000 L was established as demonstration units (Fig. 2.4). Public were reluctant to visit or to obtain residential medical treatment at these primary

Fig. 2.3 School RWH system

Fig. 2.4 RWH system in rural hospital

Fig. 2.5 Direct groundwater recharging system

health care units since there was no water access for drinking or sanitation use. These centres provide health care daily to pregnant mothers, infants and patients with Chronic Kidney Disease of unknown aetiology reported in the area.

2.3 Impact

It has been reported that during the drought and floods in recent years, these households and institutes installed with rainwater harvesting systems were able to cope better. During the drought experience over the last 2 years, the households that had rainwater collected and stored were better off since they had readily available water at home. During 2018 floods in Kilinochchi district, when the surface water sources and well water were contaminated, the only clean water source that was available to the community was the rainwater harvesting tank. Mrs Selwarasa Inparasi from Musurampitti village in Divisional Secretariate Division of Kandawalai explained her experience, "*When we encountered a flood disaster on the 22nd December 2018, we used the RWH tank water for entire water needs for months afterwards, since the tubewells were not accessible and dug wells were contaminated with flood water*". More than 17,900 persons are provided with access to safe drinking water through RWH during this project.

RWHS built schools, which functioned as flood shelters and provided water to the people during the floods in 2018 until the government aid reached them. RWHS maintenance groups are set up with students, teachers and parents to ensure long-term sustainability. A manual for operation and maintenance of RWHS in schools was published by LRWHF with collaboration of PLAN SL and National Water Supply &

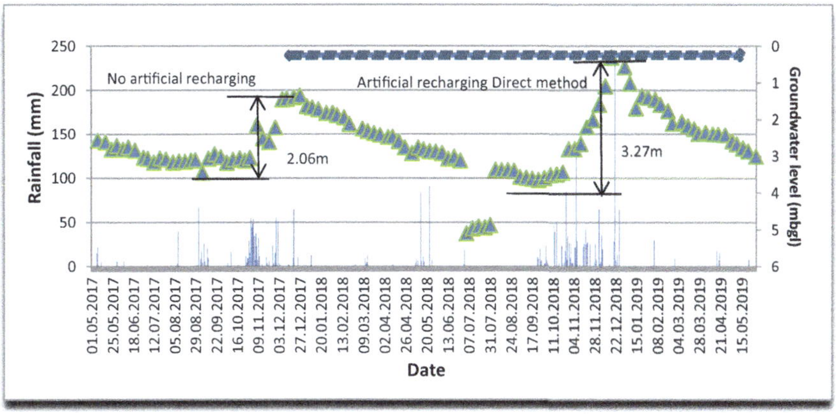

Fig. 2.6 Well Water Level and Rainfall at Madduvill Nadu GTMS

Drainage Board (NWS&DB) to provide practical solutions and guidance for school teachers, parents, caretakers and students for operation and maintenance of RWHS.

Rainwater collected is the main drinking water source for many households living in Chronic Kidney Disease of unknown aetiology (CKDu) prevalent areas. CKDu reported in many provinces in the North and Uva Province has caused great fear among people living in these areas on the quality of the available groundwater due to water hardness and contamination from agrochemicals. A recent survey has recorded that 52.8% of households with CKDu patients drink rainwater (LRWHF Survey 2018). From a water quality survey in the affected district, it was found that rainwater harvesting tank had the best quality drinking water compared with other available water sources.

In the roof water recharge system installed at Madduvil Nadu Government Tamil Maha School, in Kilinochchi Sri Lanka, the water level in the well rose by 3.27 m in the year 2018, which is 1.21 m higher than the non-recharging year of 2017 (Fig. 2.6). Also, it was observed that the water level in the well was almost near to the ground surface (0.3 mbgl), which facilitated the availability of groundwater during the dry season.

2.4 Conclusion

1. Rainwater is a valuable resource, which should be exploited in the most efficient way. Governments should explicitly recognize rainwater harvesting in their water supply policy to encourage and promote rainwater harvesting for human consumption and domestic use, to protect the people's health and livelihood, to ensure good utilization and conservation of water resources and to facilitate guidance and capacity building.

Beyond domestic use, rainwater harvesting can be used for agriculture, soil conservation and environmental management. Run-of collection structures are used for improving the health of watersheds through retention and slowing down the run-off of water to encourage restoration of aquifers, thus improving both quality and quantity of well water.

Rainwater is a valuable resource, which should be exploited in the most efficient way. In order to achieve SGD6 of universal access to safely managed water by 2030, rainwater harvesting is a must in Sri Lanka and many other South Asian and African countries

References

Eckstein D, Hutfils ML, Winges M, Germanwatch e.V (2017) Global Climate Risk Index 2019. https://www.germanwatch.org/en/17307. Accessed 17 Feb 2020

Hazard profile of Sri Lanka (2012) Disaster Management Centre, Ministry of Disaster Management, Colombo Sri Lanka and UNDP. (file:///D:/RAIN% 20WATER/My%20Documents/REPORTS/Hazard%20Profiles%20of%20Sri%20Lanka.pdf. Accessed 17 Feb 2020

Jayasinghearachchi DA (2008) Impacts of climate change on water resources and adaptation measures, Department of Meteorology Sri Lanka. http://www.meteo.gov.lk. Accessed 17 Feb 2020

Mansure U (1999) Quality of Rain Water in storage at Badulla District, Sri Lanka. Report submitted to Lanka Rain Water Harvesting Forum, Sri Lanka

National Water Supply & Drainage Board, Regional Support Centre, Uva Province. http://www.waterboard.lk/web/index.php?option=com_content&view=article&id=51&Itemid=211&lang=en#key-facts-figures-and-targets. Accessed 17 Feb 2020

Sri Lanka Water Vision 2025 (2000) Sri Lanka National Water Partnership & LIFE-WRU & WRS & IWMI, 2000. "Water vision 2025 Sri Lanka," IWMI Books, Reports H025883, International Water Management Institute

Sri Lanka Rain Water Policy and Strategy (2005) Government of Sri Lanka. http://lankarainwater.org/wp/?page_id=70. Accessed 17 Feb 2020

Survey on the Perception of Users of Rain Water Harvesting Systems in the Seven Project Districts (2018) Climate Resilient Integrated Water Management Project. LRWHF, Sri Lanka

The Spatial Distribution of Poverty in Sri Lanka, Report of Department of Census and Statistics - Sri Lanka Poverty Global Practice, World Bank Group August 2015. http://www.statistics.gov.lk/poverty/SpatialDistributionOfPoverty2012_13.pdf. Accessed 17 Feb 2020

Chapter 3
Empowering Poor Farmers for Climate Smart Water Management and Agriculture Production: Lessons from Andhra Pradesh

S. P. Tucker, Joseph L. Plakkoottam, and R. Ratnakar

Abstract This paper explores how participatory groundwater management initiatives undertaken in the erstwhile Andhra Pradesh in the last three decades have brought prosperity to small and marginal farmers while addressing environmental and climate change concerns. The outcomes of the pioneering Indo-Dutch APWELL Project (Andhra Pradesh Groundwater Borewell Irrigation Schemes, 1995–2003), in community-based groundwater management practices implemented in seven districts of Andhra Pradesh, which commissioned about 3500 group-owned borewell schemes benefitting nearly 15,000 families of small and marginal farmers owning 14,000 hectares of dryland are first discussed The paper further outlines the adoption and upscaling of lessons learned in sustainable water management practices piloted in the APWELL Project in subsequent projects such as the AP Farmer Managed Groundwater Systems (APFAMGS) Project and the Participatory Groundwater Management (PGM) component in the AP Community Based Tank Management Project (APCBTMP) (www.apmitanks.org). Similar initiatives under the APDAI Project and a number of NGO-sponsored community management of groundwater resources are also discussed. It further identifies the tasks ahead in harnessing participatory methodologies in increasing productivity of land and water through appropriate practices such as precision farming to rise to the challenges posed by climate change and water scarcity. The authors draw upon their vast experience in the design, implementation and management of natural resource management projects and use project documents and field visits to support the discussion.

S. P. Tucker
Former Chief Secretary Andhra Pradesh, and Chairman Poverty Learning Foundation, Hyderabad, India

J. L. Plakkoottam (✉)
Water Natural Resources and Climate Change Poverty Learning Foundation, Hyderabad, India
e-mail: joseph.plakkoottam@gmail.com

R. Ratnakar
Extension Education Institute (S. Region), ANGRAU, Hyderabad, India

R. K. Mishra et al. (eds.), *Sustainable Climate Action and Water Management*, Advances in Geographical and Environmental Sciences,
https://doi.org/10.1007/978-981-15-8237-0_3

Keywords Climate smart water management and agricultural production · Community based water management initiatives · Lessons in participatory groundwater management

3.1 Introduction

The Indo-Dutch APWELL Project was implemented in seven drought-prone districts of Andhra Pradesh from 1995 to 2003. The long-term objective of the project was to improve the living conditions of small and marginal farmers, through sustainable and environmentally sound interventions. The project also strove to make women farmers as equal partners with male farmers in agriculture and related activities. The immediate objective of the project was to provide groundwater irrigation facilities for small and marginal men and women farmers. They own and maintain these borewell irrigation systems, with a potential to increase agricultural production. The farmers formed Water User Groups (WUGs) for construction, operation, and maintenance of the groundwater irrigation systems. Clusters of WUGs form Borewell User Associations (BUAs), which were legally registered, for training, inputs, agro-processing, and generation of profit. Important components of the project were the following: Groundwater resource development where feasible, land-and-water management by the users, extension and training, activities for gender integration and institutional development, environment management, and monitoring and evaluation (Plakkoottam, J.L., van der Wal. J.W.K., Reddy, N.S.B., Prasad, J. 2007).

Implementing agencies: The project operated in seven drought-prone districts of Andhra Pradesh—Anantapur, Chittoor, Kadapa, Kurnool, Mahbubnagar, Nalgonda and Prakasam. Physical activities such as groundwater prospecting, drilling, yield testing, and construction of the distribution systems, were done through the AP State Irrigation Development Corporation (IDC), with its technical staff under the Executive Engineer in each district. The social, institutional, gender, groundwater management, agricultural and eco-development aspects of the project were implemented by the field staff employed through a state level or district level agency/NGO. A District Field Coordinator (DFC), part of the consultant's team, supervised the work in each district. A technical assistance team consisting of national and international experts on various disciplines advised and coordinated project activities through the NGOs. The expatriate consultants were organized through ARCADIS Euroconsult and the local consultants through an Indian agency.

Project implementation process: The WUG conducted a Planning and Budgeting meeting, together with the NGO and an IDC representative present, to agree on the layout of the system, costing, and how the farmers would contribute their share of 15% of the total cost (which might include labor). After construction and commissioning of the system, IDC provided a warranty for 1 year, and then handed it over completely to the WUG. Meanwhile, women self-help groups were organized first for thrift and credit activities leading toward undertaking income generation activities. Once the WUGs began operation and management of their own groundwater irrigation systems, Borewell Users' Associations (BUAs) were formed and registered under the Mutually Aided Societies Act. Various capacity building activities were organized for the community-based organizations, such as assistance in group organization and management, leadership and dispute resolution, gender activities and investments, water management training, sustainable agricultural practices, environment management, eco-farming, agro-processing and marketing.

Environmental sustainability: The APWELL project adequately addressed likely environmental problems. The Project conducted a novel Environment viability assessments (EVAs) before borewells were approved only in areas where groundwater was assessed to be sufficient and was not overdeveloped by government guidelines. Soil and water quality analysis were also done for each farm. Crop plans were made for sustainable land and water use based on borewell yield. Farmers were trained in land and soil management methods. The BUA took the responsibility for overall water management including recharge measures (Irrigation & CAD Department 2008; Witjes et al. 1999).

Achievements of APWELL: APWELL Project provided groundwater irrigation sources to 3,462 Borewell User Groups (BUGs) comprising 13,878 families of small and marginal farmers, cultivating 34,489 acres.

APWELL Cost sharing	**Rs. Million**
Netherlands Government Grant	411.72
Borewell User Groups	60.86
Government of India	58.43
TOTAL	**531.01**

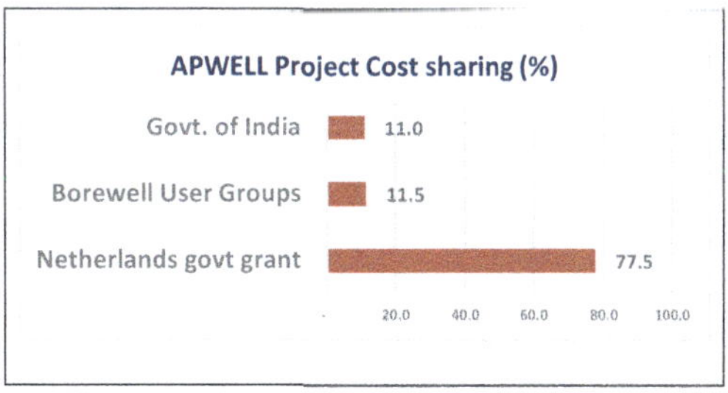

The APWELL project also organized vibrant community-based organizations: 603 Self Help Groups (SHGs), and 246 Borewell User Associations. Some of the several pilot activities initiated during the final years (1999–2003) of the project, viz., (i) Participatory Hydrological Monitoring and Crop Water Budgeting, (ii) Artificial Recharge, and (iii) Ecofarming, showed potential for developing as key areas toward sustainable agriculture and natural resource management.

APWELL Project approach and benefits: The philosophy and approach of the APWELL Project differed from the traditional public service attempts to develop borewells in several aspects. (i) Borewell development was fully controlled by a public service unit, specifically set up for the purpose; (ii) Agencies drilled, operated and managed the borewells for an indefinite period thus restricting the role of farmers to that of water users only; and (iii) The criteria adopted for selection of farmers for subsidy suffered from many drawbacks. In most cases, extraneous factors, including local politics, were more decisive than the socioeconomic status of the selected farmers (Plakkoottam et al. 2007).

In the model adopted by the APWELL project, the small and marginal farmers were initially organized into small groups and they made a limited contribution of 15% of the investment cost. The unique features of this approach were: (i) Social mobilization was an integral part of the program; (ii) Farmers became the owners of the borewells after a warranty period of 1 year; (iii) Participatory hydrological monitoring and supportive agricultural extension activities were integral of the program; (iv) operation, maintenance, management of borewells and equitable sharing of water by members of borewell user groups (BUGs); and (v) NGOs became the custodians of the selection criteria and they ensured meticulous application of the pre-determined guidelines (ARCADIS EUROCONSULT. 2003).

District	Villages	BUGs	Total N of Families	Average Family/BUG	Total Ayacut (acres)	Average Ayacut (Acre)		GPH	Average (Rs)	
						Per BUG	Per family (acre)	Average Yield	Cost per borewell	BUG contribution
Anantapur	39	415	1,396	3.4	4,410	10.6	3.2	4,009	131,724	16,159
Chittoor	110	419	2,076	5.0	3,481	8.3	1.7	3,109	141,242	17,171
Kadapa	59	415	2,160	5.2	3,978	9.6	1.8	2,995	150,625	18,167
Kurnool	78	518	2,013	3.9	5,299	10.2	2.6	4,557	143,036	16,765
Mahbubnagar	55	821	2,741	3.3	8,605	10.5	3.1	2,604	129,987	15,610
Nalgonda	42	299	1,439	4.8	3,018	10.1	2.1	3,569	153,300	18,796
Prakasam	87	575	2,053	3.6	5,698	9.9	2.8	3,523	142,660	16,635
Total	**470**	**3,462**	**13,878**	**4.0**	**34,489**	**10.0**	**2.5**	**3,523**	**140,102**	**19,790**

Source APWELL Project: Final Report. ARCADIS EUROCONSULT (2003, pp. 76–77)

The gains from the APWELL project have been two-fold, viz. *process* related and *product* specific. They are briefly described below (Plakkoottam et al. 2007).

Participatory hydrological monitoring: The belief and philosophy of APWELL that efficient and effective management of ground water resources is possible only when there is an active involvement of people at the grass root level, stood the test of time through the concept was to put into practice only during the later part of the project implementation. Extensive participatory hydrological monitoring was made an integral part of the project through the constitution of village-level committees. The committee members were provided hands-on training in simple tools and techniques of groundwater level measurements and awareness was created on the implications and undesirable consequences of improper groundwater management.

Linkages with watershed development programs: The collaborative working mechanisms established and sustained with the on-going watershed development programs and schemes (e.g. DPAP, DPEP, APRLP) by the project proved very productive.

NGO participation: Establishment and nurturing of micro-level institutions such as Water User Groups (WUGs), Borewell Users Associations (BUAs), and Self-Help Groups (SHGs) under the project were anchored through credible field level NGOs.

Active farmer involvement beginning from project formulation to project implementation and management stage was the prime reason for the success of the project. Farmers were actively involved in the selection process of marginal and small farmers, in the system design, crop mix decisions, and water sharing mechanisms fostering the spirit of co-operation to take appropriate decisions.

Gender integration: Traditions and conventions dictated women in the project area to be confined to home and on-farm work only. The training and exposure visits under the project enabled women farmers to look beyond the narrow confines of farm and home. Women gained greater access to financial assistance through banks, other developmental funding organizations and government departments. The most intangible process related benefit, however, has been in the form of enhanced self-concept of women and greater confidence gained by women groups to assert their needs, aspirations and priorities: e.g., visiting banks to discuss their financial needs, negotiating loans on their own, greater participation in group activities and village affairs.

Concept of water equity: The concept of collective and equitable water sharing has been the central focus of the project. Achievement of collaboration even under highly demanding circumstances and working in a co-operative spirit was time-intense and involved the following steps: (i) discussions initiated in WUG forums and members entering into an agreement to own and operate borewells collectively and share water equitably, (ii) the concept of water sharing, crop planning, sharing of operation and maintenance expenditure and other modalities are discussed, and (iii) agreement is reached on equitable sharing of electricity charges.

Increased agricultural output: Three major factors viz., changes in the cropping pattern, enhanced farm productivity and increased livestock production resulted in increased agricultural output, though variance between districts and between farmers were visible due to differing soil characteristics, varying rainfall patterns and cultural practices.

Increased family income has been substantial during the project period. In addition, indirect benefits such as 200% increase in land value, higher income accruals through sale of livestock products, non-farming activities etc. have been noteworthy.

Improved living conditions: The most visible impact of increased family income has been in the improvement of living conditions of farmers. Food supplies became more assured, either because of higher production of food crops or because of higher purchasing power of farmers. The diets of the farmers were increasingly becoming enriched with increased intake of superior food products, including meat products and vegetables. Increased income also helped the farmers to obtain better education, better housing, and better farm implements.

Reduced migration: In the pre-project days, the only employment avenue available for the farmers was in the faraway towns and cities, during lean seasons. The gainful employment opportunities generated by the project altered the scenario, which resulted in the dramatic reduction of off-farm migration.

3.1.1 Upscaling Lessons of APWELL Project

In July 2001, RNE started funding the AP State Water Conservation Mission (WCM). In its interim work plan, the WCM had proposed that the APWELL PHM activity be expanded and transferred to the Water Conservation Mission Programme. From April 1, 2002, the APWELL project was brought under the Panchayat Raj and Rural Development (PR&RD) department in order to share the APWELL experience in natural resource management and sustainable agriculture by communities, to larger areas in the state. In November 2002, the Rural Development Department of the Government of Andhra Pradesh submitted a proposal called "APWELL and beyond" to the Central Government for seeking funds from the Royal Netherlands Government for the next phase of the project. However, due to sudden changes in bilateral funding arrangements, this proposal was not funded and the APWELL Project was closed in 2003 (ARCADIS EUROCONSULT 2003).

3.2 APFAMGS Project (2003–2009)

The experience gained from the groundwater management and PHM in APWELL Project guided the design and implementation arrangements of the Andhra Pradesh Farmer managed Groundwater Systems Project (APFAMGS). It focused on capacity development of water users in managing their groundwater resource for crop production in a commonly agreed and sustainable way. In 2004, the Netherlands Embassy in India handed over the role of donor to the UN-FAO with a budget ceiling of USD 6.785 million for 51 months of implementation. The institutional arrangement comprised a Nodal NGO, as the implementing agency and eight Partner NGOs, with the support of a Technical Support Team (www.apfamgs.org; Das, S.V.G. & Burke, J. 2013).

The APFAMGS Project largely achieved the expected results. Through a number of training sessions, the farmers began to understand the seasonal occurrence and distribution of groundwater in their habitations as well as in the whole Hydrological Unit they belonged to. Farmers became capable of estimating seasonal groundwater recharge, draft and balance. Farmers collect and record rainfall and associated groundwater data. They have grasped the concept of groundwater as a common property resource and are ready to wisely manage it for collective benefit. APFAMGS started Farmer Water Schools (FWS) to facilitate experiential learning of different cultivation techniques and cropping patterns for optimal use of scarce groundwater resource using adult learning principles. A key element in the FWS is the Crop Water Budget session before the Rabi season starts, as a community decision-making exercise to adopt suitable agricultural practices, based on groundwater availability. Other aspects that contributed include: Training on Low External Input Agriculture, inputs from the Bio-agents production center, documentation of best agricultural practices and Farmer Training Teams, which led to external inputs reduction, with environmental and health benefits to stakeholders (www.slidesharenet.indiawaterportal.apfamgs).

Building on the experience of effective gender mainstreaming and women empowerment from the APWELL Project contributed to APFAMGS' success. The local institutions and platforms for common decision-making at various levels contributed to social capital development at different levels.

The final evaluation team recommended that:

> APFAMGS experience is a breakthrough in the management of groundwater and in securing livelihoods of poor farmers in India: since both are key concerns of the central Government of India and of many State Governments, the approach should be adopted and mainstreamed in the Government's policy and development work (www.globalissues.org; IBRD/WB 2010).

3.2.1 Participatory Groundwater Management (PGM) in APCBTMP

The experiences and lessons of APWELL and APFAMGS Projects provided the strategic directions for making future investments in sustainable groundwater management. The PGM component of the AP Community Based Tank Management Project APCBTMP, 2007–2016 was designed incorporating the processes, methods and procedures developed under the successful pilots to organize and train groundwater users to collect the hydrological data necessary for crop water budgeting, foster social regulations, promote social equity and improve water use efficiency and productivity (www.apmitanks.in). PGM in APCBTMP aimed at empowering groundwater users in the tank influence zone to wisely manage the dynamic groundwater resources, replenished through rainfall, surface water sources and return circulation from irrigated areas. (www.ap.gov.in). The Participatory Groundwater

Management (PGM) component of APCBTMP was declared as the second ranked finalist for "Water for Life" UN-Water Best Practices award 2014 edition (I&CAD 2014; Paul 2014).

3.3 The Way Forward

The early success of the APWELL Project (Witjes et al. 1999) inspired adoption in similar initiatives under the APDAI Pilot and a number of NGO-sponsored community management of groundwater resources during the first decade of 2000. These experiences also have contributed to the design of on-going projects such as the Andhra Pradesh Drought Management Project (APDM 2017) funded by the IFAD, and the World Bank Assisted APIIATP (2019–2026).

In a recent rapid assessment of the APWELL Project, it was observed that most of the small and marginal farmers who got equitable access to groundwater for nearly two decades have made substantial socioeconomic progress, even when vagaries of climate change have affected them in terms of groundwater depletion. Those farmers whose groundwater sources have sustained and have adopted climate smart agricultural practices through micro-irrigation and permaculture are thriving. Surplus creation by small and marginal farmer beneficiaries was a major legacy of the APWELL Project. However, new projects designed taking APWELL lessons have mostly focused on knowledge dissemination rather than on providing access to groundwater to poor farmers through group borewell schemes.

There is a need to systematically study the resilience of these project interventions and document these experiences for future policy directions and lesson learning.

Bibliography

Arcadis Euroconsult (2003) APWELL Project: Final Report

Das SVG, Burke J (2013) Small holders and sustainable wells: a retrospect—participatory groundwater management in Andhra Pradesh (India). FAO of UN, Rome

International Bank for Reconstruction and Development/The World Bank (2010) Deep Wells and Prudence: towards pragmatic action for addressing groundwater overexploitation in India. The World Bank, Washington, DC. www.worldbank.org

Irrigation & CAD Department (2008) Participatory groundwater management in Andhra Pradesh community based tank management project. Government of Andhra Pradesh, Hyderabad

Paul S (2014).And Not a Drop to Waste. IPS-Inter Press Service. http://www.ipsnews.net/2014/05/drop-waste

Plakkoottam JL, Raje S, Reddy NE, Sharma SNN (2014) Participatory groundwater management in APCBTMP: impact assessment study, irrigation & CAD Department. Government of Andhra Pradesh, Hyderabad

Plakkoottam JL, van der Wal JWK, Reddy NSB, Prasad J (2007) Groundwater Management Experiences in Andhra Pradesh: Strategies and Prospects. Hyderabad: Irrigation & CAD Department Irrigation & CAD Department, Government of Andhra Pradesh

Tucker SP, Gupta S (2010) improving irrigation performance management: navigating through Andhra Pradesh experiences, irrigation & CAD Department. Govt. of Andhra Pradesh, Hyderabad
Witjes B, van Raalten DW, Plakkoottam JL (eds) (1999) Farmer managed borewell irrigation systems. ARCADIS Euroconsult, BKH, IRDAS, Hyderabad, India
www.apfamgs.org
www.ap.gov.in
www.apmitanks.in
www.globalissues.org
www.slidesharenet.indiawaterportal.apfamgs

Chapter 4
Watershed Prioritization in GIS: A Comparison of Multiple Approaches

Stutee Gupta, Ramraj Pendkar, C. Sarala, and K. M. Reddy

Abstract Various Soil and Water Conservation (SWC) measures are widely implemented to reduce the surface runoff in the degraded and drought-prone watersheds. However, their implementation requires thorough assessment of morphometric and hydrological parameters prevailing in the watershed. This research was conducted to characterize the sub-watersheds for SWC in a drought-prone watershed of Ahmednagar district, Maharashtra using morpho-metric analysis in GIS domain. Their prioritization has been carried out using two different Multi-Criteria Decision Analysis (MCDA) approaches viz. Analytical Hierarchy Process (AHP) and Technique for Order Preference by Similarity to an Ideal Solution (TOPSIS) for a comparative assessment. The findings were further supplemented based on the severity of surface processes, quantitative estimation of soil erosion and surface runoff derived using Soil and Water Assessment Tool (SWAT). The priority value obtained using the AHP and TOPSIS ranged from 1.452–2.658 to 0.215–0.586, respectively. The comparison based on normalized values from the two approaches, shows similar results for the 73% of sub-watersheds and slight variation in the remaining. However, surface runoff and soil erosion-based prioritization carried out using SWAT have been found to have slightly better semblance with the TOPSIS model as compared to that of AHP. The

S. Gupta (✉) · R. Pendkar · K. M. Reddy
Rural Development and Watershed Management Division, National Remote Sensing Centre, Balanagar, Hyderabad, Telangana, India
e-mail: stutee_g@nrsc.gov.in

R. Pendkar
e-mail: ramrajpendkar@gmail.com

K. M. Reddy
e-mail: reddy_km@nrsc.gov.in

C. Sarala
Centre for Water Resources, Institute of Science and Technology, Jawaharlal Nehru Technological University, Hyderabad, Telangana, India
e-mail: hodcwr@gmail.com

R. K. Mishra et al. (eds.), *Sustainable Climate Action and Water Management*, Advances in Geographical and Environmental Sciences,
https://doi.org/10.1007/978-981-15-8237-0_4

study concludes that though the morphometric analysis and MCDA is and continue to remain the most widely used tool for watershed characterization, supplementing them with the hydrological inputs derived from SWAT provides a more effective approach for planning SWC measures at a local level.

Keywords Watershed analysis · MCDA · SWAT · Soil and water conservation · GIS · Morphometric analysis

4.1 Introduction

Water is the most widespread substance that plays a very important role both for the environment and in human life (Stec and Zeleňáková 2019). Around 1.2 billion people, or almost one-fifth of the world's population, live in the areas of physical scarcity of water and 500 million people are approaching this situation. Another 1.6 billion people, or almost one-quarter of the world's population, face economic water shortage, where countries lack the necessary infrastructure to take water from rivers and aquifers (UN-Water, FAO 2007). The problem is further exacerbated due to lack of proper water resource management and over exploitation to the extent that losses that may occur are believed to be more alarming than those caused by weapons of mass destruction in warfare. Excess flooding of water and drought situations are prevailing in the Indian subcontinent regions with short seasonal variations mainly due to the poor watershed management practices. Similarly, in spite of the average annual rainfall of 300 mm–650 mm received in the major part of the country, most of the places suffer with unavailability of water throughout the summer as the stormwater drain to the downstream watershed due to lack of surface water features. The excessive surface runoff gradually erodes away the top soil into the downstream. To overcome such situations, implementation of various Soil and Water Conservation (SWC) measures has been widely recommended by planners and resource managers. However, cautious prioritization of the watershed based on the availability of natural resources and the nature of the physical processes affecting them are necessary for identification of the critical areas of erosion, which produce maximum sediment (Chowdary et al. 2013).

The prioritization of sub-watersheds is of paramount importance in developing catchment area treatment plan and implementing watershed management activities. It can help in taking necessary precautionary SWC measures a priori to ensure effective development (Jaiswal et al. 2015). Thus, watershed characterization and management require detail information for topography, drainage network, water divide, channel length, geomorphologic and geological setup of the area for watershed management and its prioritization (Javed et al. 2009). Use of toposheets and field surveys has been conventionally very popular among the reserchers for extracting the topography and the various morphometric parameters. However, over the years, there has been a significant development in the variety of digital elevation models (DEMs) available from the satellite imaging and also their processing techniques (Jenson 1991). Currently, DEMs available from various sources at different spatial resolutions and majority of them are free of cost. As a result, DEMs have become a basic input to derive hydrological features for obtaining watershed information using computer-based programmes in a fast, precise, updated and inexpensive manner (Li and Wong 2010). There has been a significant development in the models to define ridges, channels, watersheds, and other hydrologic features from them using Geographical Information System (GIS). Quantitative morphometric analysis performed using GIS provides significant know how about the nature of the watershed.

As watershed prioritization depends on several interdependent and spatially distributed parameters, the Multi-criteria Decision Analysis (MCDA) tools are considered to be very useful for an effective decision-making (Chung and Lee 2009). MCDA helps in indexing potential sub-watersheds in an area by assigning suitable weightages to all the parameters based on the priority index values, i.e. the one with the higher values indicates higher priority. Simple additive weighting (SAW), Saaty's AHP (AHP), compound factor (CF) and Technique for Order Preference by Similarity to Ideal Solution (TOPSIS) etc. are some of the common MCDA techniques used by the researchers.

Though the morphology of a watershed exhibits a strong control on its overall functioning, further integrating it with land use change assessments using hydrological model in GIS provides valuable insights for land and water management at different spatial and temporal scales (Parupalli et al. 2019). Several physically based distributed parameter models have been developed to predict runoff, erosion, sediment and nutrient transport from rural and agricultural watersheds under various management regimes (Patel and Nandhakumar 2016). Among them, the Soil and Water Assessment Tool (SWAT) has been found to be very promising and most widely used. It incorporates spatially and physically distributed watershed inputs to simulate surface and sub-surface flow, sediment generation and deposit, and nutrient movement and fate through the watershed system (Gassman et al. 2010).

Whereas, a variety of methods as presented in this study are widely used commonly by the researchers the decision-makers are often concerned about their effectiveness and suitability over a watershed. The present study demonstrates the efficacy of MCDA over morphometric analysis and that of SWAT over the MCDA in a guided manner. Multi-Criteria Decision Analysis (MCDA) tools viz. AHP and TOPSIS were used to prioritize sub-watersheds and their further comparative assessment was carried out. The findings were also substantiated with the severity of surface processes based on the quantitative estimation of the soil erosion and surface runoff derived using SWAT.

4.2 Study Area

The study represents a part of Sina river sub-basin which is a tributary of Bhima river in the Ahmednagar district of Maharashtra. It comprises of three watersheds namely 4D7I5, 4D7I4 and 4D7I1 (as per Watershed Atlas of India). It extends from 18° 27′ 49.92″N to 19° 15′ 23.80″N latitudes and 74° 28′ 56.26″ E to 75° 21′ 55.17″E longitudes (Fig. 4.1), having an area of 2,225 km sq. The district has general slopes from

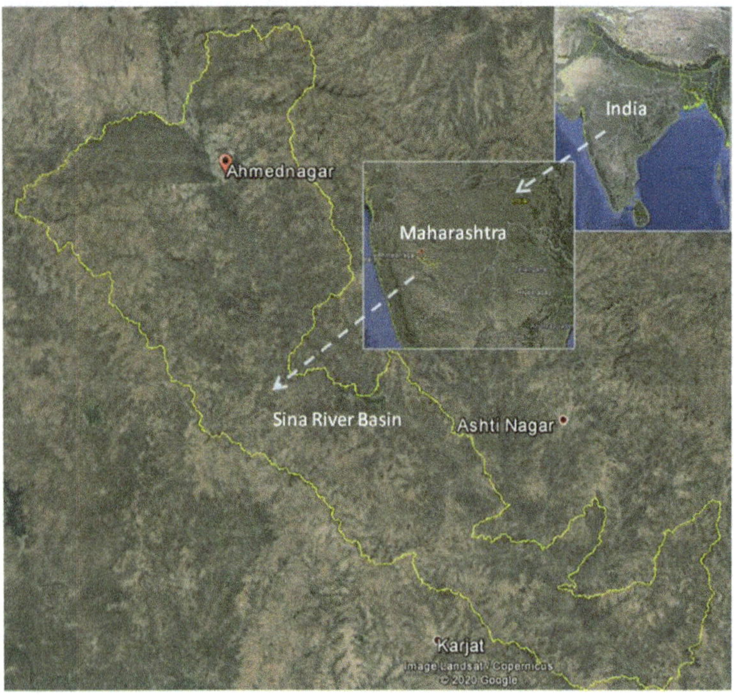

Fig. 4.1 Location map of Study Area (Sina river basin, Ahmednagar, Maharashtra)

West to East. The western part, the Akola on the Sahyadris, it is the highest part of the district. Altitude is more than 2,500 feet. The extreme eastern part Shevgaon taluka is a plain, which is not 1,500 feet above mean sea level. Climatologically, the study area is classified as semi-arid. An uneven rainfall distribution in the region further makes it a drought-prone area. Nearly 80% of the study area is under agricultural area and the forests cover around 5%, water bodies and built-up area constitutes 2% and 10%, respectively.

4.3 Methodology

The methodology for the study constitutes characterization of sub-watersheds and derivation of morphometric parameters, their further prioritization using MCDA. Towards the end, hydrological analysis was carried out to further compare and substantiate the findings based on runoff and sediment yield values derived using SWAT model.

4.4 Sub-watersheds Characterization

The characterization of watersheds and delineation of sub-watersheds is an important aspect for planning and management for water harvesting as well as soil conservation. The influence of topography on the morphology of surface hydrologic networks has been well documented in the literature (McMaster 2002), leading to the widespread use of DEMs. For this study, we used Cartosat-2 DEM (30 m × 30 m spatial resolution) obtained from ISRO Bhuvan portal. The detailed morphometric analysis of the watershed including linear, and shape parameters as well as relief aspects (Table 4.1) has been carried out to generate the sub-watershed prioritization map (Julius et al. 2017).

Table 4.1 Formulae adopted for computation of Morphometric parameters

S. no.	Morphometric Parameters	Formula	Reference
Drainage network analysis			
1	Stream Order (N_u)	Hierarchical rank	Strahler (1952)
2	Number of Stream Segments	GIS analysis	–
3	Total Stream length (Lu)	GIS analysis	–
4	Bifurcation Ratio (Rb)	$Rb = Nu/N_{u+1}$	Strahler (1964)
Basin Geometry			
5	Basin Area (A)	GIS analysis	–
6	Basin perimeter (P)	GIS analysis	–
7	Basin Length (L_b) Kms	$L_b = 1.312*A^{0.568}$	Nooka Ratnam (2005)
8	Form Factor Ratio (R_f)	$R_f = A/Lb^2$	Horton (1932)
9	Elongation Ratio (R_e)	$R_e = 2/Lb * (A/\pi)^{0.5}$	Schumm (1956)
10	Circularity Ratio (R_c)	$R_c = 12.57 * (A/P^2)$	Miller (1953)
11	Texture Ratio (R_t)	$R_t = N_1/P$	Schumm (1956)
12	Compactness Coefficient (C_c)	$C_c = (0.2841 * P)/A^{0.5}$	Gravelius (1914)
Drainage analysis			
13	Stream Frequency (F_s)	$F_s = Nu/A$	Horton (1932)
14	Drainage Density (D_d) Km/Kms^2	$D_d = Lu/A$	Horton (1932)
15	Length of overland flow (L_g)	$L_g = 1/D_d*2$	Horton (1942)
Relief Characteristics			
16	Height of Basin Mouth (z) m	GIS Analysis/DEM	–
17	Maximum Height of the Basin (Z) m	GIS Analysis/DEM	–
18	Total Basin Relief (H) m	$H = Z - z$	Strahler (1952)
19	Relief Ratio (R_{hl})	$R_{hl} = H/Lb$	Schumm (1956)
20	Ruggedness Number (R_n)	$R_n = Dd * (H/1000)$	Patton and Baker (1976)

4.5 Multi-criteria Decision Analysis

The sub-watershed prioritization was carried out by applying MCDA (Gopinath et al. 2016) techniques viz. AHP and TOPSIS that are commonly used in watershed prioritization studies.

- Box 1
- Constructing the Pair-wise comparison matrix.
- Synthesizing the pair-wise comparison matrix.
- Calculating the priority vector.

- Calculating the consistency ratio.
- Calculating the λ_{max}.
- Calculating the Consistency index.
- Higher the consistency index represents higher the priority level.

- Box 2
- Construct a normalized decision matrix (rij).
- Construct a normalized weighted decision matrix (Vij = W* rij).
- Determine the Ideal (A*) and Negative Ideal (A') Solutions.
- Calculate the separation measures for each alternative Si*,Si'.
- Calculate the relative closeness to the ideal solution (Ci*), 0 < Ci* < 1.
- Give the highest priority to the option which contains Ci* closest to 1.

AHP technique uses hierarchical structures to represent a problem, which is further assigned priorities for the alternatives, based on the user decision (Saaty 1980). In this approach, a matrix of pair-wise comparisons (ratios) is constructed between the factors responsible for the erosion. These factors are scaled as 1 to 9 where 1 indicates that the two factors are equally important and 9 indicates that the one factor is more important than the other. Reciprocal of 1–9 (1/1 and 1/9) shows that 1 is less important than the other. Steps followed in AHP are summarized in Box 1.

TOPSIS is another well-known method for classical MCDA. It is based on the principle of defining the ideal and the anti-ideal solution (Hwang and Yoon 1981). It is followed by defining the alternative solution that should have the shortest distance from the ideal solution and the longest distance from an anti-ideal solution. TOPSIS assumes that we have m alternatives and n attributes/criteria. Steps followed in applying TOPSIS are given in Box 2.

Normalization of the consistency index and relative closeness values are done to prioritize the sub-watersheds accordingly using Eq. 4.1. Based on the normalization of Priority Index and Relative Closeness values, prioritization has been done using natural breaks method.

$$\text{Normalized value} = (R - Rmin)/(Rmax - Rmin) \tag{4.1}$$

4.6 Hydrological Analysis

To understand the severity of surface processes, quantitative estimation of soil erosion and surface runoff was carried out using ArcSWAT 2012. A step-by-step procedure as outlined in the SWAT user guide (Winchell et al. 2010) has been followed for the analysis. The input thematic datasets used included land use/land cover (LU/LC), soil and DEM. The land use is one of the most important factors that affect runoff, evapotranspiration and surface erosion in a watershed. The LU/LC map was obtained from the SIS-DP, NRSC and further reclassified corresponding to the model parameterization (Fig. 4.2). The study area was found to contain five soil texture classes based on the soil map acquired from National Remote sensing Centre (Fig. 4.3). The DEM was used to delineate the catchment and analyze the drainage patterns as well as derive slope parameter (Fig. 4.4). The meteorological data required for this study were obtained from SWAT Global Weather Data (https://globalweather.tamu.edu/), which includes daily weather data (Rainfall, Temperature, Relative Humidity, Wind, and Solar Radiation) from the year 1979 to 2014. The land use and the soil data in a projected Grid file format were loaded into the Arc SWAT interface to determine the area and hydrologic parameters of each land-soil category simulated within each sub-watershed. The slope classification into four classes (0–3, 3–10, 10–20 and >20) were applied and the slope grids were reclassified. The gridded layers of land use, soil and slope were overlaid to derive Hydrological Response Units (HRUs), which represent the smallest spatial unit of the model. The standard HRU definition approach clumps all similar land uses, soils, and slopes within a sub-watershed based upon

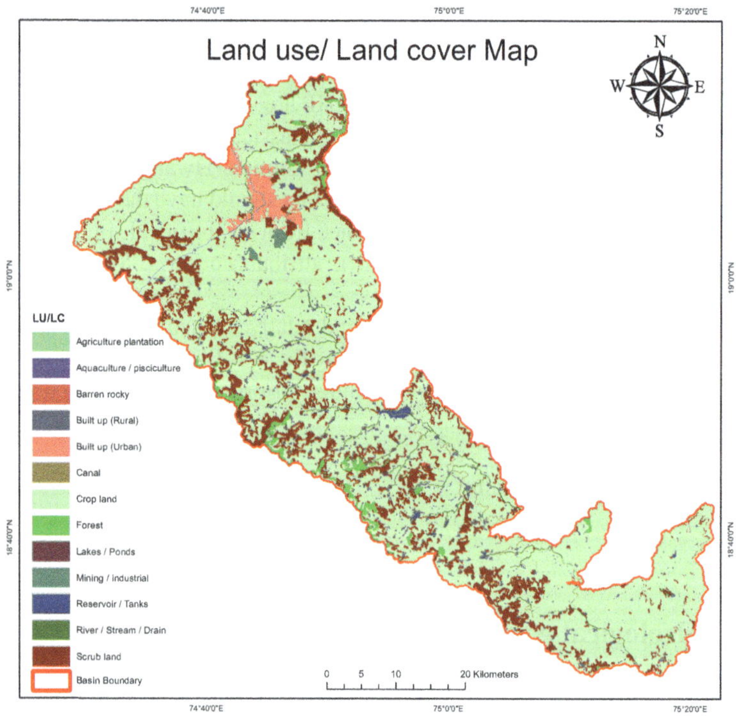

Fig. 4.2 Land Use/Land Cover map (*Source* NNRMS, NRSC; Scale 1:50 000)

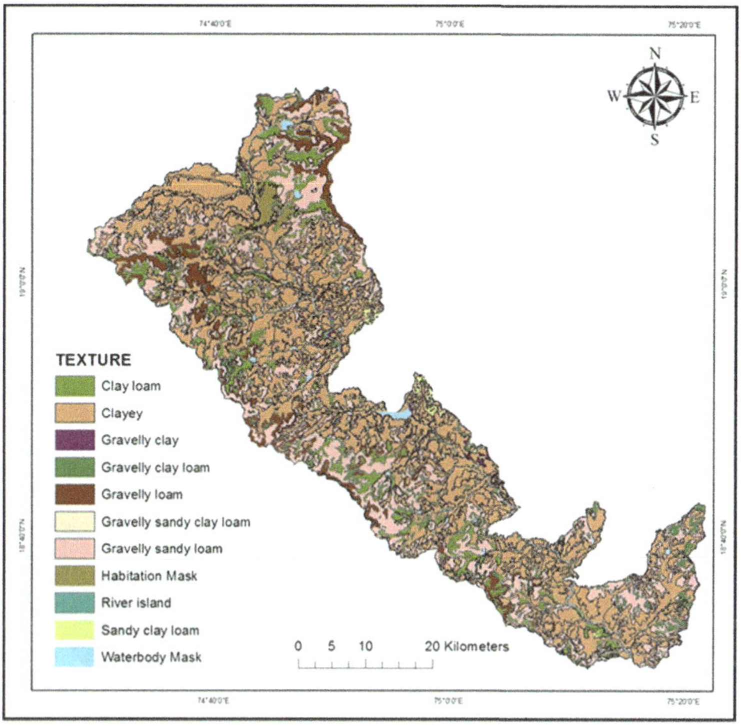

Fig. 4.3 Soil map (*Source* NNRMS, NRSC; Scale: 1:50 000)

user-defined thresholds. Using daily or sub-daily rainfall values, SWAT simulates surface runoff volumes and peak runoff rates for each HRU. SWAT incorporates two methods for estimating surface runoff: The Soil Conservation Service (SCS) Curve Number (CN) procedure and the Green and Ampt-infiltration method. In this study, the SCS curve number method was used since it is less data demanding in terms of sub-daily data. Sediment yield is estimated using Modified Universal Soil Loss Equation (MUSLE).

4.7 Results

4.7.1 Sub-watershed Characterization

The morphometric parameters for basin geometry, drainage analysis and relief characteristics etc. were calculated and evaluated for each sub-watershed (Table 4.2) based on which, 30 sub-watersheds were characterized in the study area (Fig. 4.5). In terms of basin geometry, there existed significant variation in basin area, perimeter,

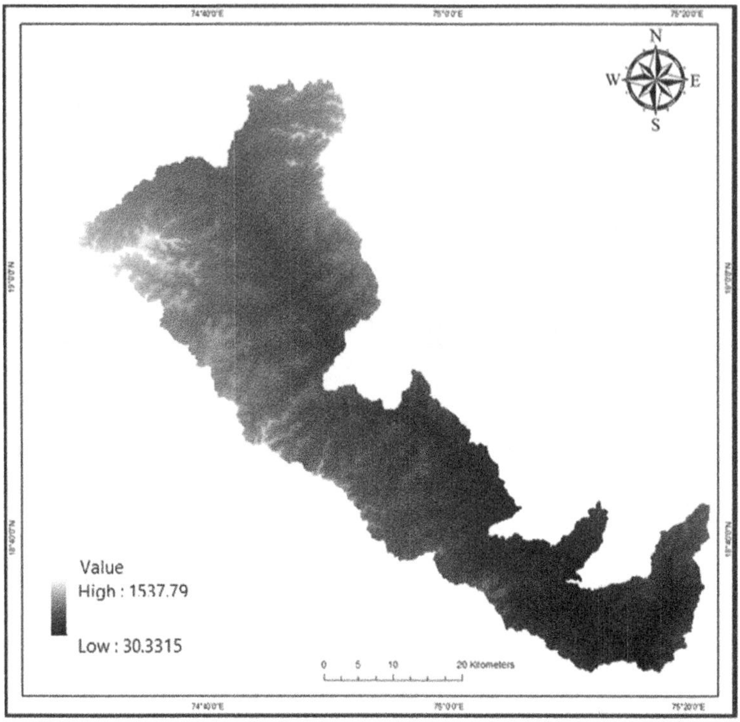

Fig. 4.4 Digital Elevation Map (*Source* Cartosat-2 v(3), 30m Spatial resolution)

basin length, mean bifurcation ratio, circularity ratio and texture ratio. Whereas, there was not much variation in terms of the elongation ratio, form factor ratio and compactness coefficient. Drainage analysis also shows a lot of variation between the sub-watersheds, which are partly dependent on basin geometry parameters. Similarly, relief characteristics were also found to have wide variations with total basin relief ranging from as high as 316 m to as low as 40 m. The relief ratio was also found to vary significantly, having maximum and minimum values as 0.025 and 0.06, respectively.

Relatively higher basin geometry values were obtained for the sub-watersheds 4D7I5-4, 4D7I1-5 and 4D7I1-6, lowest for the sub-watersheds 4D7I4-11, 4D7I5-6 and 4D7I5-3. Higher values of drainage analysis were observed for the sub-watersheds 4D7I5-3, 4D7I1-1 and 4D7I5-6, lowest values are observed in the sub-watersheds 4D7I5-10, 4D7I5-13 and 4D7I1-6. High-relief characteristic values are observed in 4D7I5-2, 4D7I5-8 and 4D7I5-1, lower values are observed in sub-watersheds 4D7I4-11, 4D7I1-4 and 4D7I5-6.

Table 4.2 Sub-watershed wise morphometric parameters

Sub watershed	Basin Geometry									Drainage Analysis			Relief Characteristics				
	Rbm	Lb Km	A km²	P Km	Rf	Re	Rt	Rc	Cc	Fs	Dd Km/km²	Lg	z m	Z m	H m	Rh	Rn
4D7I5-1	1.74	18.73	107.87	88.10	0.01	0.63	1.27	1.22	2.41	2.08	1.38	0.36	586.00	895.00	309.00	0.02	0.47
4D7I5-2	1.57	12.34	51.70	50.52	0.02	0.66	0.93	1.02	2.00	1.76	1.29	0.39	585.00	896.00	311.00	0.03	0.42
4D7I5-3	1.19	5.58	12.77	35.66	0.01	0.72	0.28	0.36	2.84	2.19	1.99	0.25	579.00	650.00	71.00	0.01	0.08
4D7I5-4	1.91	23.50	160.75	98.90	0.02	0.61	1.60	1.63	2.22	1.97	1.40	0.36	579.00	895.00	316.00	0.01	0.45
4D7I5-5	2.18	10.63	39.76	53.14	0.01	0.67	0.75	0.75	2.39	1.94	1.31	0.38	570.00	806.00	236.00	0.02	0.35
4D7I5-6	1.05	6.61	17.25	34.88	0.01	0.71	0.57	0.49	2.39	2.26	1.66	0.30	571.00	625.00	54.00	0.01	0.07
4D7I5-7	1.88	9.66	33.61	51.08	0.01	0.68	0.57	0.66	2.50	1.87	1.39	0.36	563.00	654.00	91.00	0.01	0.12
4D7I5-8	2.09	12.42	52.29	54.28	0.02	0.66	0.96	0.96	2.13	1.91	1.42	0.35	564.00	868.00	304.00	0.02	0.41
4D7I5-9	1.93	14.49	68.58	63.04	0.02	0.65	1.02	1.09	2.16	1.91	1.33	0.38	549.00	823.00	274.00	0.02	0.39
4D7I5-10	1.72	16.84	89.39	91.88	0.01	0.63	0.80	0.97	2.76	1.63	1.24	0.40	548.00	857.00	309.00	0.02	0.41
4D7I5-11	1.95	14.90	72.05	59.72	0.02	0.64	1.21	1.21	2.00	1.87	1.30	0.39	546.00	838.00	292.00	0.02	0.42
4D7I5-12	1.52	15.38	76.25	77.00	0.01	0.64	0.83	0.99	2.51	1.67	1.23	0.41	588.00	895.00	307.00	0.02	0.42
4D7I5-13	2.16	14.55	69.15	69.94	0.01	0.65	0.96	0.99	2.39	1.95	1.31	0.38	549.00	701.00	152.00	0.01	0.23
4D7I4-1	1.96	12.82	55.32	55.12	0.02	0.66	1.09	1.00	2.11	2.13	1.28	0.39	589.00	730.00	141.00	0.01	0.24
4D7I4-2	1.92	17.09	91.80	67.20	0.02	0.63	1.34	1.37	1.99	1.96	1.21	0.41	527.00	670.00	143.00	0.01	0.23
4D7I4-3	1.73	17.81	98.66	87.56	0.01	0.63	1.10	1.13	2.50	1.93	1.23	0.41	524.00	739.00	215.00	0.01	0.34
4D7I4-4	1.75	16.32	84.64	72.84	0.02	0.64	1.11	1.16	2.25	1.89	1.26	0.40	525.00	781.00	256.00	0.02	0.38
4D7I4-5	2.09	18.59	106.37	87.86	0.01	0.63	1.22	1.21	2.42	2.03	1.29	0.39	511.00	766.00	255.00	0.01	0.40
4D7I4-6	1.63	12.70	54.40	55.06	0.02	0.66	0.93	0.99	2.12	1.78	1.32	0.38	510.00	698.00	188.00	0.02	0.26
4D7I4-7	1.44	13.47	60.32	69.04	0.01	0.65	0.90	0.87	2.53	2.12	1.36	0.37	491.00	605.00	114.00	0.01	0.18
4D7I4-8	1.95	14.08	65.23	65.96	0.02	0.65	1.08	0.99	2.32	2.16	1.32	0.38	494.00	672.00	178.00	0.01	0.29
4D7I4-9	1.49	11.17	43.40	52.34	0.02	0.67	0.84	0.83	2.26	2.05	1.24	0.41	482.00	578.00	96.00	0.01	0.16
4D7I4-10	1.83	10.35	37.94	47.82	0.02	0.67	0.80	0.79	2.21	1.98	1.25	0.40	482.00	584.00	102.00	0.01	0.16
4D7I4-11	1.68	4.99	10.51	25.44	0.02	0.73	0.59	0.41	2.23	2.66	1.63	0.31	479.00	519.00	40.00	0.01	0.07
4D7I1-1	4.86	12.39	52.10	57.14	0.02	0.66	0.88	0.91	2.25	1.96	1.44	0.35	472.00	623.00	151.00	0.01	0.21
4D7I1-2	1.97	18.42	104.68	83.90	0.02	0.63	1.04	1.25	2.33	1.65	1.33	0.38	479.00	657.00	178.00	0.01	0.22
4D7I1-3	2.44	18.06	101.15	72.10	0.02	0.63	1.22	1.40	2.04	1.72	1.34	0.38	459.00	623.00	164.00	0.01	0.21
4D7I1-4	2.67	12.33	51.62	58.24	0.02	0.66	0.84	0.89	2.30	1.86	1.28	0.39	454.00	529.00	75.00	0.01	0.11
4D7I1-5	1.91	26.69	201.15	109.04	0.02	0.60	1.78	1.85	2.18	1.92	1.34	0.37	441.00	598.00	157.00	0.01	0.23
4D7I1-6	1.73	23.35	159.00	99.97	0.02	0.6	1.43	1.59	2.25	1.79	1.22	0.41	441.00	600.00	159.00	0.01	0.23

Icon
- ● Value >= 80%
- ◕ Value < 80 and >= 60%
- ◑ Value < 60 and >= 40%
- ◔ Value < 40 and >= 20%
- ○ Value < 20%

Fig. 4.5 Sub-watersheds map of study area

4.8 Analytical Hierarchy Process

The results obtained from the Analytical Hierarchy Process were analyzed in terms of priority index value. Originally ranging from 1.452 to 2.658, these values were normalized from 0 to 1 for further comparisons. Percentage values were derived for final grouping of the sub-watershed into five classes viz. very low (<10% value), low (10–40% value), medium (40–60% value), high (60–90% value) and very high (>90% value) as shown in Table 4.3. It is found that the number of sub-watersheds in low to very low priority is 12 whereas in high-very high priority is 13, thus showing almost equal distribution. The number of sub-watershed in the medium priority category was found to be six. However, it is found that sub-watersheds 4D7I5-4, 4D7I5-8, 4D7I5-3 and 4D7I5-1 are under very high priority and need more concerted efforts for effective planning SWC measures (Figs. 4.6 and 4.7).

Table 4.3 Sub-watershed wise Priority Index (AHP), Normalized value and %value

Sub-watershed	Priority Index	Normalized value	Normalized %	Prioritization category
4D7I1-6	1.452	0	0	Very low
4D7I4-2	1.572	0.0995	9.9502	Very low
4D7I4-9	1.599	0.1219	12.1891	Low
4D7I4-10	1.599	0.1219	12.1891	Low
4D7I4-6	1.728	0.2289	22.8856	Low
4D7I1-4	1.742	0.2405	24.0464	Low
4D7I4-4	1.746	0.2438	24.3781	Low
4D7I4-1	1.771	0.2645	26.4511	Low
4D7I1-5	1.85	0.3300	33.0017	Low
4D7I5-2	1.889	0.3624	36.2355	Low
4D7I1-3	1.902	0.3731	37.3134	Low
4D7I5-12	1.915	0.3839	38.3914	Low
4D7I4-3	1.954	0.4163	41.6252	Medium
4D7I1-2	1.968	0.4279	42.7861	Medium
4D7I5-11	1.993	0.4486	44.8590	Medium
4D7I4-7	2.026	0.4760	47.5954	Medium
4D7I5-7	2.04	0.4876	48.7562	Medium
4D7I5-10	2.052	0.4975	49.7512	Medium
4D7I4-5	2.206	0.6252	62.5207	High
4D7I5-13	2.217	0.6343	63.4328	High
4D7I5-9	2.26	0.6700	66.9983	High
4D7I4-8	2.321	0.7206	72.0564	High
4D7I5-6	2.384	0.7728	77.2803	High
4D7I1-1	2.414	0.7977	79.7678	High
4D7I5-5	2.479	0.8516	85.1575	High
4D7I4-11	2.505	0.8731	87.3134	High
4D7I5-4	2.545	0.9063	90.6302	Very high
4D7I5-8	2.598	0.9502	95.0249	Very high
4D7I5-3	2.601	0.9527	95.2736	Very high
4D7I5-1	2.658	1.0000	100	Very high

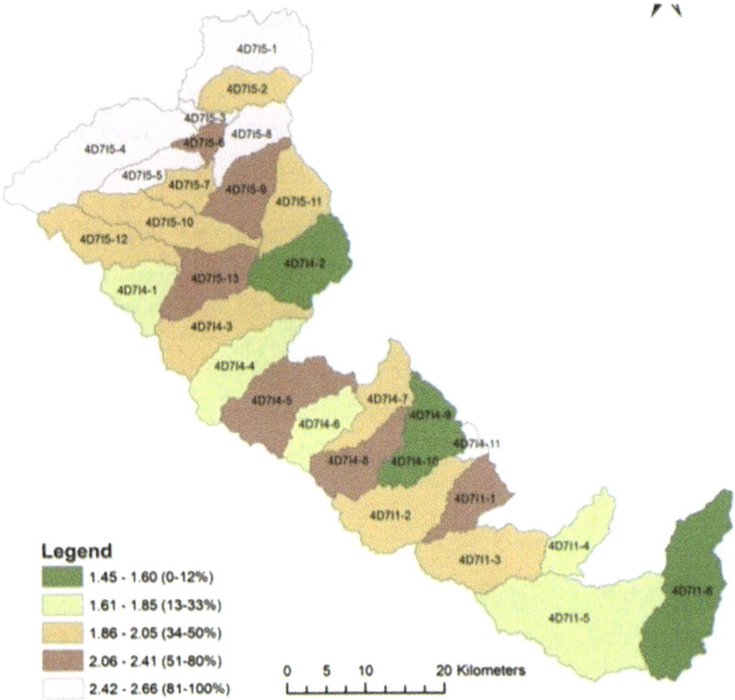

Fig. 4.6 Sub-watershed wise MCDA Analysis (Priority index and Percentage values)—AHP

4.9 Technique for Order Preference by Similarity to an Ideal Solution

The spatial distribution of sub-watersheds based on original priority index values and percentage normalized values along with prioritization classes are shown in Figs. 4.8 and 4.9, respectively. The ranking of the sub-watersheds using TOPSIS resulted in the relative closeness values ranging from 0.215 to 0.56. These values were normalized from 0 to 1 and further percentage values were derived for final grouping of the sub-watersheds into five classes (Table 4.4). It is found that 15 sub-watersheds are in very low to low priority whereas 12 are in the very high priority category. The number of sub-watersheds in the medium priority category has been found to be 4. Very high priority has been obtained only for one sub-watershed 4D7I5-1.

4.10 Soil Erosion Analysis

Soil erosion values for the sub-watersheds as obtained using SWAT model were found to range from 4.75 tonnes/ha/yr in 4D7I1-6 sub-watershed to a maximum value of

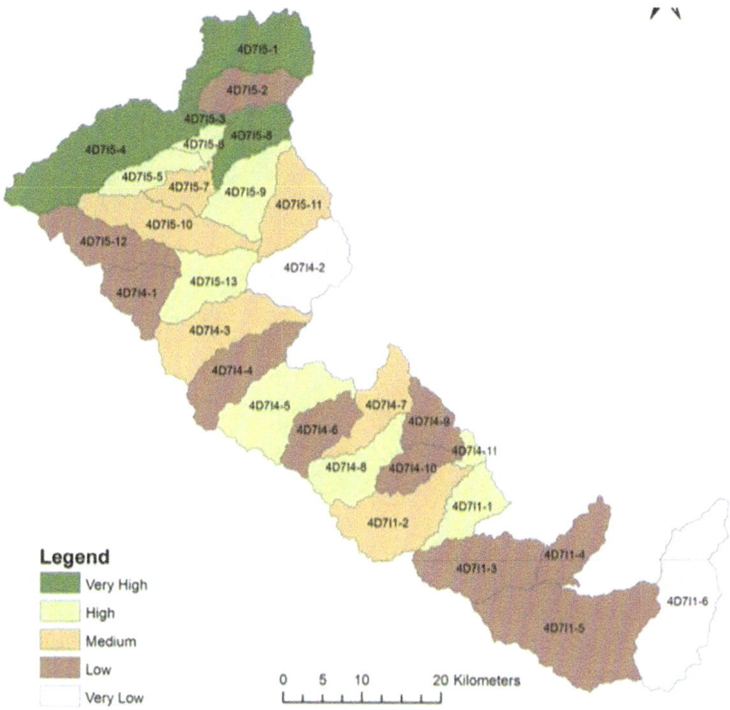

Fig. 4.7 Sub-watershed Prioritization based on AHP

273.619 tonnes/ha/yr in 4D7I5-1; these values are normalized and percentile distribution was obtained for final prioritization of sub-watersheds (Table 4.5, Fig. 4.10) and a sub-watershed prioritization map was obtained (Fig. 4.11). Most of the sub-watersheds were prioritized into medium and high categories. Eight sub-watersheds are found in very low priority, 17 in low priority, 4 in medium and only 1, i.e. 4D7I5-1 in very high priority.

4.11 Surface Runoff Analysis

Surface runoff values for the sub-watersheds ranged from 159.78 mm/yr in 4D7I1-6 subwatershed to 285.73 mm/yr in 4D7I5-1. These values are normalized (Fig. 4.12) and percentile values were calculated (Fig. 4.13) generating for comparative analysis prioritization of the sub-watershed and (Table 4.6). Three sub-watersheds are found in very low priority, 5 in low priority and 14 in the medium priority. Only one watershed, i.e. 4D7I5-1 was found in the very high priority.

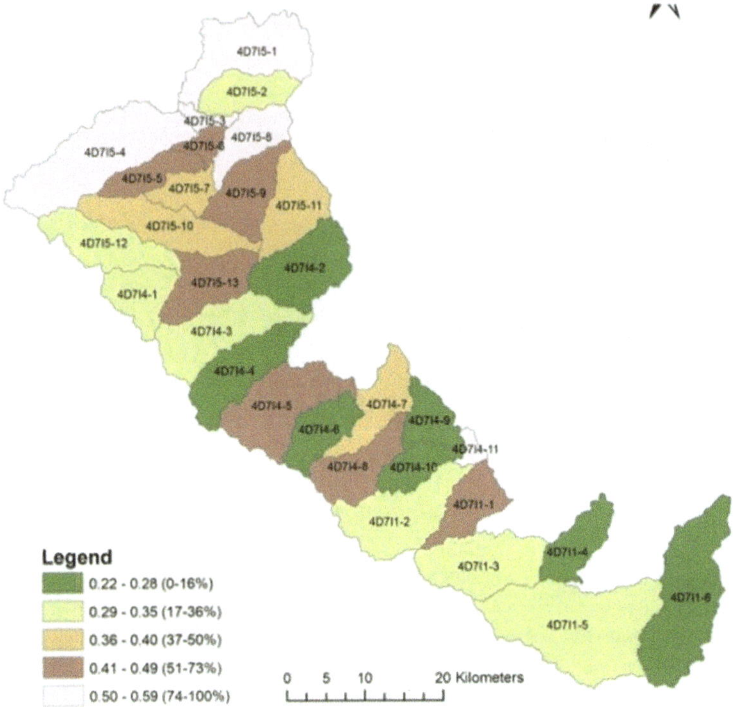

Fig. 4.8 Sub-watershed wise Relative closeness and Percentage values

4.12 Discussions

The discussions below compare the results obtained from the four popularly used methods to understand their effectiveness as demonstrated in the study area.

AHP vis-à-vis TOPSIS

Based on the priority index and relative closeness values, sub-watersheds 4D7I5-1, 4D7I5-4, 4D7I1-1 and 4D7I5-8 were found to have higher priorities in both the cases. The lower priorities are assigned to sub-watersheds 4D7I4-9, 4D7I4-10 and 4D7I1-4. Remaining sub-watersheds were moderately prioritized according to both criteria. However, in case of sub-watersheds, 4D7I5-3, 4D7I4-11 (areas of these two sub-watersheds are very less when compared to other sub-watersheds) show dissimilar priorities in two MCDA approaches implying the need for further investigations. Overall, comparison shows 46.7% match in the AHP and TOPSIS results.

AHP vis-à-vis **SWAT**

The comparison of the sub-watersheds with respect to the priority index and surface runoff and soil erosion results show that the very high and high preference zones of AHP are overrated. In case of medium, low and very low preferences sub-watersheds

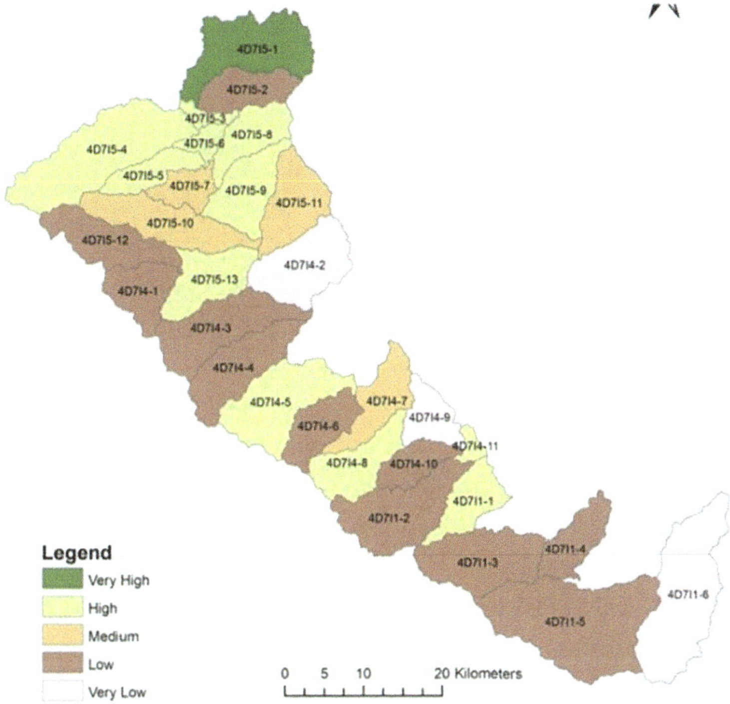

Fig. 4.9 Sub-watershed Prioritization based on TOPSIS

of AHP, we found high uncertainty in terms of the corresponding surface runoff preferences. The comparison of soil erosion preferences obtained from SWAT and AHP priority index ranges shows matching prioritization value in 33% sub-watersheds.

Table 4.4 Sub-watershed wise Relative Index Values (TOPSIS), Normalized value and %value

Sub watershed	Relative Closeness (Ci)	Normalized value	Normalized %	Prioritization category
4D7I1-6	0.215	0.0000	0.0000	Very low
4D7I4-2	0.237	0.0593	5.9299	Very low
4D7I4-9	0.25	0.0943	9.4340	Very low
4D7I4-10	0.256	0.1105	11.0512	Low
4D7I4-6	0.258	0.1159	11.5903	Low
4D7I1-4	0.264	0.1321	13.2075	Low
4D7I4-4	0.276	0.1644	16.4420	Low
4D7I4-1	0.313	0.2642	26.4151	Low
4D7I1-5	0.324	0.2938	29.3801	Low
4D7I5-2	0.33	0.3100	30.9973	Low
4D7I1-3	0.341	0.3396	33.9623	Low
4D7I5-12	0.341	0.3396	33.9623	Low
4D7I4-3	0.347	0.3558	35.5795	Low
4D7I1-2	0.348	0.3585	35.8491	Low
4D7I5-11	0.368	0.4124	41.2399	Medium
4D7I4-7	0.371	0.4205	42.0485	Medium
4D7I5-7	0.395	0.4852	48.5175	Medium
4D7I5-10	0.401	0.5013	50.1348	Medium
4D7I4-5	0.448	0.6280	62.8032	High
4D7I5-13	0.455	0.6469	64.6900	High
4D7I5-9	0.457	0.6523	65.2291	High
4D7I4-8	0.48	0.7143	71.4286	High
4D7I5-6	0.482	0.7197	71.9677	High
4D7I1-1	0.484	0.7251	72.5067	High
4D7I5-5	0.486	0.7305	73.0458	High
4D7I4-11	0.502	0.7736	77.3585	High
4D7I5-4	0.506	0.7844	78.4367	High
4D7I5-8	0.506	0.7844	78.4367	High
4D7I5-3	0.525	0.8356	83.5580	High
4D7I5-1	0.586	1	100	Very high

TOPSIS vis-à-vis **SWAT**

The evaluation of the sub-watersheds in terms of TOPSIS and SWAT preferences resulted in comparable values in 56.7% sub-watersheds. The very high and high preferences sub-watersheds assigned by TOPSIS were also assigned to high priority

Table 4.5 SWAT-based Soil Erosion Values

Sub-watershed	Soil Erosion (tonnes/ha/yr)	Normalized value	Normalized %	Prioritization category
4D7I1-6	4.78	0.00	0.00	Very low
4D7I4-2	5.06	0.00	0.11	Very low
4D7I4-9	13.30	0.03	3.17	Very low
4D7I4-10	18.20	0.05	5.00	Very low
4D7I4-6	19.16	0.05	5.35	Very low
4D7I1-4	20.00	0.06	5.66	Very low
4D7I4-4	21.11	0.06	6.08	Very low
4D7I4-1	28.14	0.09	8.69	Very low
4D7I1-5	38.99	0.13	12.73	Low
4D7I5-2	40.87	0.13	13.43	Low
4D7I1-3	52.27	0.18	17.67	Low
4D7I5-12	54.26	0.18	18.41	Low
4D7I4-3	54.80	0.19	18.61	Low
4D7I1-2	59.63	0.20	20.40	Low
4D7I5-11	66.32	0.23	22.89	Low
4D7I4-7	67.16	0.23	23.20	Low
4D7I5-7	70.26	0.24	24.36	Low
4D7I5-10	72.71	0.25	25.27	Low
4D7I4-5	83.33	0.29	29.22	Low
4D7I5-13	88.09	0.31	30.99	Low
4D7I5-9	88.61	0.31	31.18	Low
4D7I4-8	93.74	0.33	33.09	Low
4D7I5-6	97.55	0.35	34.51	Low
4D7I1-1	101.18	0.36	35.86	Low
4D7I5-5	110.74	0.39	39.42	Low
4D7I4-11	114.64	0.41	40.87	Medium
4D7I5-4	116.23	0.41	41.46	Medium
4D7I5-8	152.61	0.55	54.99	Medium
4D7I5-3	189.29	0.69	68.63	Medium
4D7I5-1	273.62	1.00	100	Very high

in the SWAT-based surface runoff preferences. The overall similarity in the prioritization of sub-watersheds using SWAT and TOPSIS is 40%, which indicates slightly better matching as compared to that of AHP.

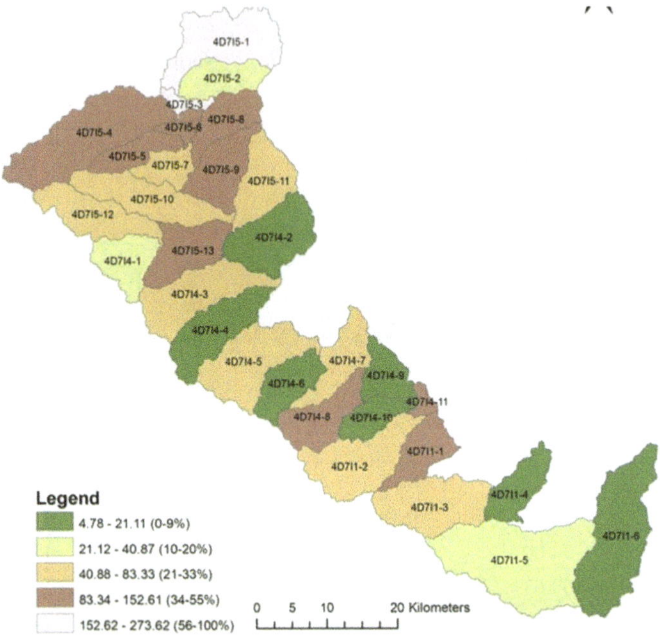

Fig. 4.10 Sub-watershed wise Soil Erosion values and percentage values

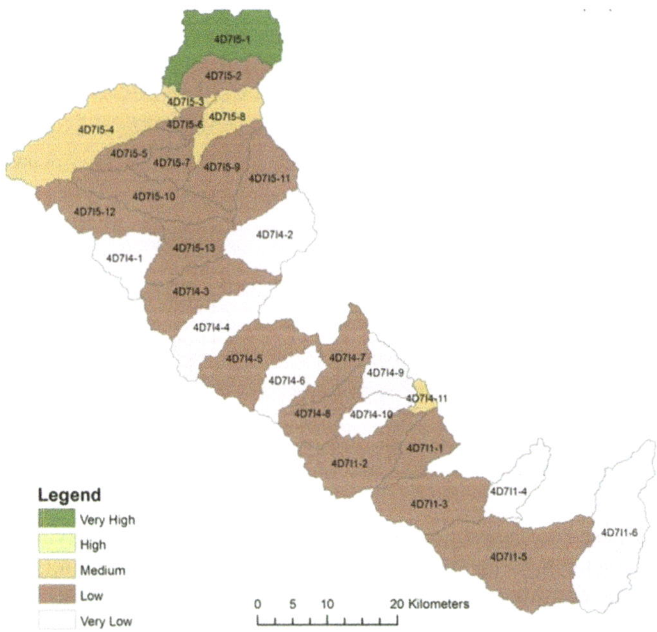

Fig. 4.11 Sub-watershed Prioritization based on Soil erosion

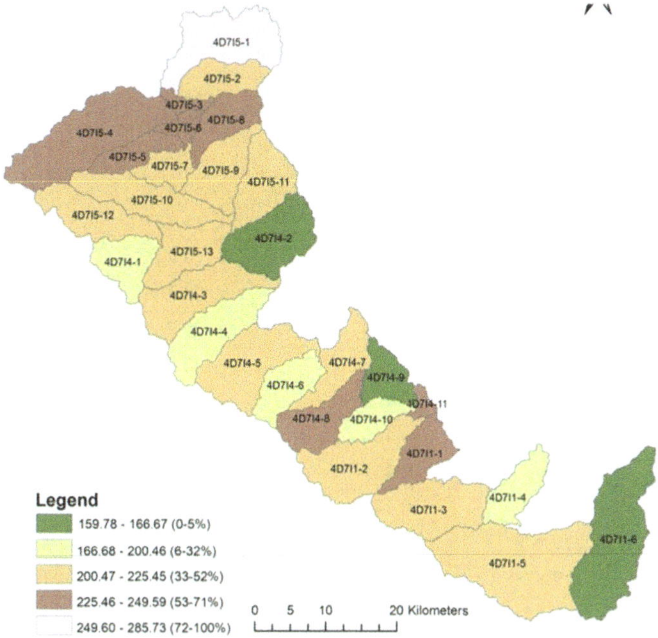

Fig. 4.12 Sub-watershed wise Surface runoff values and percentage values

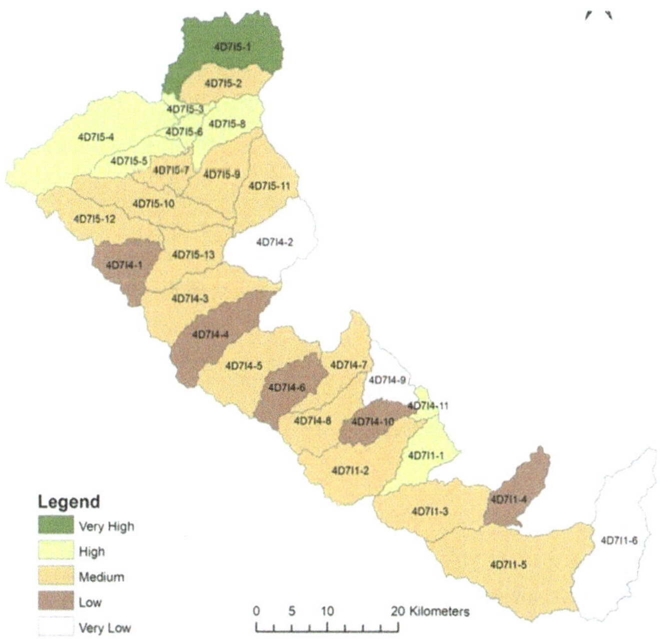

Fig. 4.13 Sub-watershed Prioritization based on Surface runoff

Table 4.6 SWAT-based Run-off Values

Sub-watershed	Surface Runoff (mm/yr)	Normalized value	Normalized %	Prioritization category
4D7I1-6	159.784	0.000	0.000	Very low
4D7I4-2	160.989	0.010	0.957	Very low
4D7I4-9	166.67	0.055	5.467	Very low
4D7I4-10	179.163	0.154	15.387	Very low
4D7I4-6	184.78	0.198	19.846	Very low
4D7I1-4	189.007	0.232	23.203	Very low
4D7I4-4	197.277	0.298	29.769	Very low
4D7I4-1	200.46	0.323	32.296	Very low
4D7I1-5	211.547	0.411	41.099	Low
4D7I5-2	212.138	0.416	41.568	Low
4D7I1-3	214.03	0.431	43.070	Low
4D7I5-12	216.347	0.449	44.910	Low
4D7I4-3	217.408	0.458	45.753	Low
4D7I1-2	217.878	0.461	46.126	Low
4D7I5-11	218.781	0.468	46.843	Low
4D7I4-7	219.212	0.472	47.185	Low
4D7I5-7	222.153	0.495	49.520	Low
4D7I5-10	222.262	0.496	49.607	Low
4D7I4-5	222.338	0.497	49.667	Low
4D7I5-13	224.531	0.514	51.408	Low
4D7I5-9	225.454	0.521	52.141	Low
4D7I4-8	234.631	0.594	59.427	Low
4D7I5-6	236.956	0.613	61.273	Low
4D7I1-1	243.152	0.662	66.193	Low
4D7I5-5	244.269	0.671	67.080	Low
4D7I4-11	246.472	0.688	68.829	Medium
4D7I5-4	247.557	0.697	69.690	Medium
4D7I5-8	247.742	0.698	69.837	Medium
4D7I5-3	249.585	0.713	71.301	Medium
4D7I5-1	285.731	1.000	100.000	Very high

4.13 Conclusion

Morphometric plays an important role in basin-level construction. The different morphometric parameters of a watershed are the best representative of underlying geology, geomorphology, relief, slope, climate as well as hydrological dynamics. Thus they are widely used for the watershed level analysis using MCDA and will continue to be so. However, hydrological models are found to be more useful for local-level planning of soil and water conservation as they also consider the existing land use land cover practices and the climatic components, therefore having much impact on behaviour of the watershed. Results also show the clear indication of surface runoff and sediment yield independent of each other. Accordingly, the soil and water conservation measures in an area should be based on hydrological analysis.

References

Chowdary VM, Chakraborthy D, Jeyaram A, Murthy YK, Sharma JR, Dadhwal VK (2013) Multi-criteria decision making approach for watershed prioritization using analytic hierarchy process technique and GIS. Water Resour Manage 27(10):3555–3571

Chung ES, Lee KS (2009) Identification of spatial ranking of hydrological vulnerability using multi-criteria decision making techniques: case study of Korea. Water Resour Manag 23(12):2395–2416

Gassman PW, Arnold JJ, Srinivasan R, Reyes M (2010) The worldwide use of the SWAT Model: Technological drivers, networking impacts, and simulation trends. In: 21st century watershed technology: improving water quality and environment conference proceedings, 21–24 February 2010, Universidad EARTH, Costa Rica. American Society of Agricultural and Biological Engineers, p 1

Gopinath G, Nair AG, Ambili GK, Swetha TV (2016) Watershed prioritization based on morphometric analysis coupled with multi criteria decision making. Arab J Geosci 9(2):129

Hwang CL, Yoon K (1981) Methods for multiple attribute decision making. In: Multiple attribute decision making. Springer, Berlin, Heidelberg, pp 58–191

Jaiswal RK, Ghosh NC, Galkate RV, Thomas T (2015) Multi criteria decision analysis (MCDA) for watershed prioritization. In: Winchell M, Srinivasan R, Di Luzio M, Arnold JG (2010) ArcSWAT Interface for SWAT 2009—User's Guide. GSWR Agricultural Research Service & Texas Agricultural Experiment Station, Temple, TX

Javed A, Khanday MY, Ahmed R (2009) Prioritization of sub-watersheds based on morphometric and land use analysis using remote sensing and GIS techniques. J Indian Soc Remote Sens 37(2):261

Jenson SK (1991) Applications of hydrologic information automatically extracted from digital elevation models. Hydrol Process 5(1):31–44

Julius JJ, Singh D, Brema J (2017, October) Morphometric analysis of linear and areal aspects of bhagirathi river basin—A GIS approach. In: 2017 2nd International Conference on Communication and Electronics Systems (ICCES). IEEE, pp. 549–554

Li J, Wong DW (2010) Effects of DEM sources on hydrologic applications. Comput Environ Urban Syst 34(3):251–261

McMaster KJ (2002) Effects of digital elevation model resolution on derived stream network positions. Water Resour Res 38(4):13–1

Parupalli S, Padma Kumari K, Ganapuram S (2019) Assessment and planning for integrated river basin management using remote sensing, SWAT model and morphometric analysis (case study: Kaddam river basin, India). Geocarto Int 34(12):1332–1362

Patel DP, Nandhakumar N (2016) Runoff potential estimation of Anjana Khadi Watershed using SWAT model in the part of lower Tapi Basin, West India. Sustain Water Resour Manag 2(1):103–118

Saaty T (1980) The analytic hierarchy process (AHP) for decision making. Kobe, Japan

Stec A, Zeleňáková M (2019) An analysis of the effectiveness of two rainwater harvesting systems located in Central Eastern Europe. Water 11(3):458

UN-Water and FAO (2007) Coping with water scarcity: challenge of the 21st century. UN-Water and FAO, Rome

Chapter 5
Efficiency Assessment of Sewage Treatment Plants for Treating Wastewater: Case Study of Narela Sewage Treatment Plants in Delhi, India

Shikha Yadav, Subhash Anand, Paramjit, and Jagbir Singh

Abstract Water pollution occurred due to many agents, discharge of wastewater is one of the large amounts of untreated sewage into the water bodies, which is due to 28 k of treatment capacity and ill maintenance of sewage treatment plants (STPs). Delhi city is the second-largest contributor of wastewater in the country after the state of Maharashtra. The extent of the wastewater treatment capacity available in Delhi with an available treatment plant varies from 103.18% (Sen Nursing Home Nala STP), 102.03% (Delhi Gate STP) and 73.88% (Rithala STP) to 17.50% (Narela STP) of the sewage generated. This micro-level, field survey study through an unstructured questionnaire analyses and tries to identify major problems, potentiality gap, flaws and suggests remedies for Narela STP in Delhi. The study observes that this plant lacks commercially viable technology due to large capital costs and complex maintenance process and also lack of adequate skilled manpower. According to the study, to enhance the efficiency of the sewage treatment plant, more resources need to be allocated for research and development along with better training of manpower.

Keywords Wastewater · Efficiency · Capacity · Assessment · Sewage treatment plant · Narela

---

S. Yadav (✉) · S. Anand
Department of Geography, Delhi School of Economics, University of Delhi, Delhi 110007, India
e-mail: shikhayadav356@gmail.com

S. Anand
e-mail: sanandpv@yahoo.co.in

Paramjit
Department of Economics, Delhi School of Economics, University of Delhi, Delhi 110007, India
e-mail: jit@econdse.org

J. Singh
Department of Geography, SSN College, University of Delhi, Alipur, Delhi 110036, India
e-mail: drjagbirs@gmail.com

R. K. Mishra et al. (eds.), *Sustainable Climate Action and Water Management*,
Advances in Geographical and Environmental Sciences,
https://doi.org/10.1007/978-981-15-8237-0_5

5.1 Introduction

The population increase and growing urbanization have been the driving force for human migration from countryside to the cities in India. Due to urbanization, a number of environmental issues and challenges are coming up, which are negatively affecting our natural resources. One such resource is freshwater, which is constantly under threat. To improve this situation of water stock, authorities decided to improve the water supply system of cities, towns as well as villages through the construction of water pipelines, well-managed sanitation structure and sewage treatment plants. Global sustainable development challenges concentrate in cities (Singh 2020) but these cities are not only sites of problems but also sites of solutions. Since the last few decades, demand for the water has been considerably increasing and side by side the water supply significantly augmented. Wastewater is used water, comprising of substances such as human waste, food scraps, oils, soaps, chemicals, domestic waste, etc. Today, several industrial plants and commercial establishments are significantly contributing to the amount of wastewater. Unfortunately, due to the absence of resources such as skilled manpower, capital, etc., sewerage or improved sanitation got minimal attention. Businesses and industries also contribute their share wastewater (Kumar and Khursheed 2010). Use of huge quantity of water leads to the generation of wastewater in the absence of its proper treatment, which further leads in the generation of tremendous amount of wastewater. Out of the total world freshwater, the country of India has 4% (Sheel 2016), but there is dearth of research focusing on water scarcity in India (Bhattacharyya 2020). Wastewater management plans do not address this increasing pace of wastewater generation (Chakraborty and Mukhopadhyay 2014). Delhi, which is already facing shortage of water, is going to face worse in near future (Rai and Nagpal 2017). In Delhi, the overall wastewater problem is tremendous with an estimated 1800 million litres per day being disposed of without treatment in the Yamuna River (Zimmer 2014).In comparison to urban areas, rural areas do not discharge much amount of sewage because of low per capita supply that is why urban areas matter a lot. Government of NCT of Delhi identified water supply and sanitation as priority section (Kumar 2020). The present paper aims to attempt to analyse the efficiency and capacity of Narela sewage treatment plant in Delhi.

5.2 Study Area

Delhi has an area of 1,483 square kilometres (Anand 2010) and support 16.75 million human population (Singh et al. 2018) and its maximum length is 51.90 km and the greatest width is 48.48 km. Two prominent features of the geography of Delhi are the Yamuna flood plains and the Delhi ridge. Study area chose present research paper is Narela in Delhi. It is located in the northern plain of India with latitudes from 28°24′17 north to 28°53′00″ north and east longitudes from 76° 50′24″ to 77° 20′37″

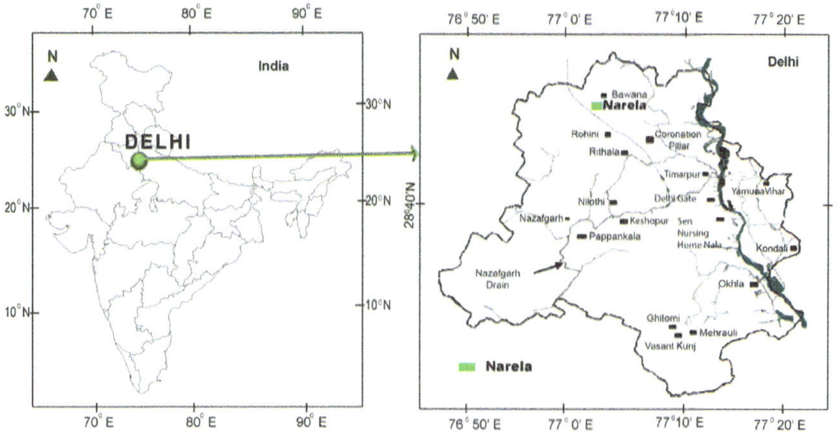

Fig. 5.1 Sewage Treatment Plants in Delhi. *Source* Compiled by Authors based on CSE, 2017

east. The Narela is a zone in North Delhi Municipal Corporation (MCD 2020) and sub-city is a tehsil, which is situated in the North Delhi district of NCT of Delhi (Fig. 5.1). It is located just off the Grand Trunk (GT) Road, its locations made it a significant market town for the encompassing areas, during the nineteenth century. It was developed as the third mega sub-city project undertaking of Delhi Development Authority (DDA) in the urban expansion project of Delhi, after Rohini Sub-city and Dwarka Sub-city. The total area of the zone is 9866 hectares of which urbanizable zone is 7365 hectares. The remaining 2501 hectares is under green belt towards Northern border of NCT of Delhi according to the Master plan Delhi-2021. The 'Narela Industrial Area', started developing in the mid-1980s, is today significant with complexes/buildings. Narela is separated into different colonies such as Rajeev Colony, Naibasti, Punjabi Colony, Gautam Colony, Swatantra Nagar, Shivaji Colony, Master Colony, Sanjay Colony, Indra Colony, Police Colony, etc.

5.3 Data Sources and Methodology

The major objective of this chapter is to analyze the problems and efficiency gap in Narela STP. In order to fulfil the objective, primary and secondary sources of data have been collected. As a primary data source, personal interviews were held with the officials at the Narela treatment plant. To capture the nuanced, inherited issues, unstructured questionnaires were used in the field and an extensive discussion was done with the executive engineers (2), junior engineers (2) and laboratory analysts (4) during the month of December 2019 to bring out the major shortcomings. Secondary data have been obtained from a number of sources such as Delhi Jal Board (DJB) reports and Economic Survey of Department of Planning, NCT Delhi. In Delhi, 20 sewage treatment plants are constructed with different capacities depending on different criteria. Out of these 20 STPs, Narela STP is chosen using purposive sampling; the focus of the study is primarily based on the analysis of the efficiency and capacity of Narela STP. In Narela STP, the lowest treatment capacity is located in Singhola Village near to Narelain North Delhi. The sewage treatment capacity of Narela sewage treatment plant is 10 MGD.

5.4 Results and Discussion

As per the United Nations, the population of Delhi will be about 43 million by the year 2035. Decadal water demand and net water supply in Delhi are given in the figure below. According to the figure, it can be seen that over the decades, all three measures, i.e., water requirement, sewage generation and net water supply, have been increasing. However, the increase is greater from 2004 to 2011 than from 2011 to 2021 for water requirement and sewage generation, whereas the increase in net water supply has been greater in the later decades (from 2011 to 2021) rather than the former (from 2004 to 2011) (Fig. 5.2). It can be seen that there are persistent gaps in the water requirement and net water supply in Delhi. In such scenario, wastewater management becomes crucial to meet the rising demand of the population in Delhi.

The Delhi Jal Board (DJB) is the organization that is responsible for the supply of potable water to a large part of Delhi. Delhi Jal Board is also providing the services of treatment and disposal of wastewater. According to Delhi Jal Board, about 60% of the total generated wastewater is treated and the rest is discharged into either different drains or the Yamuna River. Presently, about 1349 million litres per day of wastewater generated in Delhi is treated by sewage treatment plants and the rest of the wastewater is being discharged into the drains without any treatment or making any use. However, the capacity of sewage treatment in Delhi has been increased from 402.40 million gallons per day (MGD) in March 2001 to 607.26 million gallons per day in March 2018. Thus, during a period of 17 years, 205 million gallons per day capacity has been increased. Among the 20 sewage treatment plants, maximum capacity has been recorded at Okhla STP followed by Rithala, Keshopur and Kondli

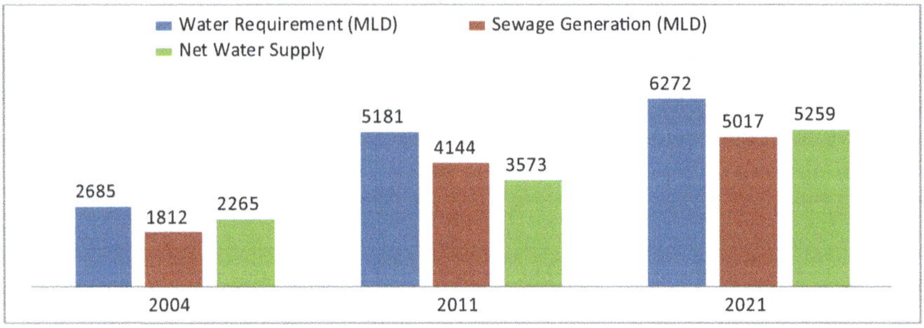

Fig. 5.2 Decadal Water Requirements in Delhi. *Source* Economic Survey of Delhi, 2018–19

I, II, III and IV in March 2018. It has been found that 166 MGD is the difference between the capacity and treatment of sewage treatment plants (Table 5.1).

The various sewerage zones served a varied number of population in their area. The population served by each of the sewerage zones is given in Fig. 5.3. It is clear that the highest population was served by Okhla Sewerage Zone followed by Trans Yamuna, Keshopur and so on and the least by Coronation Pillar at 7.72 lakhs (Table 5.2). However, the proportion of population served in relation to the total population remains the same at 75% for all the Sewerage Zones.

Table 5.1 Sewage Treatment Plants (STPs) in Delhi **(in MGD)**

S. No.	Name of Sewerage Treatment Plants (STPs)	Capacity till 31.3.2001	Capacity till 31.3.2018	Actual Treatment till 31.3.2018	% of Utilisation
1	Okhla	140.00	140.00	101.76	72.68
2	Keshopur	72.00	72.00	64.36	89.38
3	Coronation Pillar with Oxidation Ponds at Timarpur	46.00	30.00	22.33	74.43
4	Rithala	40.00	80.00	58.71	73.38
5	Kondli I, II, III, IV	45.00	70.00	63.01	90.01
6	Yamuna Vihar I, II	10.00	35.00	17.77	50.77
7	Vasant Kunj	5.00	5.20	2.66	51.15
8	Ghitorni	5.00	5.00	00	00
9	Pappankalan	20.00	40.00	33.81	84.52
10	Narela	10.00	10.00	1.75	17.50
11	Najafgarh	5.00	5.00	2.10	42
12	Delhi Gate	2.20	17.20	17.55	102.03
13	Sen Nursing Home Nala	2.20	2.20	2.27	103.18
14	Rohini	–	15.00	5.37	35.80
15	Nilothi	–	60.00	32.36	53.93
16	Mehrauli	–	5.00	3.48	69.60
17	CWG Village	–	1.00	0.18	18
18	Molarbad	–	0.66	0.54	81.81
19	Kapeshera	–	5.00	1.92	38.40
20	Chilla	–	9.00	9.07	100.77
	Total	**402.40**	**607.26**	**441**	**72.62**

Source Economic Survey of Delhi, 2018–19

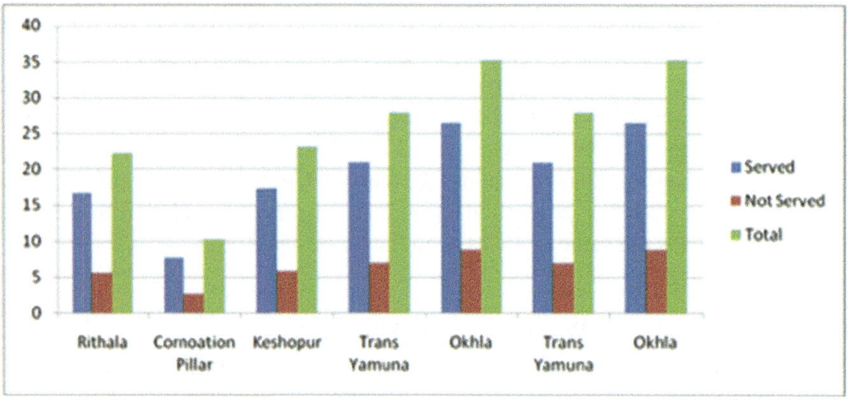

Fig. 5.3 Served and Not Served Population by Different Sewage Treatment Plants (Population in Lakhs)

Table 5.2 Status of Population Coverage in Different Sewerage Zones

S. No.	Sewerage zone	Population (Lakhs)		Total
		Served	Not served	
1.	Rithala	16.70	5.56	22.26
2.	Coronation Pillar	7.72	2.57	10.29
3.	Keshopur	17.39	5.78	23.14
4.	Trans Yamuna	20.99	6.99	27.98
5.	Okhla	26.51	8.84	35.35
6.	**Total**	**89.24**	**29.74**	**119.02**

Source Delhi Jal Board, 2017

5.4.1 Locational Aspects

Narela STP is located in North Delhi with medium density of population living in unauthorised colonies with minimal sewerage facility. The population is a crucial factor, which affects the consumption of water and generation of wastewater. Another factor can be a sewerage system of Delhi, which made various categories as per DJB;

(i) Planned colonies: about 50% of the population live in planned colonies are fully sewered;
(ii) Unauthorised/regularised colonies;
(iii) Resettlement colonies and
(iv) Urban villages.

Rural villages, unauthorised colonies and JJ Cluster (City Sewerage System 2006). The sewerage facility is provided in planned colonies unauthorised/regularised colonies, JJ Resettlement colonies and urban villages, but sewer facilities lack in rural area, unauthorised colonies, slums and JJ clusters (CPCB 2004).

5.4.2 Understanding Efficiency Determination of STP

Efficiency of STPs depends on many factors like biological oxygen demand (BOD), chemical oxygen demand (COD), TSS, FC and turbidity removal, and STPs exhibit different physical, chemical and microbiological efficiencies depending upon characteristics of influent sewage. There is a need to define one common parameter, which could determine the total efficiency of plant in terms of physical, bio-chemical and micro-biological removal efficiencies. The parameter will also help in taking a decision for efficient reuse of effluents. Colmenajero et al. (2006) determined the general efficiency indicator to compare overall performances of the different plants.

General efficiency was an average of TSS, COD, BOD and ammonia removal efficiencies.

$$\mathbf{IEa = 1/3\,[E_{TUR} + E_{BOD} + E_{Fc}]}$$

Where **IEa** is the actual integrated efficiency in (%),

E_{TUR} is average efficiency of turbidity removal (%),

E_{BOD} is average efficiency of BOD removal (%),

E_{FC} is average efficiency of FC removal (%) and

IEs is the standard integrated efficiency.

The IEa was found to be lesser than IEs for Narela STP, as indicated in Table 5.3. There is a need for tertiary treatment for effluent from Narela STP so that its treated water may be utilized in a better way for horticulture, agriculture, construction, irrigation, industries for cooling machine etc. In Narela STP, employing conventional ASP, BOD removal efficiencies were measured at 84%. The treatment capacity of Narela STP remained the same at 10 MGD during the period of 17 years (2001–2018), as shown in Fig. 5.4.

In order to evaluate the performance and identify the main problems, technical and administrative parameters were used. Based on the findings, recommendation measures were proposed for the same. This paper assesses the performance of the sewage treatment plant and brings forth the reasons for technology, location aspects, maintenance and cost, operation and management staff and design and construction.

Table 5.3 Comparison of Efficiencies: Actual and Standard of Narela STP

STP	Actual Integrated Efficiency (IEa)	Standard Integrated Efficiency (IEs)
Narela	80.88	93.50

Source Delhi Jal Board, 2017

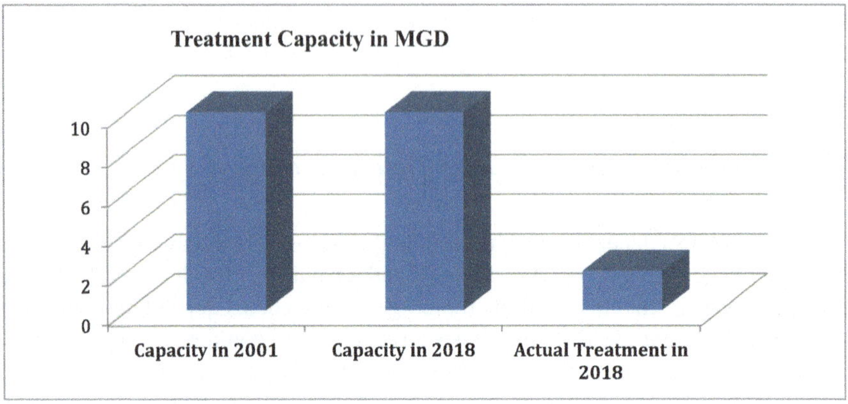

Fig. 5.4 Capacity and Actual Treatment in Narela STP. *Source* Economic Survey of Delhi, 2018–19

5.4.3 Operation and Management

The authorities made an extensive investment in construction and establishment of sewage treatment plants but not on the operational and maintenance aspects and consequently (Sinha and Nazimuddin 2008). A sewage treatment plant once constructed needs to have operation and maintenance through qualified and experienced personnel to serve efficient and effective operation of STP, qualified operator and supervisor are being provided by Delhi Jal Board, which maintains all the records of the operation and attends and records day to day problems. The manpower requirement for STP operation and maintenance can be classified into plant manager, plant operator and laboratory analyst. According to the field survey at Narela STP, only 15 labourers, 1 junior engineer, 1 executive engineer and 1 assistant engineer were employed, While Suez India Pvt. Ltd., a France based company, privately managing Sen Nursing Home Nala STP, Delhi Gate STP and Okhla STP, hired their own staff who works in day and night shifts, handling STPs.

5.4.4 Technological Aspects

Depending on the type of technology adopted, it is possible to profitably reuse the effluents from the STP. The high microbiological quality of wastewater and sludge permit them to be used for crop irrigation or cultivation of vegetables (Guidelines of Sewage Treatment Plant 2006). With 10 MGD sewerage capacity, Narela STP was constructed in the year 2003. The process of treatment is called SAF (Submerged Aerated Filters) technology and is used to reduce the organic loads of domestic and commercial wastewater, which in turn decreases biological oxygen demand and a

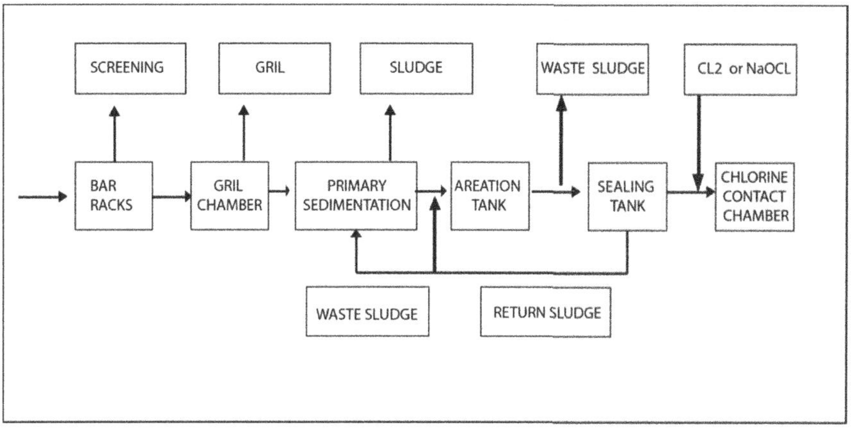

Fig. 5.5 Flow Diagram (ASP) Activated Sludge Process at Narela STP. *Source* Complied/Designed by the Authors from Feedback Received during the Field Survey

significant quantity of TSS, while the Chlorination is being done for the removal of faecal coliform (FC).

In this case study, the activated sludge process used at Narela STP is a type of wastewater treatment process for treating sewage or industrial wastewaters using aeration and a biological stuff composed of bacteria and protozoa (Fig. 5.5).

5.5 Post-treatment Water Quality

It is a matter of fact the ultimate goal of a STP is to treat the sewage water and convert it into a usable form through different processes of removing contaminants from the wastewater. Besides that, another major purpose is to produce environmentally safe sewage water, called effluent, and solid waste, called sludge or bio-solids, suitable for disposal or reuse. Reuse is often for agricultural purposes, but more recently, sludge is used as a fuel source. There are various water quality parameters, such as, BOD, COD, TSS, PH, etc., which are depicted in the following tables for the Narela STP (Table 5.4).

At Narela sewage treatment plant, all the parameters reduced after treatment of sewage. Biological oxygen demand decreased by 93%, whereas chemical oxygen demand by 57% and TSS by 87%. Here too, there is a marginal reduction in pH value of the treated water (Fig. 5.6).

From the survey, it comes out that the overall performance of Narela STP is not adequate. Various problems came to light, after an extensive discussion with operation and management staff. This plant is based on conventional technology called Activated Sludge Process (ASP), which is cost-effective but is outdated with distressing results. A major cause of low utilisation capacity (17%) of this plant is

Table 5.4 Narela Sewage Treatment Plant Performance in December, 2019

Total flow	1.75 MGD		
	Raw sewage parameters (Inlets)	Treated sewage parameters (Outlets)	Reduction after treatment
BOD (Mg/l)	200	15	93%
COD (Mg/l)	350	150	57%
TSS (Mg/l)	220	28	87%
pH	7.40	7.25	2%

Source Field Survey, 2019

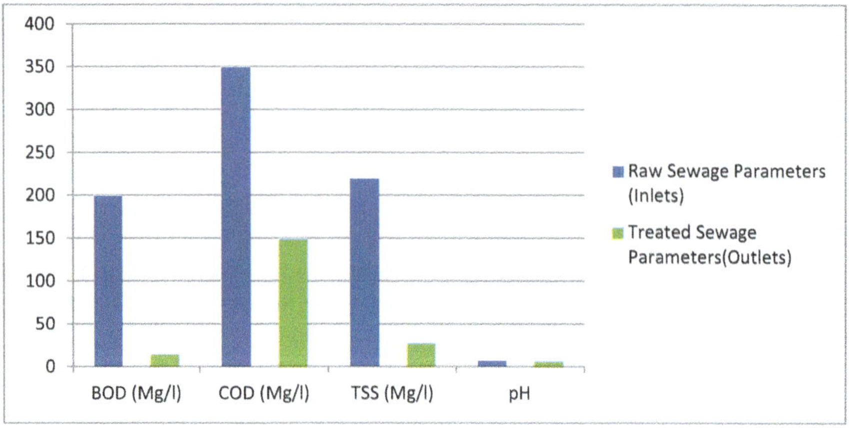

Fig. 5.6 Raw sewage parameters and treated sewage parameters for Narela STP. *Source* Compiled by authors based on fieldwork, 2019

non-availability of sewage water, which is due to lack of interceptors and pump houses in Narela region; once wastewater is treated, it is discharged into the supplementary drain, except 30% of treated wastewater that is sent for horticulture, construction, irrigation of fields etc. Additionally, waste stabilisation does not exist here viz, the treated effluents are not used to produce biogas, bio-energy, bio-fertilizers, etc.

It was found that everyday status report with regards to condition and functioning of various parts, units, equipment etc. is not prepared and maintained properly at site. Operation and management of the plant are not satisfactory due to poor housekeeping as well as poor preventive maintenance (Table 5.5). At the site, some very crucial issues were found. More trained and skilled human resource is required at Narela STP.

Table 5.5 A Snapshot of Narela STP

Narela STP
Area:8000 square metres
Total Capacity: 10 MGD
Sources of Sewage Water: Narela city, Tigri village, Supplementary Drain No:6, Police Colony, DDA colony, Industrial Area etc.
Technology: Activated Sludge Process, Conventional Technology for Primary, Secondary and Tertiary Treatment
Usage of Treated Wastewater: Horticulture, agriculture, construction, irrigation, industries for cooling machines, etc.
Byproducts: None
Management: Government-owned (DELHI JAL BOARD) STP
Utilization: Low (17%), because with low population density, there is a lack of interceptors and pump houses; low flow, open drains, lack of sludge pipes, non-availability of water etc.
Human Resources: Lack of skilled labourers; only 10–12 men; 1 junior engineer, 1 assistant engineer and 1 executive engineer
Problems: 1. Underutilised due to non-availability of sewage water. 2. Lack of skilled labourers 3. Obsolete technology 4. Water quality not satisfactory

Source Field Survey, December, 2019

5.6 Conclusion

The Narela sewage treatment plant lags in the matters of better technology, skilled human resources and better management of sludge parameters. The sludge remains of in Sen Nursing Home Nala STP, Okhla STP and Delhi Gate STP are used for production of bio-gas, bio-fuel, electricity and bio-land fillings, but the treated water of Narela STP is not of potable standards. At Narela STP, the treated sewage water quality is not at good standard. To rectify this, it requires technological upgradation and more trained workers. Further, the better catchment of sewage water is necessary to fully utilize the sewage treatment capacity of Narela STP. The establishment of a sewage treatment plant needs a huge investment and updated technology together with skilled manpower. Therefore, it becomes immensely important for the authorities to spend on the operation and maintenance of treatment plants; otherwise the entire practice becomes futile. Keeping in mind the utility of wastewater, Narela sewage treatment plant should be operated, functional and updated with the good working style, full capacity and state of art technology. Wastewater generated from the northern part of Delhi should be properly treated, reused and managed for maintaining the urban water sustainability. Science advisory and science–policy interface will be very effective in this regard.

References

Anand S (2010) Solid waste management. Mittal Publications, New Delhi, p 9
Bhattacharyya R, Prasad S (2020) Water Scarcity in Delhi: Mapping for Solutions and the Way Forward. In: Singh RB, Srinagesh B, Anand S (eds) Urban health risk and resilience in Asian Cities. Advances in geographical and environmental sciences. Springer, Singapore, pp 421–444; https://doi.org/10.1007/978-981-15-1205-6_24
Chakraborty D, Mukhopadhyay K (2014) Introduction. Water pollution and abatement policy in India. Springer, Dordrecht, pp 1–21
City Sewerage System (2006) Jawaharlal Nehru Urban Rural Mission. CPD, Delhi
Colmenarejo M et al (2006) Evaluation of municipal wastewater treatment plants with different technologies at Las Rozas, Madrid (Spain). J Environ Manage 81(4):399–404
CPCB (2004) Status of sewerage and sewage treatment plants in Delhi, Control of Urban Pollution Series: CUPS 2003–2004. Central Pollution Control Board, New Delhi
CSE (2017) A waste water recycling manual with case studies, Centre for Science and Environment, Ministry of Housing and Urban Affairs. New Delhi, Centre for Science and Environment
Delhi Jal Board (2017) Pre-feasibility Study Report on Rehabilitation of Sewer System, Delhi Jal Board, Government of NCT, New Delhi. http://www.delhijalboard.nic.in/djbdocs/whats_new/news/pdf/DFR3Sewerage%20Vol%20I%20-%2017%20Nov%202017.pdf. Accessed 11 Aug 2020
Economic Survey of Delhi (2018–2019). Water supply and sewage, Chapter-13, Planning Department, Government of the National Capital Territory of Delhi, New Delhi, pp 217–237
Guidelines for Siting of Sewage Treatment Plant (2006) Urban Development and Housing Department. Support Programme for Urban Reforms in Bihar and DFID, United Kingdom, Government of Bihar Initiative
Jamwal P, Mittal AK, Mouchel, J-M (2009) Efficiency evaluation of sewage treatment plants with different Technologies in Delhi (India). Environ Monitor Assess 153(1-4):293–305
Kumar A, Khursheed S (2017) Utilization of treated effluent of STP's in Delhi. IJESC 7(6):256–262
Kumar H et al (2020) understanding drinking water quality problems in Shahdara Region of Delhi. J Interdiscipl Cycle Res XII (VII), July:1311–1323
Kumar PR, Pinto LB, Somashekar R (2010) Assessment of the Efficiency of Sewage Treatment Plants: a Comparative Study between Nagasandra and Mailasandra Sewage Treatment Plants. Kathmandu Univ J Sci Eng Technol 6(2):115–125
Municipal Corporation of Delhi (MCD) (2020) North Delhi Municipal Corporation Zones. https://mcdonline.nic.in/ndmcportal/. Accessed 14 Aug 2020
Rai SC, Nagpal J (2017) An assessment of domestic water use practices in Delhi. In: Poonam S, Swati R (eds) Sustainable smart cities in India. Springer, pp 445–458
Sheel A (2020) Managing India's freshwater. https://www.livemint.com/. Accessed 12 Aug 2020
Singh RB, Anand S, Saluja V (2018) Low carbon resilient Delhi megacity for sustainable future Earth In: Himiyama Y (ed) Exploring sustainable land use in Monsoon Asia. Springer, Singapore, pp 137–155
Singh R, Srinagesh B, Anand S (eds) (2020) Urban health risk and Resilience in Asian Cities. Advances in geographical and environmental sciences. Springer, Singapore, pp 421–444. https://doi.org/10.1007/978-981-15-1205-6_24
Sinha AK, Nazimuddin M (2008) Evaluation of operation and maintenance of sewage treatment plants in India—2007. Central Pollution Control Board, Ministry of Environment and Forests, Government of India, Delhi
Zimmer A (2014) Urban political ecology in megacities: The Case of Delhi's wastewater in Singh. In: Singh RB (ed) Urban development challenges, risks and resilience in Asian Mega Cities. Springer, Tokyo, pp 119–139

Chapter 6
The Hierarchical Segregation of Groundwater Potential Zones Using Weighted Overlay Technique for the Discernment of Site Suitable Artificial Recharge Structures in the Development of Avunoor Watershed

Lokesh Kumar Chilukuri, Sathish Kumar Vasa, and Akhter Ali Mohd

Abstract Since a decade, geospatial technologies found to be the best alternative solution for quick and cost-effective development of spatio-temporal information likewise, they are also effective in spatial data analysis and prediction especially in groundwater development and management studies in various watersheds. This study involves the use of GIS and image processing techniques for the processing and interpretation of SOI toposheets and Resourcesat-2A LISS-IV satellite imageries to produce the final composite map showing varying groundwater potential zones. Avunoor Watershed of Telangana State is being chosen as the study area. The objective of this study is to delineate the varying groundwater potential zones and evaluate the sites suitable for artificial recharge in order to suggest the site suitable artificial recharge structures. The Weighted Overlay technique is being employed to segregate the heterogeneous groundwater potential zones. The results showed Avunoor Watershed has 27.27% Poor, 51.76% Moderate and 17.29% Good groundwater potential zones. Finally, after determining the sites suitable for artificial recharge, it is found that Check Dam, Recharge well, Recharge pit are the artificial recharge structures suitable to develop the groundwater resources of Avunoor Watershed.

Keywords Groundwater · GIS · Weighted overlay · Remote sensing and watershed

L. K. Chilukuri (✉) · A. A. Mohd
Department of Geography, Osmania University, Telangana, India
e-mail: lokesshhh@gmail.com

S. K. Vasa
Department of Geo-Informatics, Telangana University, Telangana, India

R. K. Mishra et al. (eds.), *Sustainable Climate Action and Water Management*, Advances in Geographical and Environmental Sciences,
https://doi.org/10.1007/978-981-15-8237-0_6

6.1 Introduction

Groundwater is a precious natural resource found in voids or spaces under subsurface of the earth. Greater than 90% of rural and approximately 30% of the urban population depend on it for drinking water (NRSA 2008). In semi-arid rural areas, hard rock terrain created less scope for the availability of surface water resources, which in turn lead to higher demand for groundwater to cater to the needs of agriculture, industries, drinking and household purposes causing its overexploitation. One such area is Avunoor watershed that is facing critical water shortage problem is being chosen as the research study area. To meet the ever-growing demand for groundwater in Avunoor watershed where agriculture is the primary activity, it is necessary to identify the groundwater potential sites for exploration and development. The above-said objective is being achieved effectively and quickly for a couple of decades by combining Satellite Remote sensing and Geographical Information System (GIS) technologies (Krishnamurthy et al. 1996a, b; Saraf and Choudhury 1998; Khan and Moharana 2002; Jaiswal et al. 2003; Lokesh et al. 2005; and Kumar et al. 2008;). Remote sensing facilitates in detecting the direct or indirect indicators of groundwater potential such as lineaments, drainage network, geomorphological and geological features (Bobba et al. 1992; Meijerink 2000) because of its synoptic view of satellite images whereas, GIS provides an environment for processing, overlay, database management and analysis of multi-thematic spatial data in deriving groundwater potential zones (Preeja et al. 2011) by weighted overlay technique (Krishnamurthy et al. 1996; Saraf and Choudhury 1998).

6.2 Study Area

Avunoor watershed falls in Rajanna Sircilla and Siddipet districts of Telangana state. The study area (Fig. 6.1) is extended between 78°37′ to 78°52′ East longitudes and 18°5′ to 18°21′ North latitudes. The Avunoor watershed area is about 370.75 km^2 and it is covered in 56 J/11, 56 J/12, 56 J/15, 56 J/16 Toposheets of SOI.

6.3 Objectives

1. To examine geomorphic features and drainage network of Avunoor watershed.
2. To identify and segregate the heterogeneous groundwater potential zones in Avunoor watershed.
3. To detect the sites for artificial recharge and recommend the site suitably artificial recharge structures in the study area.

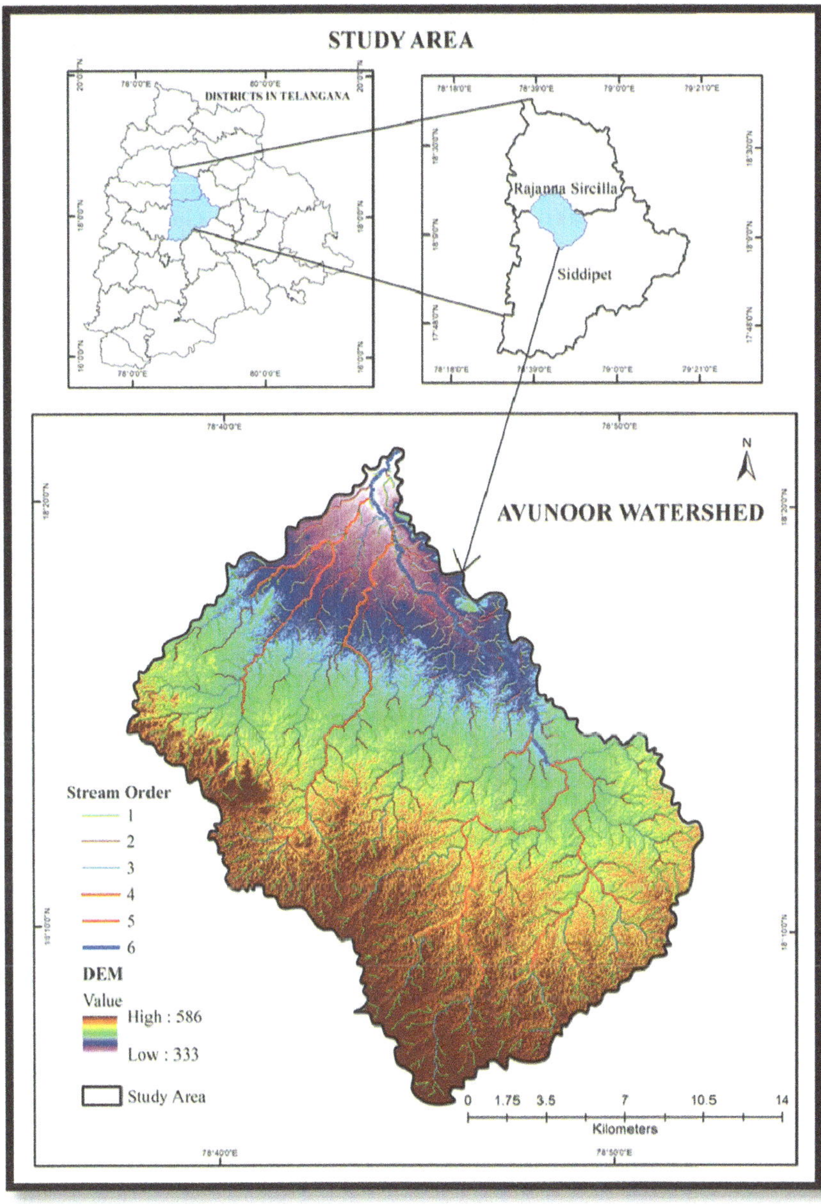

Fig. 6.1 Location of Avunoor Watershed

6.4 Data Used and Methodology

Resourcesat-2ALISS-IV satellite imagery acquired from NRSC was used for identifying geomorphology and lineaments using Image processing techniques of ERDAS IMAGINE 2015 software. SRTM DEM data obtained from USGS have been utilized to prepare drainage network and slope themes. SOI Toposheets 56 J/11, 56 J/12, 56 J/15, 56 J/16 of 1:50,000 scale were used as base maps. Groundwater potential zones in this research study are segregated by the Weighted Overlay technique of ArcGIS 10.5 software. GIS enabled the integration of geomorphology, lineaments, slope and drainage network themes to prepare the final composite map. The weights were assigned to each component, in each theme, based on their occurrence and contribution to groundwater recharge. Based on the distribution of lineaments, geomorphological features, slope and stream order artificial recharge structures were suggested.

6.5 Results and Discussion

6.5.1 Slope of the Study Area

In the slope map (Fig. 6.2), the slope is of three types.

They are 1. Low. 2. Average. 3. High.

Slope is one of the crucial indicators for groundwater prospect since it checks the infiltration of water depending on the degree of slope.

In the study area, high slope is present in the north and west side parts and the majority of the portion is occupied by low slope.

6.5.2 Drainage Network of the Study Area

Drainage network of an area depends on the topography and subsurface formations. Strahler stream ordering method is adopted in this study. There are six stream orders identified (Fig. 6.3) in the study area. The stream order is increasing when moving to the northern side of the study area. The sixth-order stream flows from east to north side of the study area.

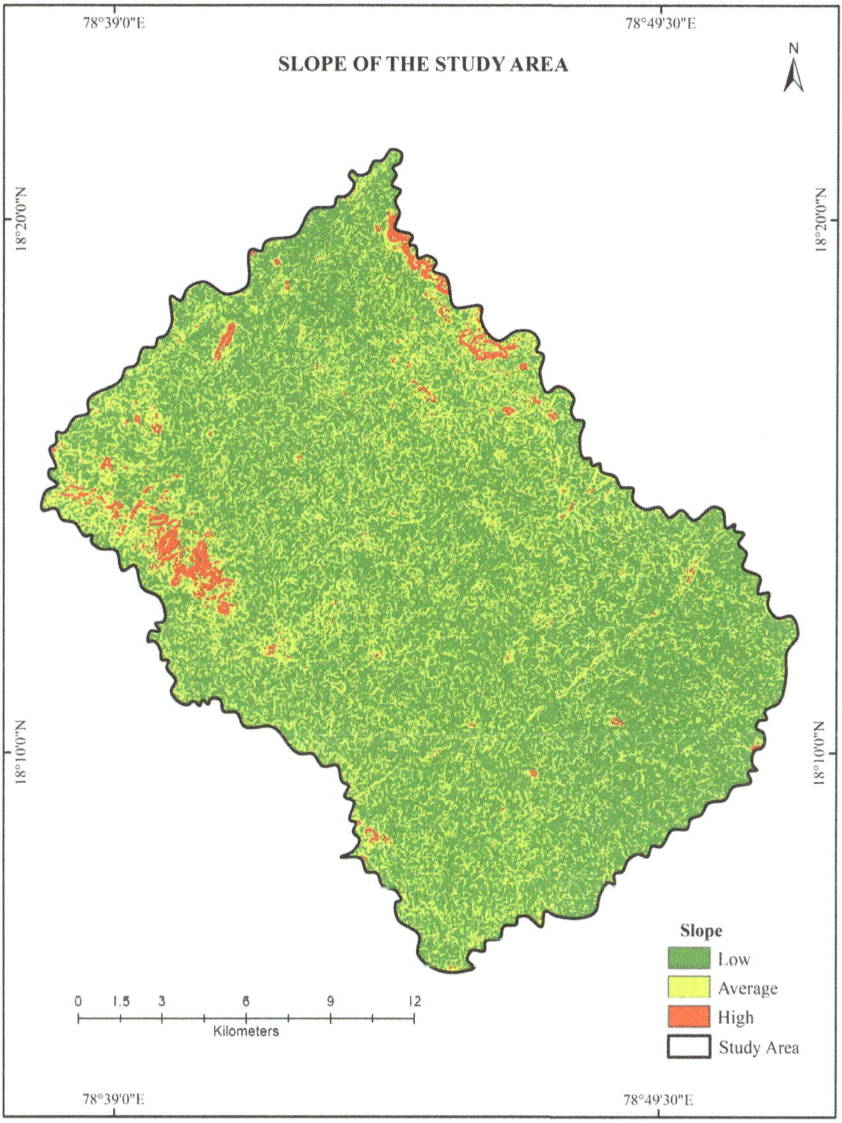

Fig. 6.2 Slope of the Study area

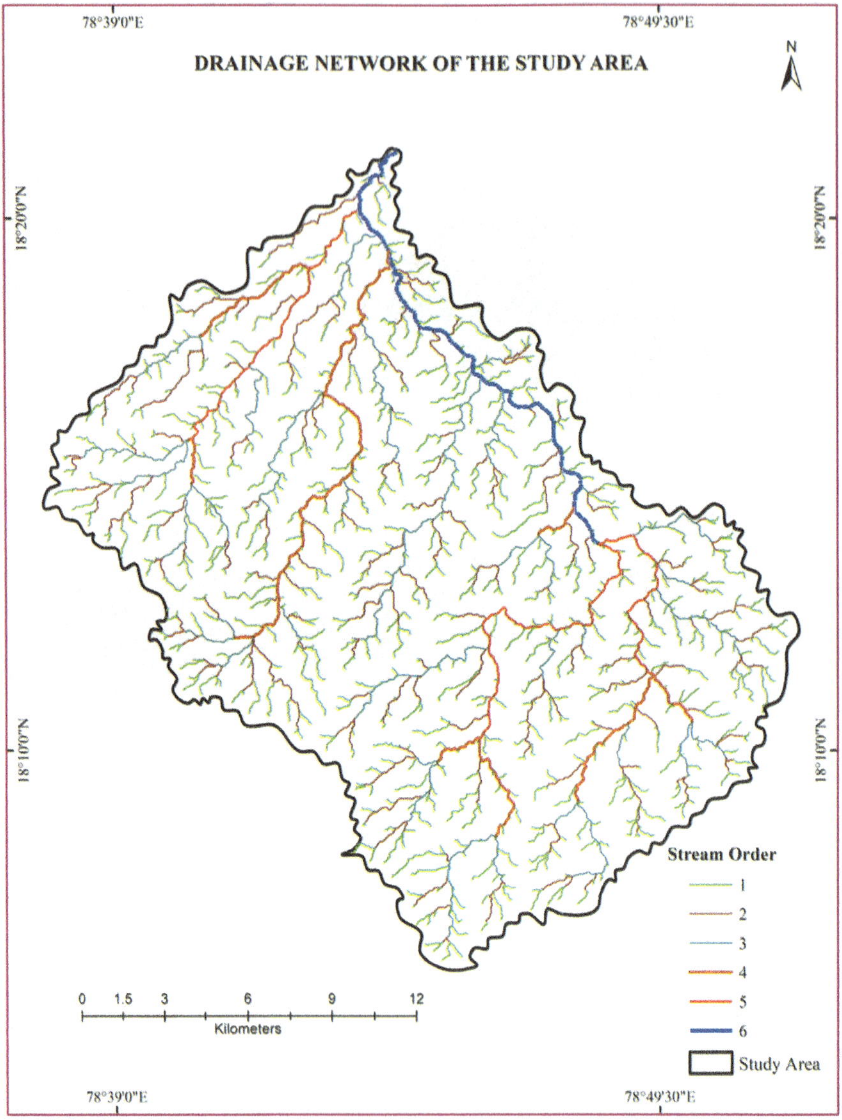

Fig. 6.3 Drainage Network of the study area

6.5.3 Lineaments of the Study Area

Lineaments are the fault/fractured zones that yield good groundwater potential. In the study area, more than 30 fractured/faults zones (Fig. 6.4) are present. Lineaments trend majorly in SW–NE and few in S–N directions. Majority of them lie in southwestern and northwestern parts of the watershed.

6.5.4 Geomorphology of the Study Area

Geomorphic units were interpreted from the Resourcesat-2A LISS-IV satellite imagery. A total of nine geomorphological features (Table 6.1) are identified (Fig. 6.5) in the present study.

6.5.4.1 Denudational Hill (DH)

These are the massive hills characterized by big boulders and sparse vegetation, which were formed due to denudation and weathering processes. It has a steep slope because of which it has poor groundwater potential. The percentage of area occupied by it is 6.93% and the area it covers is 25.69 sq km.

6.5.4.2 Dyke Ridge (D)

Wall-like structure formed by the erosion of softer material surrounding the dyke. These landforms generally act as barriers to groundwater movement. The percentage of area occupied by it is 0.57% and the area it covers 2.13 sq km.

6.5.4.3 Inselberg (I)

Isolated hill characterized by hard volcanic rock that is rising abruptly above the surrounding dry region due to erosion. They have low groundwater potential because of steep sides creating low infiltration of water and high runoff. The percentage of area occupied by it is 0.85% and the area it covers 3.16 sq km.

6.5.4.4 Pediment (PD)

Gently sloping eroded rock surface dotted with a number of small outcrops, which is formed between the hill and plain surface. It has moderate to poor groundwater

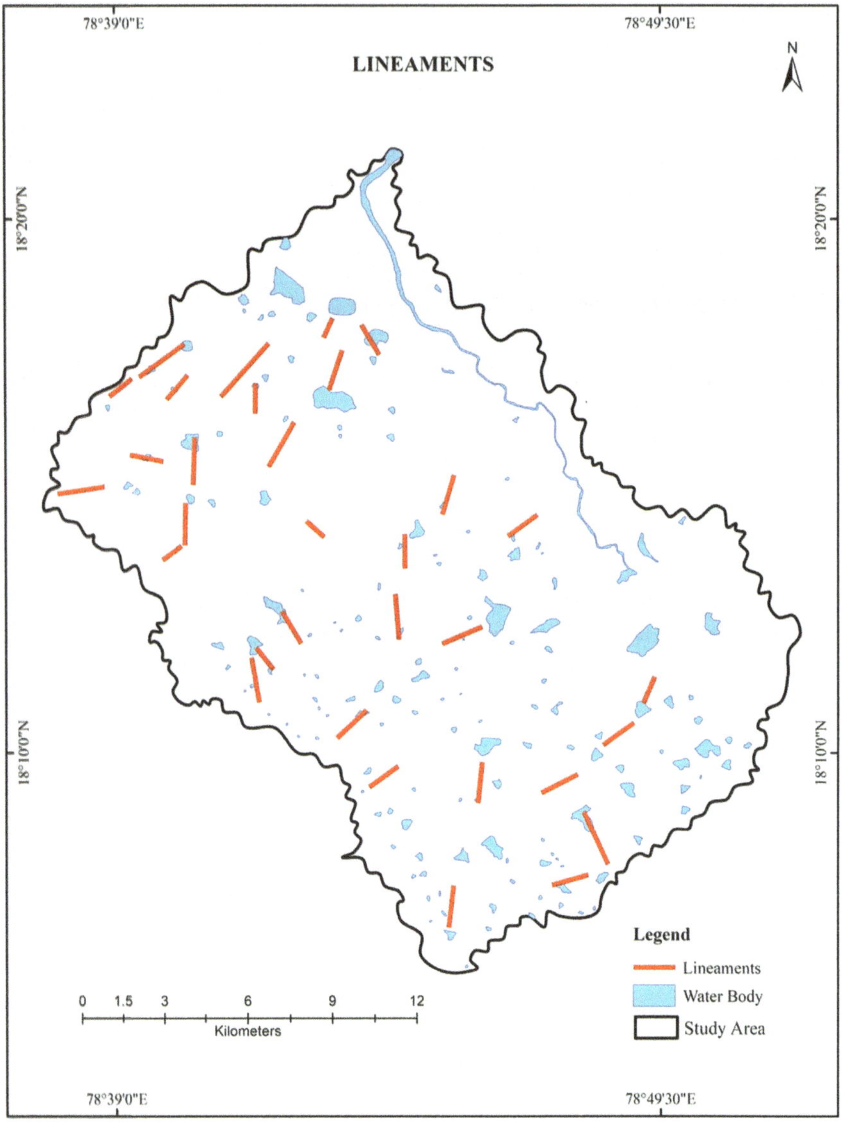

Fig. 6.4 Lineaments of the study area

Table 6.1 Geomorphic Units identified in study area

S. no	Geomorphological features	Area in sq.km	Percentage of area
1	Denudational Hill (DH)	25.69	6.93
2	Dyke Ridge (D)	2.13	0.57
3	Inselberg (I)	3.16	0.85
4	Pediment (PD)	70.11	18.91
5	Pediment Inselberg Complex (PIC)	31.35	8.45
6	Plateau Slightly Dissected (PLS)	12.97	3.50
7	Pediplain Moderate (PPM)	64.11	17.29
8	Pediplain Shallow (PPS)	147.58	39.81
9	Water Body	13.65	3.68
Total Study Area		**370.75**	**100**

potential. The percentage of area occupied by it is 18.91% and the area it covers 70.11 sq km.

6.5.4.5 Pediment Inselberg Complex (PIC)

Pediments dotted by multiple numbers of low relief inselbergs in a gently sloping pediplain, which cannot be distinguished as separate features constitute Pediment Inselberg Complex. They have moderate to poor groundwater potential because of moderate to high slope. The percentage of area occupied by it is 8.45% and the area it covers 31.35 sq km.

6.5.4.6 Plateau Slightly Dissected (PLS)

The region of a plateau has been separated into numerous smaller pieces because of intense erosion by rivers. Intense erosion created steep sides causing high run-off and low infiltration of water except at lineaments. The percentage of area occupied by it is 3.50% and the area it covers 12.97 sq km.

6.5.4.7 Pediplain Moderate (PPM)

Gently sloping buried pediplain of granite gneiss with thicker weathered material higher than Pediplain shallow. This feature has good to moderate groundwater potential. The percentage of area occupied by it is 17.29% and the area it covers 64.11 sq km.

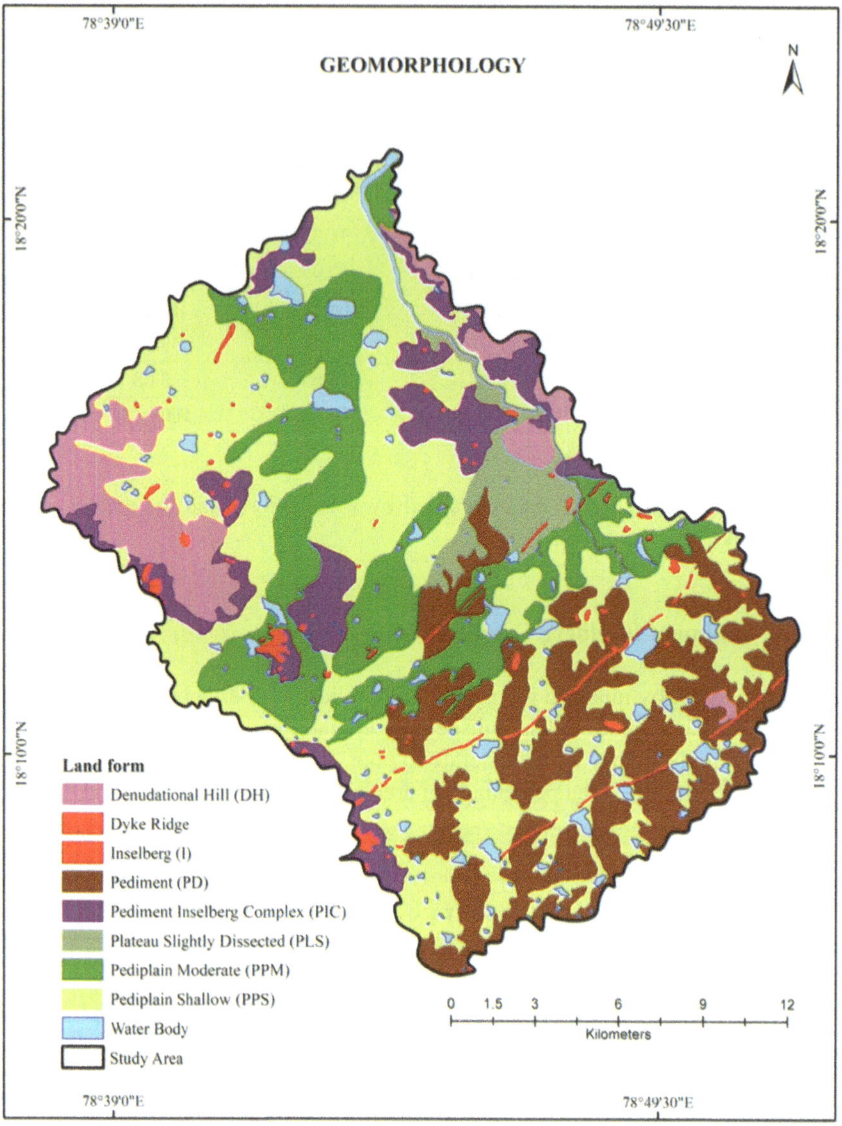

Fig. 6.5 Geomorphological features in study area

6.5.4.8 Pediplain Shallow (PPS)

Gently sloping pediplain of granite gneiss formed by coalescence of buried pediments, with a shallow overburden of weathered materials. This feature has moderate to poor groundwater potential because of less weathered material. The percentage of area occupied by it is 39.81% and the area it covers 147.58 sq km.

6.5.5 Groundwater Potential Zones

Groundwater potential zones (Fig. 6.6), Good, Moderate and Poor (Table 6.2), are derived by assigning weights to components of geomorphology, slope, drainage network and lineaments themes and integrating using weighted overlay technique.

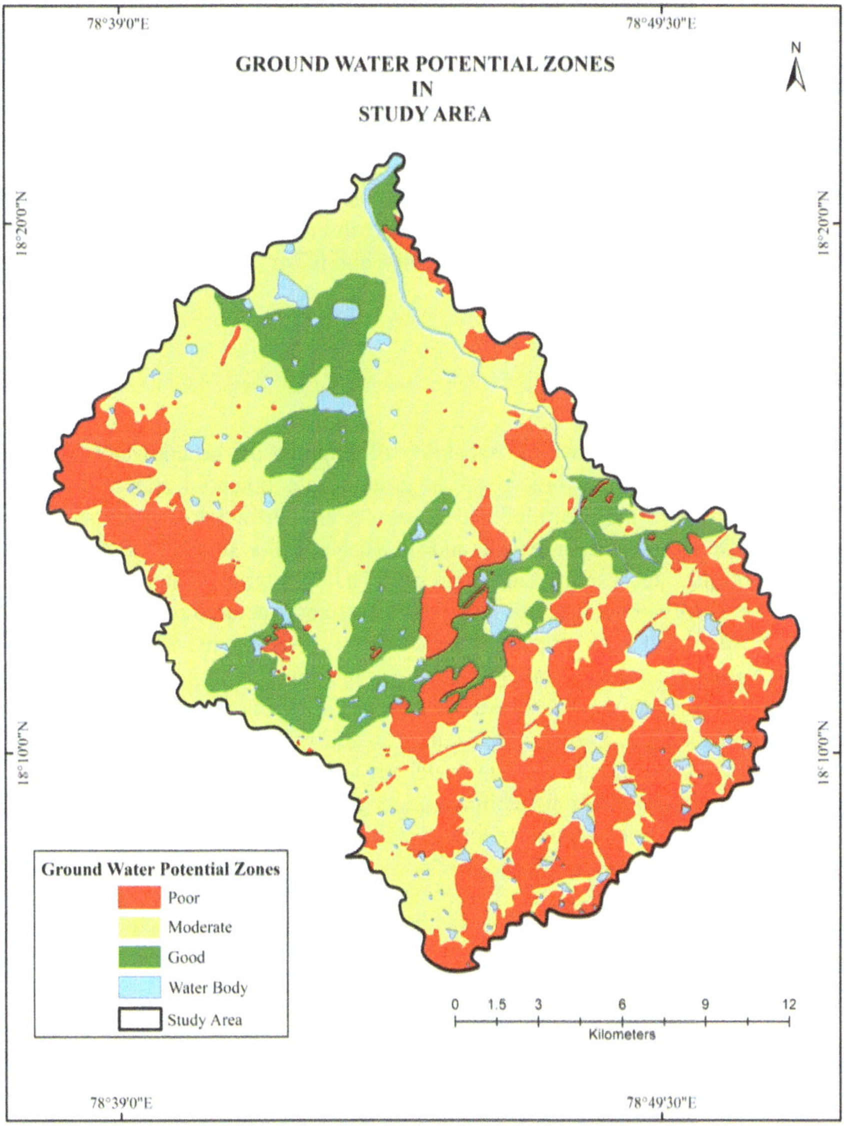

Fig. 6.6 Groundwater potential zones in the study area

Table 6.2 Groundwater Potential Zones in the Study Area

S. No	Ground water potential zones	Area in sq.km	Percentage of area
1	Poor	101.09	27.27
2	Moderate	191.89	51.76
3	Good	64.11	17.29
4	Water body	13.65	3.68
Total study area		**370.75**	**100**

Good groundwater potential zone covers 64.11 sq.km; moderate groundwater potential zone covers 191.89 sq.km and poor groundwater potential zone covers 101.09 sq.km of the Avunoor watershed. Highest groundwater potential zone observed in the study area was moderate groundwater potential zone, which covers 51.76%. The lowest groundwater potential zone observed in the study area was Good groundwater potential zone covering 17.29% (Fig. 6.7).

6.6 Conclusion

Geospatial technologies have the capability of mapping the groundwater potential zones. In this study area, three types of groundwater potential zones, i.e., Poor, Moderate and Good are identified and 22 site-specific artificial recharge structures (9—Check Dams, 5—Recharge wells and 8—Recharge pits) suitable for the augmentation of groundwater resources are suggested. The Poor groundwater potential zone has hard rock, Moderate zone has loose sediments and fractured rock and the Good zone has weathered rock, fractured rock and loses sediments. Thus, because of the above reason, in the Avunoor Watershed cultivation is more in the Moderate groundwater prospector zone whereas the wasteland is more in the Poor groundwater prospector zone. This research study can be used in the future for efficacious planning and management of groundwater resources. The weighted overlay technique adopted in this study for delineation of groundwater potential zones can be applied to similar hydrogeomorphological watersheds.

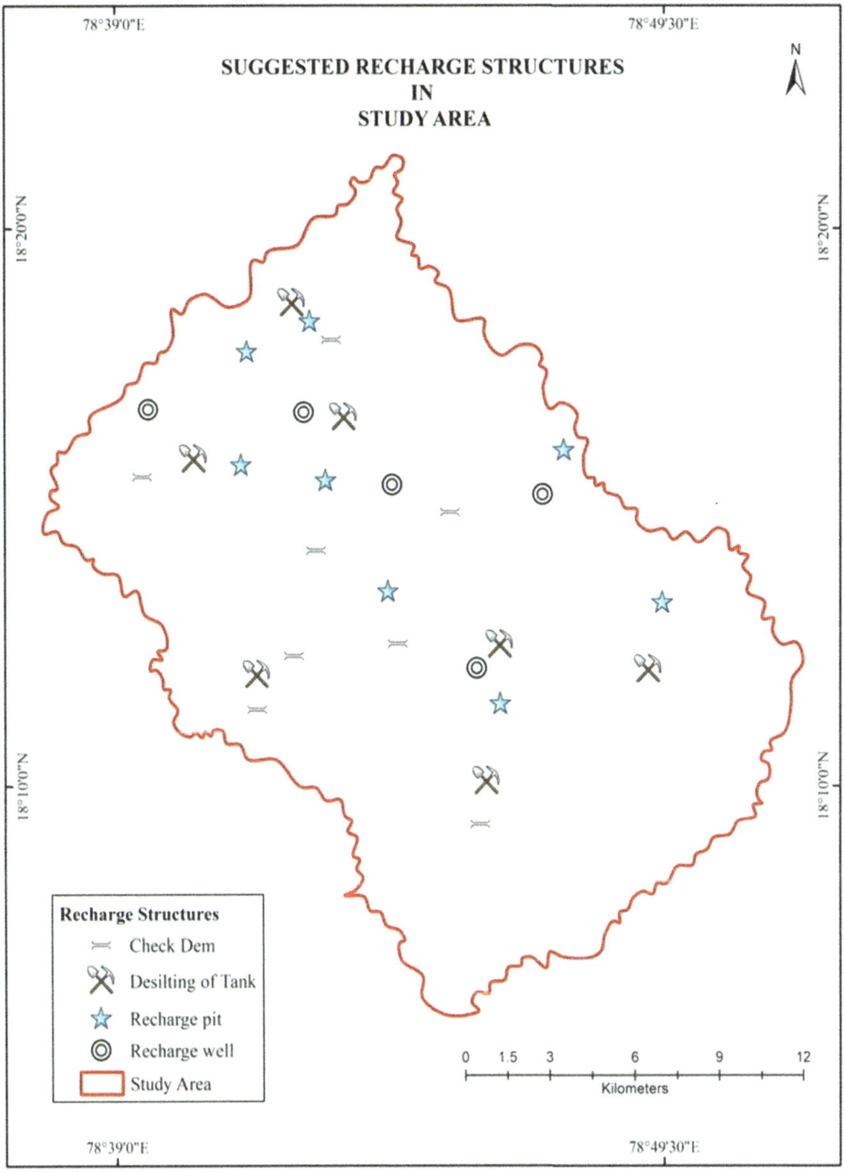

Fig. 6.7 Suggested recharge structures in study area

References

Bobba A, Bukata R, Jerome J (1992) Digitally processed satellite data as a tool in detecting potential groundwater flow systems. J Hydrol 131(1–4):25–62. https://doi.org/10.1016/0022-1694(92)90212-e

Jaiswal RK, Mukherjee S, Krishnamurthy J, Saxena R (2003) Role of remote sensing and GIS techniques for generation of groundwater prospect zones towards rural development–an approach. Int J Remote Sens 24(5):993–1008. https://doi.org/10.1080/01431160210144543

Khan MA, Narain P, Moharana PC (2006) Prospecting ground water resources using RS-GIS—A case study from arid western rajasthan of India. J Indian Soc Remote Sens 34(2):171–179. https://doi.org/10.1007/bf02991822

Krishnamurthy J, Kumar NV, Jayaraman V, Manivel M (1996a) An approach to demarcate ground water potential zones through remote sensing and a geographical information system. Int J Remote Sens 17(10):1867–1884. https://doi.org/10.1080/01431169608948744

Krishnamurthy J, Srinivas G, Jayaraman V, Chandrasekhar MG (1996b) Influence of rock types and structures in the development of drainage networks in typical hardrock terrain. ITC J 3–4:252–259

Kumar MG, Agarwal AK, Bali R (2008) Delineation of potential sites for water harvesting structures using remote sensing and GIS. J Indian Soc Remote Sens 36(4):323–334. https://doi.org/10.1007/s12524-008-0033-z

Lokesha N, Gopalakrishna GS, Gowda HH, Gupta AK (2005) Delineation of ground water potential zones in a hard rock terrain of Mysore district, Karnataka using IRS data and GIS techniques. J Indian Soc Remote Sens 33(3):405–412. https://doi.org/10.1007/bf02990011

Meijerink AMJ (2000) Groundwater. In: Schultz GA, Engman ET (eds) Remote sensing in hydrology and water management. Springer, Berlin, pp 305–325

NRSA (2008) Groundwater prospect mapping using remote sensing and GIS. Rajiv Gandhi National Drinking Water Mission Project Manual, National Remote Sensing Agency, Hyderabad, India

Joseph RPKS, Thomas JHV (2011) Identification of groundwater potential zones of a tropical river Basin (Kerala, India) using remote sensing and GIS techniques. J Indian Soc Remote Sens 39(1):83–94. https://doi.org/10.1007/s12524-011-0075-5

Saraf AK, Choudhury PR (1998) Integrated remote sensing and GIS for groundwater exploration and identification of artificial recharge sites. Int J Remote Sens 19(10):1825–1841. https://doi.org/10.1080/014311698215018

Chapter 7
Delineation of Groundwater Potential Zones on Lower and Middle Beas Basin Using Remote Sensing and Geographical Information System

Ashutosh Mohanty, Arunashis Chandra, and Pritam Ghosh

Abstract Nowadays, remote sensing and geographical information system (GIS) are effective comprehensive systemic tools for the analysis and assessment of water resources. An attempt has been made in this study to delineate groundwater potential zones through remote sensing and geographical system on the middle and lower part of the Beas River basin. The main objective of the study is to delineate different groundwater potential zones based on some selected indicators in the study area. Groundwater potential zone has been created here through the integration of thematic maps of the selected indicators using remote sensing and GIS techniques by Arc-GIS 10.1 software. Good groundwater potential zones are distributed along the river from north to southwest portion of the study area. A major portion of the study area has come out as poor or very poor groundwater potential zone. Therefore, this study can be a guideline for the exploration of groundwater in the basin area. On the other hand, this demarcation can help to make proper policy and proper plan for sustainable management of the groundwater in this area.

Keywords Groundwater potential · River basin · Water resource management

A. Mohanty (✉)
Disaster Management and Climate Change, Shoolini University, Himachal Pradesh, India
e-mail: drashutoship@gmail.com

A. Chandra
Department of Geography, Vidyasagar University, Midnapore, India
e-mail: mr.arunashischandra@gmail.com

P. Ghosh
Department of Geography, University of Calcutta, Kolkata, India
e-mail: mr.ghosh.pritam@gmail.com

R. K. Mishra et al. (eds.), *Sustainable Climate Action and Water Management*,
Advances in Geographical and Environmental Sciences,
https://doi.org/10.1007/978-981-15-8237-0_7

7.1 Introduction

Groundwater is one of the most important, precious and renewable dynamic natural resources (Fashae et al. 2014), which plays a great role in survival (Das 2017) and different developmental activities of the human being (Subba et al. 2001; Hutti and Nijagunappa 2011; Biswas et al. 2012). To ensure continuous supply of drinking water, agriculture and industrial needs, identification of groundwater potential zone and sustainable management of groundwater are essential for any region as well as any nation. Assessment of groundwater resources is extremely significant for this sustainable management.

In previous times, different researchers have used different geological, hydrogeological, geophysical and photogeological conventional techniques to determine groundwater potential zones. However, in recent days, demarcation of groundwater potential zone through the application of RS and GIS has become significant. Many researchers (Murthy 2000; Jaiswal et al. 2003; Sener et al. 2005; Srivastava et al. 2006; Chowdhury et al. 2009; Dar et al. 2011; Nag 2012; Fashae 2014; Das 2017; Sadaf 2018) have tried to delineate groundwater potential zone through these techniques on different areas or different river basins. In recent days, extensive use of satellite data along with conventional maps has made a new effective way to demarcate groundwater potential zones (Das et al. 1997; Thomas et al. 1999; Harinarayana et al. 2000; Muralidhar et al. 2000; Chowdhury et al. 2010; Bagyaraj et al. 2013; Rahmati et al. 2015; Roy et al. 2016; Das et al. 2018). Here an attempt has been made to delineate groundwater potential zones through remote sensing and geographical system on middle and lower part of Beas River basin. Beas is an important river in Himachal Pradesh and Punjab. The basin area is covering some districts of Himachal Pradesh as well as Panjab. A major part of the basin area is situated on plain land and in a few portions of Himalayan mountain region. Therefore, a detailed study on sustainable management of groundwater is necessary for this region.

The availability of groundwater directly or indirectly depends upon different factors, like geology, geomorphology, slope, soil texture, drainage density, lineament concentration, land use and land cover pattern of an area. (Nag 2005; Sener et al. 2005; Solomon et al. 2006; Ganapuram et al. 2009; Magesh et al. 2012; Mukherjee et al. 2012; Das et al. 2017, 2018).

The main objective of the study is to delineate different groundwater potential zones based on some selected indicators in the study area. There is also an attempt to search the pattern of distribution of different indicators of groundwater potential in the study area.

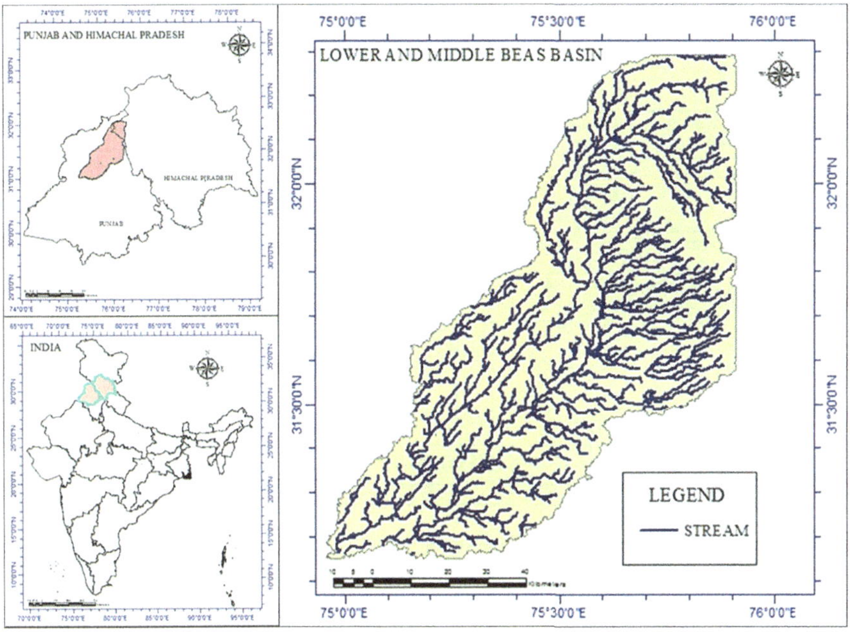

Fig. 7.1 Location of the study area (lower and middle Beas River basin)

7.2 Study Area

The river Beas originating from the lap of western Himalaya in central Himachal Pradesh flows 470 km in southwest direction to Sutlej River in Punjab. Total area of the basin of the river is approximately 20.303 square kilometers. Here, lower and middle parts of this basin have been considered as the study area, which extends from 31°8′55″ N to 32°17′24″ N and 74°57′18″ E to 75°54′41″ E and covers a total area of 5875 square kilometers (Fig. 7.1).

7.3 Database and Methodology

In the present study, the groundwater potential zone mapping is accomplished by using GIS tools. High-resolution satellite imagery, Landsat VII Enhanced Thematic Mapper Plus and SRTM Dem have been used for the preparation of the thematic maps. 'All the data are georectified and projected to Geographic Co-ordinate System 1984 (CGS-WGS) Universal Transverse Mercator (UTM) zone 43 north for easy handling in GIS environment' (Ibrahim-Bathis and Ahmed 2016).

After being rectified by radiometric correction method through Envi 5.0 software, Landsat 7 ETM+ was used for the extraction of land use and land cover map

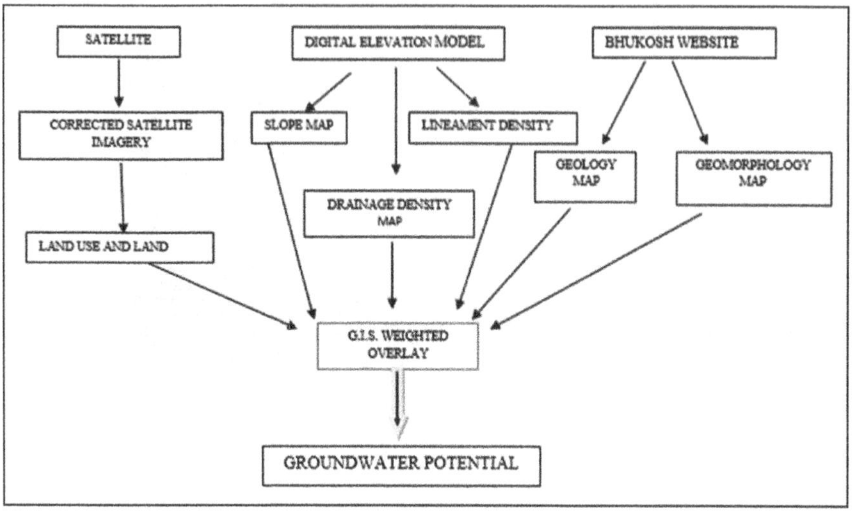

Fig. 7.2 Methodological framework of the study

by ERDAS 9. Slope map, drainage density map and lineament density map were extracted by SRTM Dem through Arc-GIS 10.1 software.

Slope map has been extracted from SRTM DEM through *Surface tool*, from 'Spatial Analyst tools' in Arc Gis10.1. Drainage map is also determined by SRTM DEM through *Hydrology tool* from 'Spatial Analyst tools' in Arc Gis10.1. Then the Drainage Density map has been accomplished by Line Density tool in Arc Gis10.1. The lineament map has been extracted by DEM from Algorithm Librarian in PCI Geometica 9.1 software, then the lineament density has been prepared by Line Density tool in Arc Gis10.1.

Different reclassed maps have been generated by assigning probable score, based on groundwater potentiality like, geology re-class map, geomorphology re-class map, slope re-class map, drainage density re-class map, lineament density re-class map, land use-land cover re-class map etc. Then these re-class maps have again been assigned by proper weighted score and at last overlay them and groundwater potential zone map is prepared (Fig. 7.2).

7.4 Results and Discussion

7.4.1 Geology

The hydrogeology of a region is highly determined by the lithology and structures of the underlying formations. Three types of geologic formation are found in the study area. Sandstone is found in some dissected areas of the eastern part of the

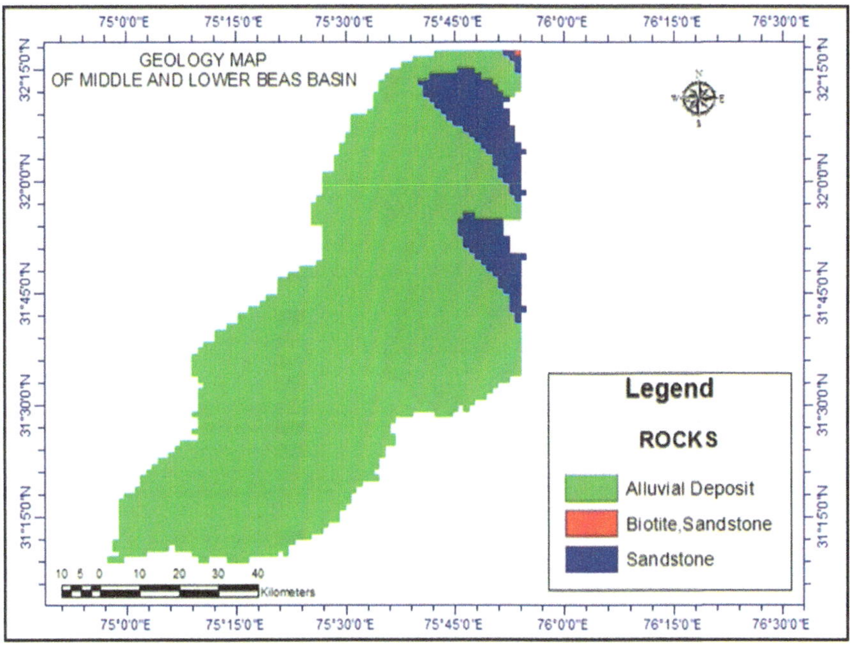

Fig. 7.3 Geology map of lower and middle Beas basin

basin (Fig. 7.3). Sandstone is considered as moderate water-bearing rocks. Rest of this place, entire western part of the basin comprises alluvial deposits. From the hydrological point of view, the area covering with alluvial deposit is characterized by good hydraulic conductivity.

7.4.2 Geomorphology

Geomorphology of a region plays a great role for groundwater prospects. Remote sensing study helps to analyze in a systemic way various hydrogeomorphic units, multispectral repetitive coverage of the terrain (Chatterjee et al. 1995; Tiwari et al. 1996). In this study, from the conventional map and Google Earth, major hydrogeological units of this region have been identified (Fig. 7.4). A major portion of the region consists of flood plain and alluvial plain. This active and older flood plain is relatively a more potential zone for groundwater than others. A highly dissected hill and valley, which has been found in eastern parts of the region, in association with more surface runoff, denotes less amount of groundwater recharge.

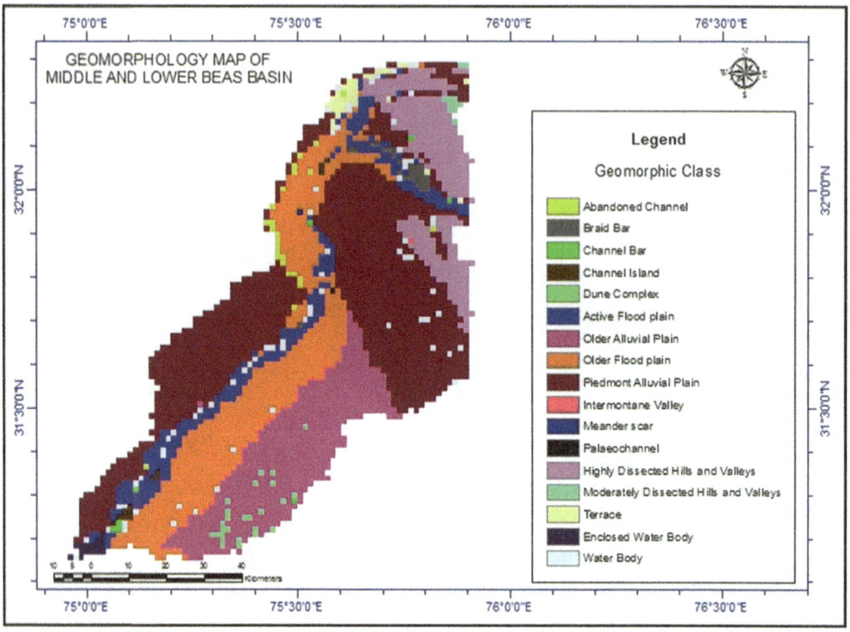

Fig. 7.4 Geomorphology of lower and middle Beas basin

7.4.2.1 Slope of the Area

Ground slope is an important factor for groundwater recharge. Steeper slope on the ground surface in association with less infiltration and rapid runoff during rainfall causes less underground water recharge (Rokade et al. 2007; Magesh et al. 2012; Gumma et al. 2012; Selvam et al. 2012). On the other hand, plane ground surface helps to store more ground water. A major portion of the basin area is found as plane land (Fig. 7.5). So a major part of the area can be considered as excellent for groundwater storage. Beside this, the area having a slope of 0–1 degree is considered as good storage. Slope of some parts of the northeastern side of the region is found greater than one degree. Hence, this part is unfavorable for groundwater storage.

7.5 Drainage Density of the Region

Drainage density of a region has a significant influence on groundwater storage. The area of high drainage density in association with great surface runoff causes less infiltration to store underground water. On the other hand, low drainage density or low surface runoff shows higher chance of groundwater recharge or potential groundwater zone. Here drainage density has been derived with length of drainage in kilometer per square kilometer area. In this study, from the placement of the Strems, a ruster map

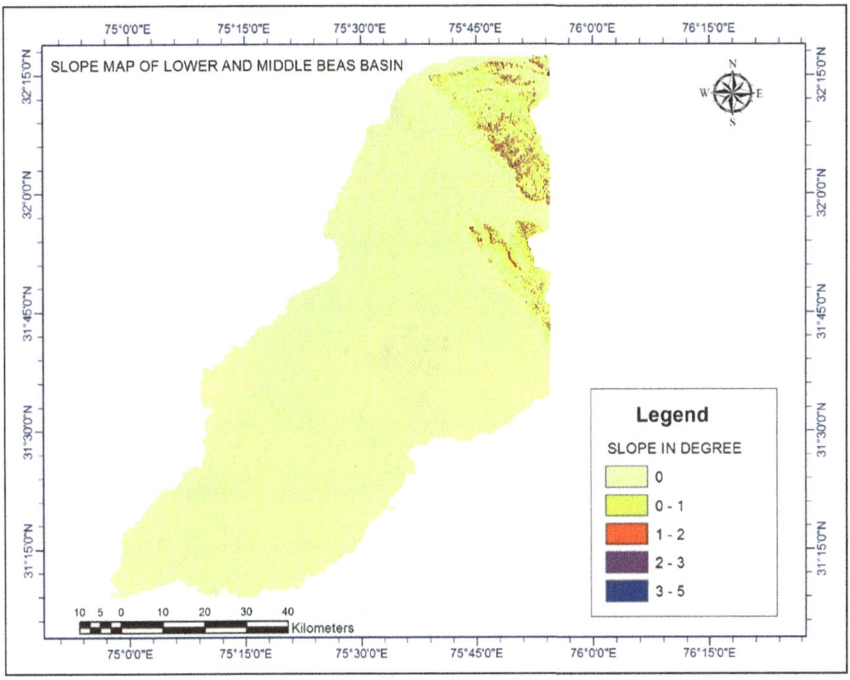

Fig. 7.5 Slope map of lower and middle Beas basin

has been prepared. Therefore, drainage density ma of the region has been prepared through line density tools of Arc Gis10.1 software. The map (Fig. 7.6) shows some discontinuous zone of lower drainage density in the eastern and northeastern parts, which probably have more potentiality to store groundwater. Southern and middle parts of the region show higher drainage density zones. It means these zones can be identified as less groundwater potential zones from this aspect. The suitability of groundwater potential zones in association with surface runoff and permeability is indirectly related to drainage density of any region.

7.6 Lineament Density of the Region

This factor is hydrogeologically significant as the presence of lineament on the ground surface in association with rapid infiltration helps to store underground water and control the movement of groundwater (Magesh et al. 2011). Therefore, it has a direct relationship with groundwater recharge. Here in this study, the lineament map has been extracted from DEM by PCI Geomatica 9.1 software. Then lineament density has been determined through line density tools of Arc-GIS software. Here lineament density has been measured with the length of lineament per unit area. The map shows

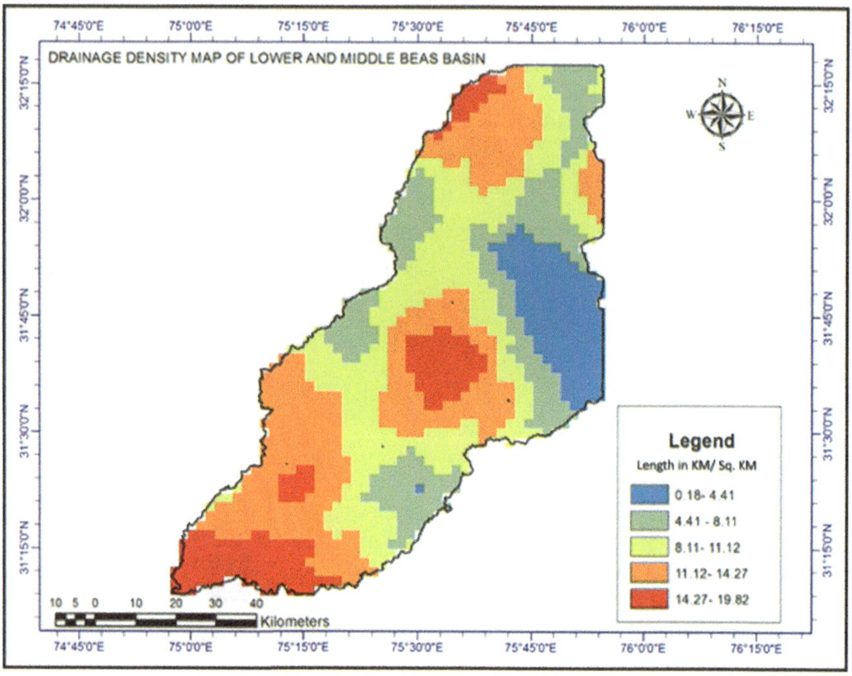

Fig. 7.6 Drainage density map of lower and middle Beas basin

(Fig. 7.7) that, the lineament is located along the river and along the marginal part of the area. These zones can be identified as more groundwater stored zones.

7.7 Land Use and Land Cover Pattern of the Area

Different types of land use characteristics are found in this region. Here Beas River is flowing from north to southwest direction (Fig. 7.8). Dense forests are found in the east and northeastern hilly region. There is a direct relationship between vegetation, waterbodies and river with groundwater recharge. Here a vast area is characterized with agricultural land and waste fallow land. They are moderately favorable for groundwater recharge and dry fallow land and settlements are inversely related to groundwater prospectiveness.

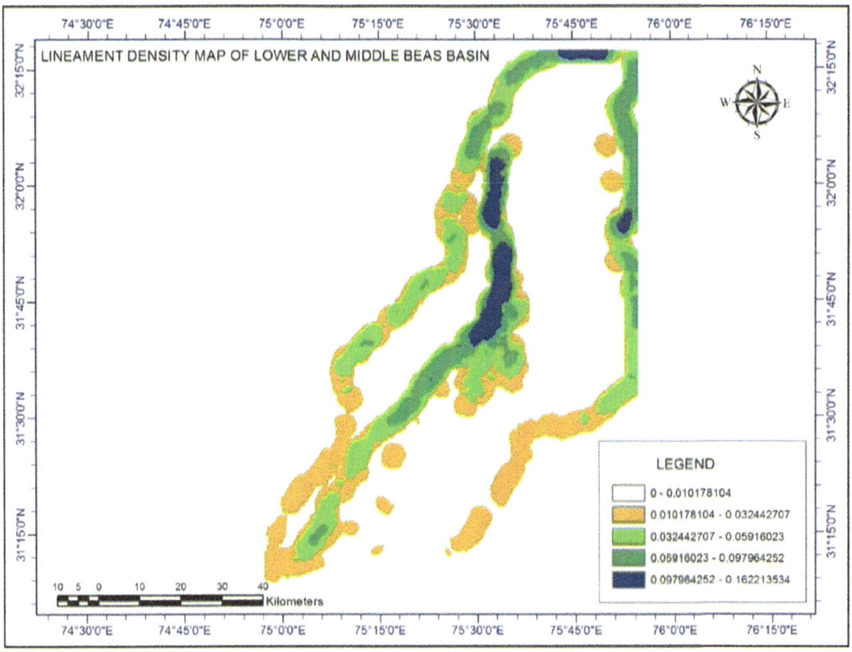

Fig. 7.7 Lineament density map of lower and middle Beas basin

7.8 Groundwater Potential Zone

Groundwater potential zone has been created here through the integration of thematic maps of the selected indicators using remote sensing and GIS techniques by Arc-GIS 10.1 software. All these factors here are being combined with certain weightage. High score has been given in case of that indicator, which has much influence on groundwater potentiality. Factors and their score for the aggregation have been given in Table 7.1. The study area has been divided into five groundwater potential zones (Fig. 7.9). The map shows that very good potential zones are found as scattered discontinuously in the middle and south parts of the region along with the river valley. Percentage of these zones is very less in the whole area.

On the other hand, good groundwater potential zones are distributed along the river from north to southwest portion. A major portion of the study area has come out as poor or very poor groundwater potential zone. Eastern and southeastern parts of the region show poor or very poor condition in groundwater potentiality.

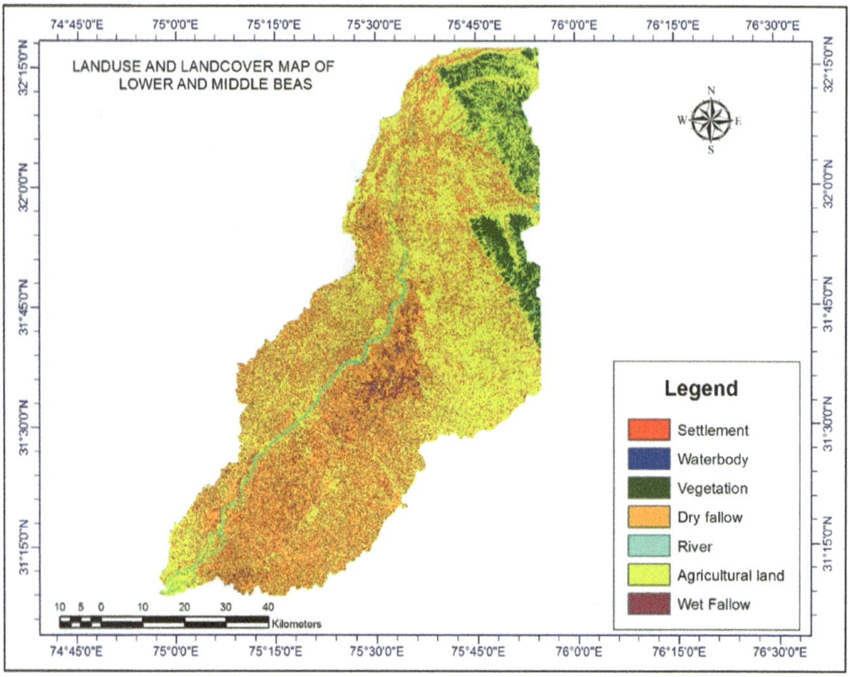

Fig. 7.8 Land use and land cover map of lower and middle Beas basin

7.9 Conclusion

In this study, with the help of satellite imagery, conventional maps and Google Earth, several types of thematic maps of some important factors that directly or indirectly associated with groundwater potential of an area have been prepared. The factor of geology, geomorphology, slope, drainage density, lineament density and land use and land cover pattern of the basin area have been considered here to delineate groundwater potential zones. Finally, with the GIS software, after integrating all these factors, the final groundwater potential map has been prepared. Here in this study, the entire basin area has been classified into five potential zones of groundwater. The very good groundwater potential zones may be selected for artificial recharge project by government. Therefore, this study can be a guideline for the exploration of groundwater in the basin area. On the other hand, this demarcation can help to make a proper policy and proper plan for sustainable management of the groundwater in this area.

Table 7.1 Level of groundwater prospect of different categories of various physical aspects in the study area

Parameter	Classes	Groundwater prospect	Weight (%) for overlay
Geology	Biotite, sandstone Sandstone Alluvial deposits	Poor Moderate Good	20
Geomorphology	Dune complex, highly dissected hill and valleys Older flood plain, piedmont alluvial plain, intermontane valley Terrace Active flood plain, meander scar, braid bar River, channel bar, waterbody	Very poor Poor Moderate Good Very good	25
Slope	0 0–1 1–2 2–3 3–5	Very good Good Moderate Poor Very poor	20
Drainage density	0.18–4.41 4.41–8.11 8.11–11.12 11.12–14.27 14.27–19.82	Very good Good Moderate Poor Very poor	10
Lineament density	0–0.00101 0.00101–0.032442 0.032442–0.05916 0.05916–0.09796 0.09796–0.16221	Very poor Poor Moderate Good very good	15
Land use and land cover	Settlement Waterbody Vegetation Dry fallow River Agricultural land Wet fallow	Very poor Very good Good Poor Very good Moderate Moderate	10

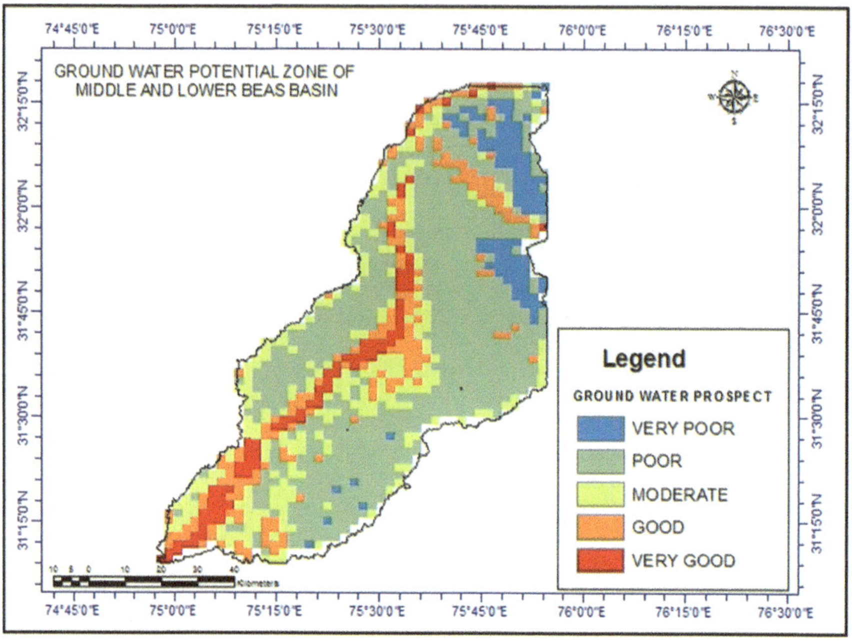

Fig. 7.9 Groundwater potential map of lower and middle Beasbasin

References

Bagyaraj M, Ramkumar T, Benkatramanan S, Gurugannam B (2013) Application of remote sensing and GIS analysis for identifying groundwater potential zones in parts of Kodaikanal Taluk, South India. Front Earth Sci 7(1):65–75. https://doi.org/10.1007/s11707-012-0347-6

Biswas A, Jana A, Sharma SP (2012) Delineation of groundwater potential zones using satellite remote sensing and geographic information system techniques: a case study from Ganjam district, Orissa, India. Res J Recent Sci 1(9):59–66

Chatterjee RS, Bhattachaya AK (1995) Delineation of drainage pattern of a coal basin related inference using satellite remote sensing techniques. Asia Pacific Remote Sens J 1:107–114

Chowdhury A, Jha MK, Chowdary VM, Mal BC (2009) Integrated remote sensing and GIS-based approach for assessing groundwater potential in West Medinipur district, West Bengal, India. Int J Remote Sens 30(1):231–250

Chowdhury A, Jha MK, Chowdary VM (2010) Delineation of groundwater recharge zones and identification of artificial recharge sites in West Medinipur district, West Bengal, using RS, GIS and MCDM techniques. Environ Earth Sci 59:1209–1222

Dar IA, Sankar K, Dar MA (2011) Deciphering groundwater potential zones in hard rock terrain using geospatial technology. Environ Monit Assess 173(1–4):597–610. https://doi.org/10.1007/s10661-010-1407-6

Das S (2017) Delineation of groundwater potential zone in hard rock terrain in Gangajalghati block, Bankura district, India using remote sensing and GIS techniques. Model Earth Syst Environ 3(4):1589–1599. https://doi.org/10.1007/s40808-017-0396-7

Das S, Pardeshi SD (2018) Integration of different influencing factors in GIS to delineate groundwater potential areas using IF and FR techniques: a study of Pravara basin, Maharashtra, India. Appl Water Sci 8(197):1–16. https://doi.org/10.1007/s13201-018-0848-x

Das S, Behera SC, Kar A, Narendra P, Guha S (1997) Hydro-geomorphological mapping in groundwater exploration using remotely sensed data—a case study in Keonjhar District, Orissa. J Indian Soc Remote Sens 25:247–259

Fashae OA, Tijani MN, Talabi AO, Adedeji OI (2014) Delineation of groundwater potential zones in the crystalline basement terrain of SW-Nigeria: an integrated GIS and remote sensing approach. Appl Water Sci 4:19–38. https://doi.org/10.1007/s13201-013-0127-9

Ganapuram S, Kumar G, Krishna I, Kahya E, Demirel M (2009) Mapping of groundwater potential zones in the Musi basin using remote sensing and GIS. Adv Eng Softw 40(7):506–518. https://doi.org/10.1016/j.advengsoft.2008.10.001

Gumma MK, Pavelic P (2012) Mapping of groundwater potential zones across Ghana using remote sensing, geographic information systems, and spatial modeling. Environ Monit Assess 185(4):3561–3579. https://doi.org/10.1007/s10661-012-2810-y

Harinarayana P, Gopalakrishna GS, Balasubramanian A (2000) Remote sensing data for groundwater development and management in Keralapura watersheds of Cauvery basin, Karnataka, India. Indian Miner 34:11–17

Hutti B, Nijagunappa R (2011) Identification of groundwater potential zone using geoinformatics in Ghataprabha basin, North Karnataka, India. Int J Geomat Geosci 2(1):91–109

Jaiswal RK, Mukherjee S, Krishnamurthy J, Saxena R (2003) Role of remote sensing and GIS techniques for generation of ground water prospect zones towards rural development: an approach. Int J Remote Sens 24(5):993–1008

Magesh NS, Chandrasekar N, Soundranayagam JP (2011) Morphometric evaluation of Papanasam and Manimuthar watersheds, parts of Western Ghats, Tirunelveli district, Tamil Nadu India: a GIS approach. Environ Earth Sci 64(2):373–381

Magesh NS, Chandrasekar N, Soundranayagam JP (2012) Delineation of groundwater potential zones in Theni district, Tamil Nadu, using remote sensing. GIS and MIF techniques. Geosci Front 3(2):189–196. https://doi.org/10.1016/j.gsf.2011.10.007

Murthy KSR (2000) Groundwater potential in a semi-arid region of Andhra Pradesh: a geographical information system approach. Int J Remote Sens 21(9):1867–1884

Muralidhar M, Raju KRK, Raju KSVP, Prasad JR (2000) Remote sensing applications for the evaluation of water resources in rain fed area, Warangal district, Andhra Pradesh. Indian Miner 34:33–40

Mukherjee P, Singh CK, Mukherjee S (2012) Delineation of groundwater potential zones in arid region of India- a remote sensing and GIS approach. Water Resour Manag 26(9):2643–2672

Nag SK, Ghosh P (2012) Delineation of groundwater potential zone in Chhatna Block, Bankura District, West Bengal, India using remote sensing and GIS techniques. Environ Earth Sci 70(5):2115–2127. https://doi.org/10.1007/s12665-012-1713-0

Nag SK (2005) Application of lineament density and hydrogeomorphology to delineate groundwater potential zones of Baghmundi block in Purulia district, West Bengal. J Indian Soc Remote Sens 33:521–529

Rahmati O, Samani AN, Mahdavi M, Pourghasemi HR, Zeinivand H (2015) Groundwater potential mapping at Kurdistan region of Iran using analytic hierarchy process and GIS. Arabian J Geosci 8(9):7059–7071

Rokade VM, Kundal P, Joshi AK (2007) Groundwater potential modeling through remote sensing and GIS: a case study of Rajura Taluka, Chandrapur District, Maharashtra. J Geol Soc India 69(5):943–948

Roy S, Sahu AS (2016) Efectiveness of basin morphometry, remote sensing, and applied geosciences on groundwater recharge potential mapping: a comparative study within a small watershed. Front Earth Sci 10(2):274–291

Sadaf R, Mahar GA, Younes I (2018) Appraisal of ground water potential through remote sensing in River Basin, Pakistan. Int J Econ Environ Geol 9(3):25–32

Selvam G, Banukumar K, Srinivasan D, Selvakumar R, Alaguraja P (2012) Identification of ground water potential zone in hard rock terrain—a case study from parts of Manapparai block Tamilnadu using remote sensing and GIS techniques. Int J Adv Remote Sens GIS 1(1):8–18

Sener E, Davraz A, Ozcelik M (2005) An integration of GIS and remote sensing in groundwater investigations: a case study in Burdur, Turkey. Hydrol J 13(5–6):826–834

Solomon S, Quiel F (2006) Groundwater study using remote sensing and geographic information system (GIS) in the central highlands of Eritrea. Hydrogeol J 14(5):729–741

Srivastava PK, Bhattacharya AK (2006) Groundwater assessment through an integrated approach using remote sensing, GIS and resistivity techniques: a case study from a hard rock terrain. Int J Remote Sens 27(20):4599–4620

Subba RN, Chakradhar GKJ, Srinivas V (2001) Identification of groundwater potential zones using remote sensing techniques in and around Guntur town, Andhra Pradesh, India. J Indian Soc Remote Sens 29(1–2):69–78

Thomas A, Sharma PK, Sharma MK, Sood A (1999) Hydro-geomorphological mapping in assessing groundwater by using remote sensing data—a case study in Lehra Gage Block, Sangrur district, Punjab. J Indian Soc Remote Sens 27:31–42

Tiwari A, Rai B (1996) Hydro-morphological mapping for groundwater prospecting using landsat-MSS images—a case study of Part of Dhanbad District, Bihar. J Indian Soc Remote Sens 24:281–285

Chapter 8
Spatial Distribution of Reactive Nitrogen in Najafgarh Lake and Adjoining Canal, Delhi

Priyanka Verma and Pamposh

Abstract Despite suffering from various anthropogenic stresses, Najafgarh Lake is considered a haven for avian fauna especially flamingoes, which visit the water body in large numbers. Input of reactive nitrogen into the system is through drainage from surrounding agricultural field, wastewater intrusion from adjacent commercialized area (Gurugram) and macrophyte litter fall. In order to understand the ecological characteristics and the impact of anthropogenic activities, this study examined spatio-seasonal variability of selected reactive nitrogen species (ammonia, nitrate, organic nitrogen, and total nitrogen) in Najafgarh Lake and surrounding canal area using geostatistical module of ArcGIS, which is based on the monitoring data collected from 2017 to 2019. Yearly, the heterogeneity of ammonia, nitrate, organic nitrogen and total nitrogen from 2017 to 2019 is 12.97 mg/l (2017–18) < 18.95 mg/l (2018–19), 13.86 mg/l (2017–18) < 21.64 mg/l (2018–19), 41.42 mg/l (2017–18) > 30.91 mg/l (2018–19), 68.49 mg/l (2017–18) > 52.36 mg/l (2018–19), respectively. Seasonally, concentration of reactive nitrogen species was found to be high in winter followed by summer and monsoon. It was observed that the spatial distribution of nitrogen species was minimum at the inlet of the canal, which gradually increase reaching maximum at outlet point of the canal considered for this study. The spatial distribution results suggest the intrusion of reactive nitrogen through local non-point sources. The seasonal variation in water quality is found to be associated with hydrological and biochemical processes. These results will help us in application of integrated water quality management via sewage treatment plants, agricultural diffuse pollution control along with monitoring and assessment of the water body.

Keywords Najafgarh canal · Reactive nitrogen · Spatial distribution · Seasonal variation · ArcGIS

P. Verma (✉) · Pamposh
University School of Environment Management, GGSIP University, Delhi, India
e-mail: priyanka221290@gmail.com

R. K. Mishra et al. (eds.), *Sustainable Climate Action and Water Management*, Advances in Geographical and Environmental Sciences,
https://doi.org/10.1007/978-981-15-8237-0_8

8.1 Introduction

Najafgarh basin is scattered over Rajasthan, Haryana and NCT Delhi and is a sub-basin of Yamuna River. Najafgarh Lake is located on the southwest part of NCT Delhi covering Najafgarh Block and northwestern part of Gurugram district, Haryana. Before the 1960s, Najafgarh jheel used to be a vast lake with an area of about 220 sq. km in southwest part of Delhi in Najafgarh District (DIFC 2014; DLF 2008). It possessed extremely rich wetland ecosystem and was considered as a haven for avian fauna and local wildlife. In 1865, the Government of North West Province drained the jheel by excavating an irregular channel from the eastern part of the jheel to Yamuna. Later, this stretch was known as Najafgarh drain, which is spread over 51 km till Yamuna. Soon after the floods of 1964 and 1977, the Flood and Irrigation Department widened the Najafgarh drain, which led to the draining of lake, which was once ecologically rich in pretext of saving Delhi from severe flooding (DLF 2017).

Presently, it is a 7-km stretch that is straddling the Delhi–Haryana border equally, from which 5 sq. km. of the area remains underwater perennially. Drainage from Bahadurgarh, Jhajjar and Rewari falls into Najafgarh jheel through outfall number 8 reaching Dhansa Barrage and finally makes its way to Yamuna (DLF 2017). Najafgarh Lake is a typically shallow polluted lake. The thick mud embankments on both sides of the lake are forested with trees and shrubs, which inhabit the remnant local wildlife. It is considered as a haven for avian fauna especially flamingos, which visit this water body in large numbers. The open water of the lake is covered by an olive-green mat of a fast-growing weed *Eichhornia crassipes.*

Apart from the contamination coming from irrigation and drainage, industrial wastewater and most of the sewage from Gurugram are entering into Najafgarh Lake, which is causing a great influence on the quality of the lake water. To summarize, the pollution status of this water body is serious (Table 8.1).

8.2 Methodology

8.2.1 Sample Collection

The water samples were collected from Najafgarh Lake and canal seasonally (summer, monsoon and winter) from 2017 to 2019. The position of the sampling points was located using GPS (Gramin eTrex 20x). The distribution of the sampling points is shown in Fig. 8.1. The water sampling points were located randomly along the fringes of the lake. There were 12 sampling points along the lake and canal sides. Three replicates were collected from each sampling point.

Table 8.1 Sampling locations and their denomination

Sampling stations	Abbreviation	Location
1	NC1	Najafgarh canal (outlet)
2	NL1	Najafgarh lake (outlet of lake)
3	NL2	Najafgarh lake (middle point of lake)
4	NL3	Najafgarh lake (inlet point of lake)
5	NLOP	Najafgarh lake opposite (point on opposite boundary)
6	NC2L	Najafgarh canal (second point on left of canal)
7	NC3L	Najafgarh canal (third point on left of canal)
8	NC2R	Najafgarh canal (second point on right of canal)
9	NC3R	Najafgarh canal (third point on right of canal)
10	DDS2	Dhansa downstream (second point on canal)
11	DDS1	Dhansa downstream (first point on canal)
12	DUS	Dhansa upstream (inlet before barrage)

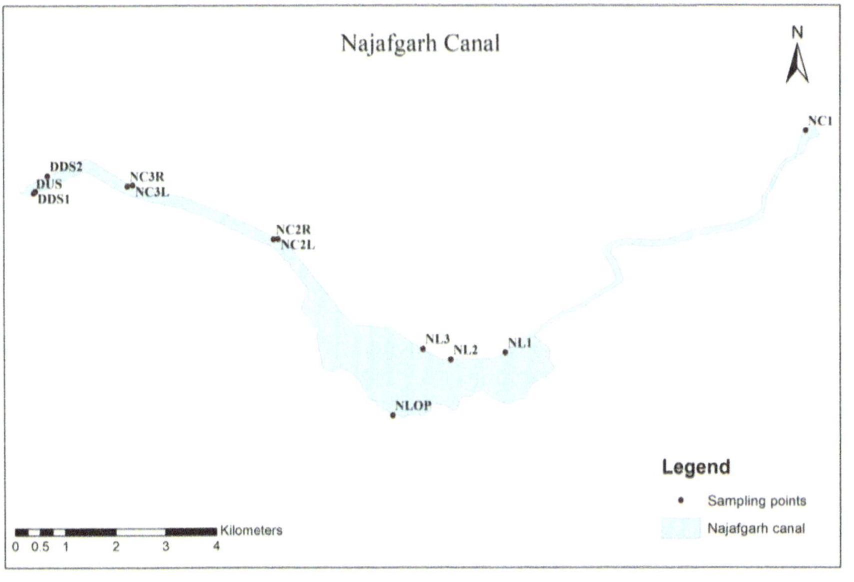

Fig. 8.1 Sampling points in Najafgarh canal

8.2.2 *Experimental and Analytical Methods*

The water samples were collected in polypropylene plastic bottles and were stored at constant temperature in an icebox. The samples were transported to the laboratory for further processing. The samples were filtered with the help of filtration

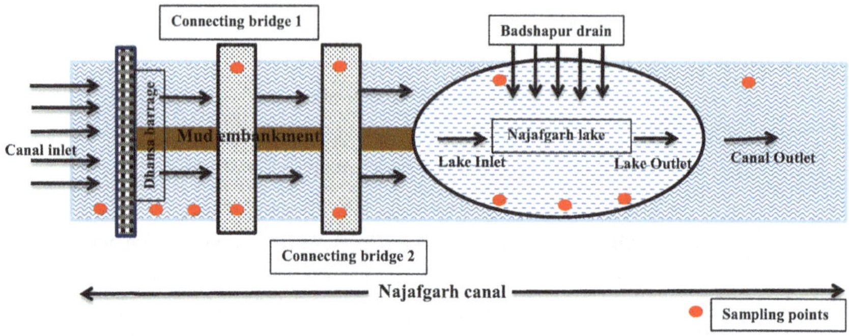

Fig. 8.2 Schematic representation of the flow of water through Dhansa barrage, outlet from Chhawla and arrangement of sampling points in Najafgarh Lake and the canal around

assembly using Grade 1 Whatman paper. The samples were analyzed for nitrate by UV spectrophotometry, nitrite by colorimetric method, ammonia by distillation method, organic nitrogen using total Kjeldahl method and total nitrogen was calculated using the sum of Kjeldahl nitrogen, nitrate and nitrite (APHA 1999; CPCB 2011). Geostatistics was calculated using the ArcGIS 10.1 software. The maps were generated with the help of inverse distance weighted (IDW) interpolation (Fig. 8.2).

8.3 Results and Discussion

8.3.1 Ammonia

Figure 8.3a–c shows that the range of ammonia is 0.82–9.32 mg/l, 1.52–10.03 mg/l and 1.60–42.62 mg/l in summer, monsoon and winter 2017–18, respectively. The summer, monsoon and winter average values are 3.40 mg/l, 2.3 mg/l, 6.26 mg/l, respectively. The IDW maps of different seasons show similar sort of spatial distribution of ammonia. Gradual increase in the concentration from inlet to outlet was observed in three seasons. The red color in the IDW map shows the maximum concentration of ammonia at NC1 (outlet canal) due to continuous sewage discharge from the nearby commercialized area and limited vegetation growth (Liu et al. 2016). Blue color shows minimum concentration that was recorded at DUS (inlet canal) in summer, monsoon and minutely high in winter season. Minimum concentration of ammonia was noted as only one drain (Badhshapur drain) joins the lake owing to low input of sewage (DLF 2018). Constant flow and dilution of the water channel decrease the concentration of ammonia (Park et al. 2011; Liu et al. 2016). Seasonally, the maximum concentration of ammonia was observed in winter (42.62 mg/l) followed by monsoon (10.03 mg/l) and summer (9.32 mg/l). The low concentration in summer and monsoon was due to high growth of planktons and aquatic plants, which are largely populating during these seasons (Abdel-Moati et al. 1990; Okbah and

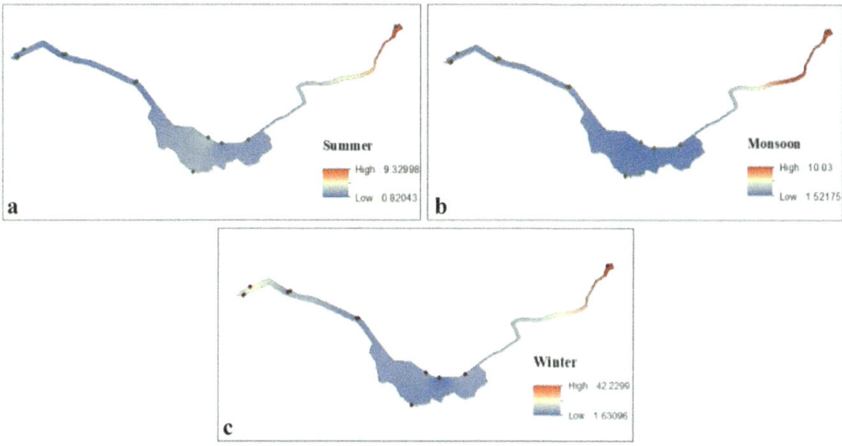

Fig. 8.3 **a**, **b**, **c** shows spatial and seasonal variation of NH_3 in 2017–18

Hussein 2006). Concentration of ammonia in the lake area was found maximum in monsoon (3.26 mg/l) followed by summer (2.99 mg/l) and winter season (2.28 mg/l).

Figure 8.4a–c shows the range of ammonia, which is 2.4–7.74 mg/l, 2.54–18.18 mg/l, 1.63–56 mg/l in summer, monsoon and winter 2018–19, respectively. The summer, monsoon and winter average values are 4.03 mg/l, 3.55 mg/l, 10.95 mg/l, respectively. The spatial distribution of ammonia in summer shows a similar pattern as previous summer season (2017–18) due to constant flow of wastewater (Fig. 8.4a). In monsoon, concentration of ammonia is maximum at the outlet point (NC1) due to large amount of flow of drainage, which is anthropogenically contaminated (Zhu et al.2011) (Fig. 8.4b). Figure 8.4c shows the maximum concentration at the outlet (NC1) and opposite boundary of lake (NLOP). It was possibly due to domestic

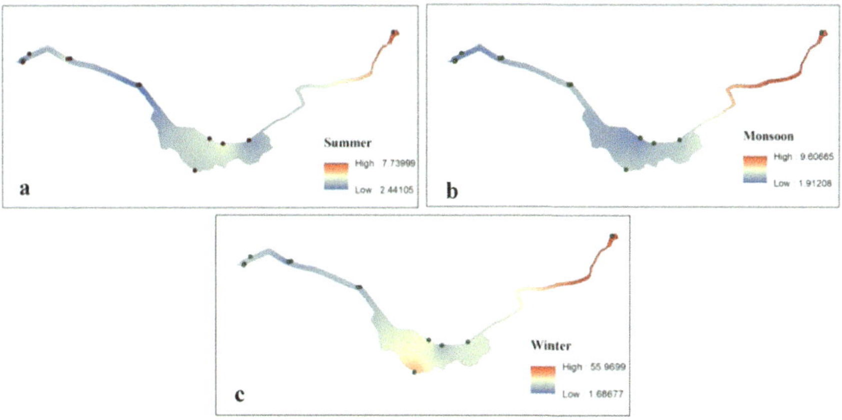

Fig. 8.4 **a**, **b**, **c** shows spatial and seasonal variation of NH_3 in 2018–19

wastewater discharge and agricultural runoff from the agricultural fields. Seasonally, the maximum concentration was observed in winter (56 mg/l) followed by monsoon (18.18 mg/l) and summer (7.74 mg/l). Concentration of ammonia in the lake was found maximum in winter (4.49 mg/l) followed by summer (2.82 mg/l) and monsoon (3.53 mg/l).

The heterogeneity of ammonia from 2017 to 2019 is 2017–18 (12.97 mg/l) < 2018–19 (18.95 mg/l), which shows an increasing pattern. Horizontally, the overall ammonia is gradually increasing from inlet to outlet of the canal.

8.3.2 *Nitrate*

Oxidative decomposition of ammonia forms nitrate as its final product (Meng et al. 2018). Figure 8.5a–c shows the range of nitrate, which is 2.87–24.15 mg/l, 0.52–0.85 mg/l and 0.58–36.71 mg/l in summer, monsoon and winter 2017–18, respectively. The summer, monsoon and winter average values are 6.58 mg/l, 0.7 mg/l and 6.62 mg/l, respectively. The spatial distribution of nitrate was maximum at outlet NC1 and NLOP in summer and winter (Fig. 8.5a and c), which may be primarily due to sewage discharge and agricultural runoff, respectively, around the sampling points (Meng et al. 2018). Seasonally, the maximum concentration was observed in winter (36.71 mg/l) followed by summer (24.15 mg/l) and monsoon (0.86 mg/l). Concentration of nitrate in the lake was found maximum in summer (6.01 mg/l) followed by winter (5.14 mg/l) and monsoon (0.68 mg/l).

The range of nitrate is 0.16–16.80 mg/l, 2.54–18.81 mg/l, 1.42–43.65 mg/l in summer, monsoon and winter 2018–19, respectively (Fig. 8.6 a, b, c). The average

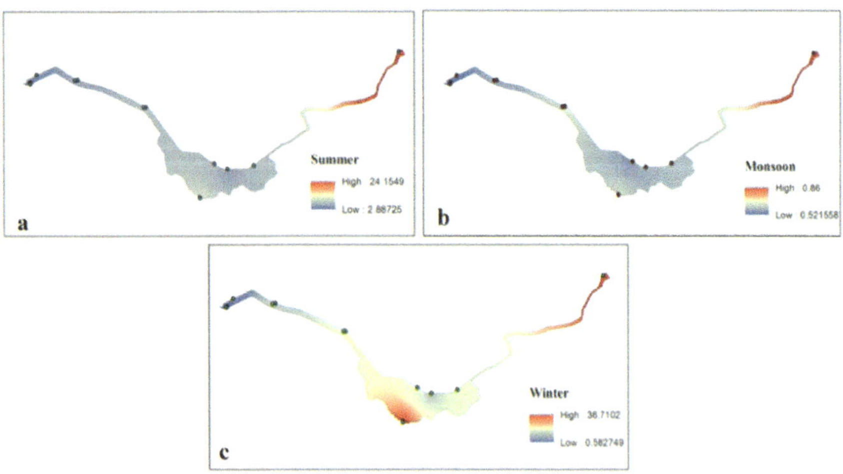

Fig. 8.5 **a**, **b**, **c** shows spatial and seasonal variation of NO_3 in 2017–18

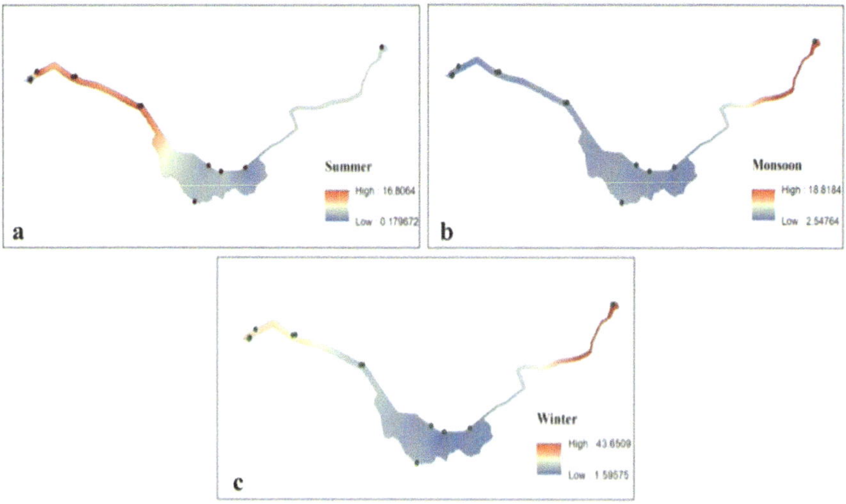

Fig. 8.6 **a**, **b**, **c** shows spatial and seasonal variation of NO_3 in 2018–19

values for summer, monsoon and winter are 7.03 mg/l, 7.14 mg/l and 7.46 mg/l respectively. Overall concentration increased in 2018–19 as compared with 2017–18 due to dredging activities taking place in the lake, which leads to the disturbance of the bottom sediment allowing it to mix with water. The spatial distribution of nitrate in summer was maximum at inlet points DDS2, NC2L, NC3L, NC2R and NC3R and minimum concentration throughout the lake (NL1, NL2, NL3) and outlet point NC1 (Fig. 8.6a). Monsoon shows the maximum concentration at the (NC1) outlet and minimum throughout the lake and canal area (Liu et al. 2016) (Fig. 8.6b). Figure 8.6c shows the maximum concentration at the inlet and outlet of the canal in winter. During the non-flood season when the barrage is closed leading to stagnation of water, which accumulates high amount of nutrients, which make it more concentrated. Seasonally, the maximum concentration was observed in winter (43.65 mg/l) followed by monsoon (18.81 mg/l) and summer (16.80 mg/l). Concentration of nitrate in the lake was found maximum in winter (4.49 mg/l) followed by monsoon (3.53 mg/l) and summer (2.82 mg/l).

The variation in nitrate concentration from 2017 to 2019 is 2017–18 (13.86 mg/l) < 2018–19 (21.64 mg/l). Nitrate gradually increases from inlet to outlet of the canal.

8.3.3 Organic Nitrogen

Dissolved organic nitrogen compounds in lakes and streams are product of photosynthetic organisms, excretion by the animals, leachate from the soil, discharge of sewage (Jørgensen 2009; Wall 2013). Figure 8.7a–c shows the range of organic

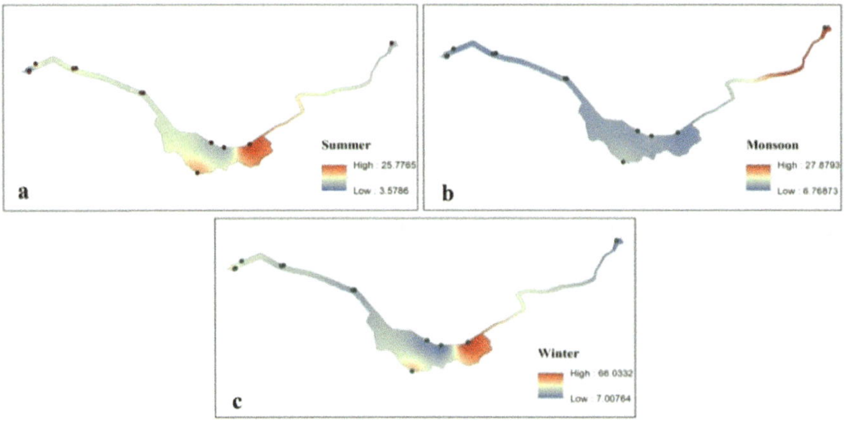

Fig. 8.7 **a**, **b**, **c** shows spatial and seasonal variation of organic nitrogen in 2017–18

nitrogen is 3.5–25.78 mg/l, 6.76–27.88 mg/l, 7.0–66.03 mg/l in summer monsoon and winter 2017–18 respectively. The average values for summer, monsoon and winter are 12.18 mg/l, 13.6 mg/l and 15.64 mg/l respectively. The spatial distribution of organic nitrogen in summer and winter was found to be maximum at NL1 (Outlet of lake) and NLOP (Fig. 8.7a and b). These points possibly have inputs from Badshapur drain and runoff from agricultural fields. Figure 8.7b shows the maximum concentration at outlet point NC1 in monsoon. Seasonally, the maximum concentration was observed in winter (66.03 mg/l) followed by monsoon (27.87 mg/l) and summer (25.77 mg/l). The average concentration organic matter at the lake was found maximum in summer (16.31 mg/l) followed by monsoon (15.89 mg/l) and winter (12.03 mg/l).

The range of organic nitrogen is 8.24–12.17 mg/l, 7.07–23.17 mg/l, 1.86–44.33 mg/l in summer, monsoon and winter 2018–19 respectively (Fig. 8.8a–c). The average values for summer, monsoon and winters are 10.40 mg/l, 13.02 mg/l and 7.47 mg/l respectively. In summer, the spatial distribution of organic nitrogen was observed maximum at inlet and subsequent points (DUS, DDS1, NC2L, NC3L, NC2R and NC3R) (Fig. 8.8a). Excess algal and plant growth at these points cause accumulation of dissolved organic nitrogen. Monsoon shows the maximum concentration at outlet point (NC1) (Fig. 8.8b). This may be due to excess discharge and flow of sewage leading to increased concentration (Wall 2013). The distribution of organic nitrogen shows high concentration at lake's outlet point (NL1), opposite boundary of lake (NLOP) and inlet point (DUS) (Fig. 8.8c). Seasonally, the maximum concentration was observed in winter (44.33 mg/l) followed by monsoon (23.17 mg/l) and summer (12.17 mg/l). The average concentration of organic nitrogen in lake is found maximum in monsoon (13.76 mg/l) followed by summer (9.39 mg/l) and winter (3.18 mg/l).

The heterogeneity of organic nitrogen from 2017 to 2019 is 2017–18 (41.42 mg/l) > 2018–19 (30.91 mg/l).

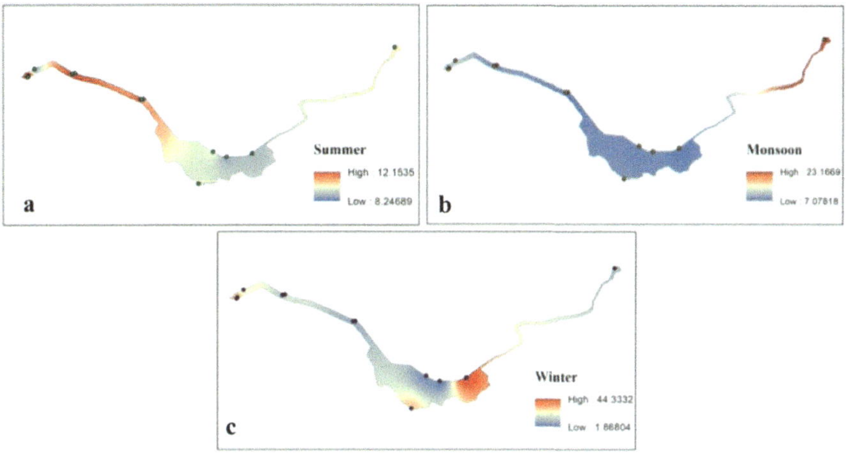

Fig. 8.8 **a**, **b**, **c** shows spatial and seasonal variation of organic nitrogen in 2018–19

8.3.4 Total Nitrogen

Total nitrogen comprises dissolved inorganic and organic nitrogen in water. It is a significant factor governing water quality. Figure 8.9a–c shows the range of total nitrogen 2.38–42.41 mg/l, 9.97–31.49 mg/l and 13.65–145.17 mg/l in summer, monsoon and winter 2017–18 respectively. The average values for summer, monsoon and winter are 22.30 mg/l, 17.59 mg/l and 28.60 mg/l respectively. The spatial distribution of total nitrogen was found to be maximum from NC1 to NL1 in summer (Fig. 8.9a). Outlet of the lake shows maximum concentration in monsoon (Fig. 8.9b).

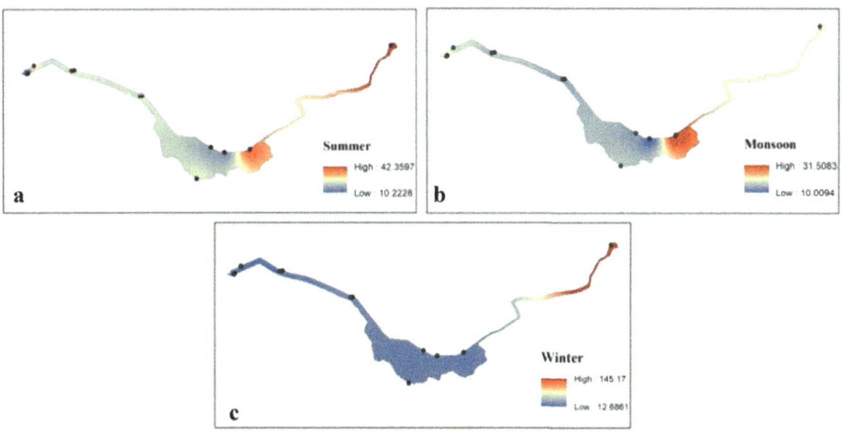

Fig. 8.9 **a**, **b**, **c** shows spatial and seasonal variation of total nitrogen in 2017–18

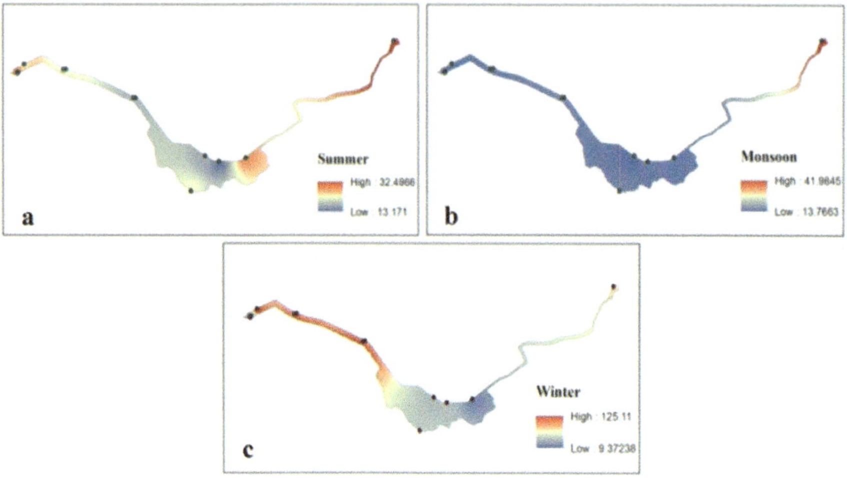

Fig. 8.10 **a**, **b**, **c** shows spatial and seasonal variation of total nitrogen in 2018–19

Figure 8.9c shows maximum concentration at the outlet point NC1 in winter. Seasonally, the maximum concentration was observed in winter (145.71 mg/l) followed by summer (42.41 mg/l) and monsoon (31.49 mg/l). In monsoon when precipitation is high, the hydrodynamic condition is increased, which decreases the retention time of the nutrient leading to decrease in concentration of total nitrogen (Goedkoop and Petersson 2000; Sun et al. 2009; Zhu et al. 2011). The average concentration of total nitrogen at the lake was observed maximum in summer (25.52 mg/l) followed by almost similar values in monsoon (19.84 mg/l) and winter (19.55 mg/l). The increased concentration of total nitrogen is the result of domestic sewage discharge, agricultural runoff and land-use change.

The range of total nitrogen is 13.11–32.49 mg/l, 12.73–38.78 mg/l and 2.28–36.27 mg/l in summer, monsoon and winters respectively (Fig. 8.10 a–c). The spatial distribution in summer was maximum at inlet points DUS, DDS1, and DDS2 (Fig. 8.10a). Monsoon season shows a high concentration at outlet point NC1 (Fig. 8.10b). Figure 8.10c shows the maximum concentration at DUS, DDS1, DDS2, NC2L, NC3L, NC2R and NC3R in winter. This may be due to the accumulation of nutrients in the absence of precipitation during non-flood season. Seasonally, the concentration of total nitrogen was approximately similar in summer (32.49 mg/l), winter (38.78 mg/) and monsoon (36.27 mg/l). The average concentration of total nitrogen in the lake was observed maximum in monsoon (21.46 mg/l) followed by summer (16.67 mg/l) and winter (4.29 mg/l).

The change in the concentration of total nitrogen from 2017 to 2019 is 2017–18 (68.49 mg/l) > 2018–19 (52.36 mg/l). Horizontally, total nitrogen was found to be increasing from inlet to outlet.

8.4 Conclusion

The concentration of ammonia, nitrate, organic nitrogen and total nitrogen differed temporally. Seasonally, the maximum reactive nitrogen was observed in winter season from 2017 to 2019. Ammonia and nitrate concentrations were discovered to be greater than categories D and A of CPCB and BIS IS: 10500 standards respectively. Standards for organic nitrogen and total nitrogen are not available in the Indian context. Spatially, the concentration of reactive nitrogen increased from inlet to outlet. Maximum concentration was observed at NC1 (outlet canal). The variables affecting the concentration of reactive nitrogen include wastewater discharge, irrigation return flow from agricultural areas and domestic wastewater from neighboring commercialized areas. Considering that the lake area has also decreased from 220 sq. km to 10 sq. km, appropriate and immediate conservation and revival measures should be taken. Since the aquifer near the lake is in a semi-critical condition, water quality improvement and restoration will assist to recharge the groundwater better. In order to implement better management methods, the point sources should be recognized. With the purpose of enhancing the water quality, stakeholders and local residents should be engaged in raising awareness. Understanding the spatial and temporal variability of physiochemical parameters and the concentration of various pollutants in water will assist in applying water quality management procedures efficiently.

References

Abdel-Moati MAR, Beltagy AI, Mamoney M (1990) Nutrient balance and biomass productivity interrelations in the coastal lagoon Burullus, Egypt. Rapp Comm Int Mer Medit 32(1):69

APHA (American Public Health Association) (2005) Standards methods for the examination of water and wastewater, 20th edn. American Public Health Association, Washington DC

Bureau of Indian Standards (BIS) (2012) Specification for drinking water. Bureau of Indian Standards IS: 10500, New Delhi, India. Available via https://www.indiawaterportal.org/articles/indian-standard-drinking-water-bis-specifications-10500-2012-second-revision. Accessed 20 Sept 2019

Central Pollution Control Board (CPCB) (1982) Surface water quality standards. Central Pollution Control Board. New Delhi, India. Available via Delhi land and finance foundation (DLF) (2017) Transformation of Najafgarh Basin under Gurgaon Rejuvenation project. Workshop was held on Transformation of Najafgarh Basin Gurgaon, India. Aug 2017 https://www.dlffoundation.in/grp/mentoring/

Department of Irrigation and Flood Control (DIFC) (2014). Flood Problem due to Sahibi River, Department of Irrigation and Flood Control, Government of NCT of Delhi, India. http://web.delhi.gov.in/wps/wcm/connect/DoIT_Irrigation/irrigation+and+flood+control/extra2/flood+problem+due+to+sahibi+river. Accessed 16 Sep 2019

Goedkoop W, Pettersson K (2000) Seasonal changes in sediment phosphorus forms in relation to sedimentation and benthic bacterial biomass in Lake Erken. Hydrobiologia 431(1):41–50. https://doi.org/10.1007/s10933-009-9403-y

Guide Manual: Water and Waste Water analysis, Central Pollution Control Board, New Delhi. https://www.indiawaterportal.org/sites/indiawaterportal.org/files/Water_and_wastewater_analysis_Guide_manual_Central_Pollution_Control_Board_2011.pdf. Accessed 13 Sep 2019

Jorgensen NOG (2009) Organic nitrogen. Encycloped Inland Waters 2:832–851. https://doi.org/10.1016/b978-012370626-3.00119-8

Liu J, Zhang X, Xia J et al (2016) Characterizing and explaining spatio-temporal variation of water quality in a highly disturbed river by multi-statistical techniques. Springer Plus 5(1):1171. https://doi.org/10.1186/s40064-016-2815-z

Meng Z, Yang Y, Qin Z et al (2018) Evaluating temporal and spatial variation in nitrogen sources along the lower reach of Fenhe River (Shanxi Province, China) using stable isotope and hydrochemical tracers. Water 10(3):231. https://doi.org/10.1371/journal.pone.0098833

Okbah MA, Hussein NR (2006) Impact of environmental conditions on the phytoplankton structure in Mediterranean Sea Lagoon, Lake Burullus, Egypt. Water Air Soil Pollut 172(1–4):129–150. https://doi.org/10.1007/s11270-005-9066-x

Park JH, Inam E, Abdullah MH et al (2011) Implications of rainfall variability for seasonality and climate-induced risks concerning surface water quality in East Asia. J Hydrol 400(3–4):323–332. https://doi.org/10.1016/j.jhydrol.2011.01.050

Sun B, Li CY, Yang ZY et al (2009) Changes of dissolved oxygen and analysis of the impact factor in main drainage channel in the Hetao irrigation district. China Rural Water Hydropower 5:68–75

Wall D (2013) Nitrogen in waters: forms and concerns. In: Wall D, Mulla D, Weiss S et al (eds) Nitrogen in Minnesota surface waters., Minnesota: Minnesota Pollution Control Agency, A2-1. https://www.pca.state.mn.us/sites/default/files/wq-s6-26a.pdf

Zhu D, Li C, Sun B et al (2011) Temporal and spatial distribution of total nitrogen, total phosphorus and COD in the largest shallow lake of the upper Yellow River. Paper presented at the International Symposium on Water Resource and Environmental Protection. Xi'an, Shannxi Province, China, 20-22 May 2011

Chapter 9
Effect of Pollution on the Sediment Bacterial Diversity and Composition in Najafgarh Lake, Delhi

Sandhya Bhat and Pamposh

Abstract Najafgarh Lake is an important bird and biodiversity area (IBA) on the Delhi Gurgaon border. It is home to many migratory as well as indigenous water birds. This wetland is under serious threat due to the adjoining drain that feeds it and carries untreated raw waste, industrial waste as well as agricultural runoffs into the lake making it highly degraded. The pollution in the lake has resulted in bioaccumulation of pesticide residues, trace metals and many toxic chemicals causing harm to aquatic flora and fauna, which in turn serves as food for the water birds. This study was undertaken for the first time in Najafgarh Lake to get an insight into the most vital component of this wetland ecosystem, i.e. bacteria using the novel approach of next-generation sequencing. The assessment of the community structure and diversity was done by sequencing the 16S rRNA amplicon. Results from sequencing showed that Najafgarh Lake harbors many phyla of bacteria like *Firmicutes*, *Proteobacteria*, *Acidobacteria* and *Bacteroidetes*. Saprophytic bacteria like *Firmicutes*, *Bacteroidetes* and *Actinobacteria* are predominant in this site while primary producers like *Nitrospira, Chloroflexi, Bacillariophyta* are less prevalent. Many bacterial species like *Sedimentibacter hydroxybenzoicus, Sulfurovum lithotrophicumuse* and *Bifidobacterium thermophilum* are the predominant members of the sediment bacterial community. The dominant presence of these bacterial species points to the fact that this lake is degraded and in dire need of restoration. Some novel bacterial communities were also found, which have the potential to be used as tools for restoration of this wetland.

Keywords Bacterial diversity · Bacterial community · 16S rRNA amplicon · Next-generation sequencing

S. Bhat (✉) · Pamposh
University School of Environment Management, GGSIP University, Sector-16C, Dwarka, New Delhi 110078, India
e-mail: sandhyabhatambardar@yahoo.com

R. K. Mishra et al. (eds.), *Sustainable Climate Action and Water Management*, Advances in Geographical and Environmental Sciences,
https://doi.org/10.1007/978-981-15-8237-0_9

9.1 Introduction

Microbes are an essential and integral part of every ecosystem (Rothschild and Mancinelli 2001; Konings et al. 2002), but their role is more predominant in case of wetlands as these are the drivers of different biogeochemical processes in wetlands (Reddy and DeLaune 2008). Bacteria play a key role in nutrient cycling and biogeochemical processes in the wetland (Reddy and DeLaune 2008) and provide critical ecosystem services to humankind (MEA 2005). The biogeochemical cycles that take place especially in wetlands are greatly influenced by microbial communities (Song et al. 2012; Liu and Liu 2013; Fang et al. 2015), and changes in the composition of these bacterial communities especially sediment microbial communities can significantly impact the biogeochemical environment of the wetlands (Van der Gucht et al. 2001; Haukka et al. 2006;). Sediment communities are more stable and there is less variation in them compared to surface and interstitial waters (Tiquia et al. 2006; Tiquia 2008; Oest et al. 2018) and thus these serve as reliable recorders of ecological responses to pollution (Nealson 1997; Li et al. 2016; Savidge et al. 2016). However, the composition and activity of the sediment bacterial communities are sensitive to environmental pollutants (Xu et al. 2014). There is a shift in the diversity, function and bacterial biomass with changes in nutrients and pollution (Flood et al. 2015). This happens by more tolerant species replacing sensitive ones and pollution degrading microbes becoming dominant members of the community (Freese et al. 2006). Information about the sediment bacterial community composition provides information about the sediment microorganisms that play key roles in nutrient cycling, heavy metal immobilization (Kandeler et al. 1996; Sandaa et al. 1999) and degradation of organic compounds (Xu et al. 2014). Bacteria serve as valuable tools for bioremediation in disturbed environments given their capabilities to degrade environmental contaminants and reduce toxicity (Parkinson and Coleman 1991; Aelion and Bradley 1991, Chaineau et al. 2005). Sediment bacterial populations can be used as a potential bioremediation strategy to overcome pollution problems associated with this wetland (Tiquia 2010; Freese et al. 2006). Microbial species in environmental samples can be identified by the comparative analysis of ribosomal RNA sequences. Among ribosomal RNA, the use of 16 s rRNA for microbial phylogenetic analysis has become reliable methodology (Woese and Fox 1977; Christen 2008; Cochrane et al. 2009; Benson et al. 2010). 16S rRNA metagenomic analysis generates sequences, which are free from PCR primer biases and thus provide good coverage (Hamady and Knight 2009). In spite of the advent of so many new methodologies, not much is known about the composition of bacterial communities in Najafgarh Lake and this study was undertaken.

9.2 Materials and Methods

9.2.1 *Study Site*

Najafgarh Lake is located on the Delhi–Gurugram border at 28.5365° N, 76.8726° E. Najafgarh Lake is already listed with Birdlife International as an Important Bird and Biodiversity Area (IBA). The ground truth about this lake is that it is in highly dilapidated condition due to apathy of people, administrative issues and majorly due to pollution. This lake is joined by Najafgarh drain, which was constructed to save Delhi from floods and which eventually drained the once huge and ecologically rich lake and polluted it severely. The Najafgarh Basin ranks 11th in a list of critically polluted industrial clusters released by CPCB in consultation with MoEF in 2009. The Najafgarh Drain basin tops the list of 33 "critically" polluted industrial hubs, scoring between 70 and 80 in the 'Comprehensive Environmental Pollution Index' (CEPI). The revival of this waterbody can bring about a lot of benefits. Due to its location, this waterbody has the potential to be developed as a tranquil public recreation place or a bird sanctuary in midst of a highly urbanized area.

9.2.2 *Sampling Strategy*

Water and sediment samples from Najafgarh Lake were collected in the summer of 2018 from five different sampling sites. From each site, both water and sediment samples were collected in triplicates from three random points, which were separated by a minimum distance of three meters from each other. A sediment corer was used and the top 15 cm of sediment was collected. The sediment was immediately sieved by using a 2 mm mesh, stored in polypropylene containers and kept in ice. Water samples were also collected in polypropylene bottles. Water pH, specific conductance and temperature were measured in the field using portable meters (WTW, Germany). Water samples (100 mL) were taken in glass vials to determine inorganic nutrient (ammonia, nitrate and nitrite) and dissolved oxygen (DO) concentrations. Both water as well as sediment samples were kept in ice, transported to laboratory and kept refrigerated (4 °C) until analysis within 24 h. In the laboratory, water was tested for DO (Modified Winkler's Method, APHA1999), TKN (Kjeldhal's method, APHA1999) Ammonia (Distillation and Volumetric analysis, APHA1995) Nitrate (UV method CPCB Manual) and Nitrite (Colorimetric method, CPCB Manual). Sediment was tested for water content (Drying method), TKN (Kjeldhal's method, APHA1999) and total carbon (Elemental Analyzer), C:N ratio was calculated. Sediment samples were subjected to DNA extraction.

9.2.3 DNA Extraction and Amplicon Library Preparation

The DNA from sediment samples was extracted by following the methodology of Mo Bio PowerSoil® DNA Isolation Kit. This purified DNA was used for 16S rRNA amplicon generation.

Bacterial and Archaeal 16S rRNA gene q PCR was done as described in Bomberg et al. (2016). V3–V4 region of the 16S rRNA genes present in the sediment metagenome was chosen for library preparation (Baker et al. 2003; Wang and Qian 2009). 25 ng of sediment metagenomic DNA was used to amplify 16S rRNA hypervariable region V3–V4 and creates a single amplicon of approximately 460 base pairs. DNA was amplified as recommended in Berry et al. (2011). The libraries were quantitated using Qubit DNA HS quantitation assay (Thermo Scientific, MA, USA) and Bioanalyzer 2100 (Agilent, CA, USA), which specifically quantitates dsDNA.

9.2.4 Sequence Processing and Analysis

Amplicon sequencing as well as processing of the data generated by the sequencing was performed on the Illumina MiSeq platform. The read data generated by the Mi Seq platform are in the form of FASTQ files. Reads were subjected to Preprocessing of reads, De-replication, Singleton removal, Operational Taxonomic Unit (OUT) Clustering, Chimera filtering. OTU picking was performed with the default UCLUST algorithm. The most abundant read in each OTU cluster was selected as the representative sequence using 97% identity. The default **Green gene reference database** (Version gg_13_5) was used to assign taxonomy to the representative sequence. Hence obtained non-redundant and representative OTUs were annotated up to species level followed by individual sample quantitation. All the above analysis was done using the bioinformatics software the **"Quantitative Insights into Microbial Ecology"** (QIIME v1.9.1) (Caporaso et al. 2010). The end product of the sequencing was the OTU table. Using the observed OTUs from OTU table, bacterial diversity, richness and taxonomic distribution of taxa were calculated. Abundance for each sample is subjected to Cumulative Sum Scaling (CSS) normalization. Diversity profiling was done at alpha and beta levels. Alpha diversity that calculates within-sample diversity was calculated using three indices namely Chao1, Shannon index and Simpson index. Beta diversity that calculates diversity for a group of samples was calculated between all the samples by using the Euclidean metric (Huson et al. 2016).

9.2.5 *Data Availability*

The sequence data from Najafgarh were submitted to NCBI sequence read archive (SRA) under accessions SRA: SRS4987034 accession number SAMN12058210, SRA: SRS4987033 accession number SAMN12058209, SRA: SRS497041 accession number SAMN12058208, SRA: SRS4987042 accession number SAMN12058208 and SRA: SRS4987045 accession number SAMN12058211.

9.3 Results and Discussion

9.3.1 *Water and Sediment Analysis*

The physio-chemical parameters for water column in the lake and the sediments are shown in Table 9.1. As the sampling was done during summer, the surface water temperature of the lake was high at around 30 °C, which influences the growth of bacteria that are more tolerant to heat. The pH of water recorded during the study was slightly alkaline (7.42). The lake waters showed low DO (4.5 mg/L), high nitrite (0.10 mg/l) and ammonia (3.17 mg/l). Low dissolved oxygen in the water column may be due to shallow and stagnant nature of the water column and the pollution. High ammonia found in natural water is regarded as indicative of sanitary pollution (Li et al. 2016), which is collaborated by the fact that this waterbody has a lot of sanitary waste flowing in it due to adjoining drains and human and animal defecation that is prevalent along its shores. Nitrite concentration was high for freshwater wetland and may be due to oxidation of excess ammonia by the ammonia-oxidizing bacteria.

Table 9.1 Physico-chemical parameters of water and sediment in Najafgarh Lake

S. No.	Water characteristics	Values
1	pH	7.8
2	Temperature (°C)	29.3
3	Specific conductance (uS/cm)	2017.23
4	Dissolved oxygen (mg/L)	4.45
5	Ammonium (mg/L)	3.17
6	Nitrates (mg/L)	0.54
7	Nitrites (mg/L)	0.10
S. No.	Sediment characteristics	Values (%)
1	Sediment carbon	3.7
2	Nitrogen content	0.069
3	Water content	3.5
4	C:N ratio	12.89

Table 9.2 Sediment bacterial community diversity index based on the Illumina Miseq sequencing data from Najafgarh Lake

Sample name	Total reads	OTUs	Chao1	Shannon index	Simpson index
NJL_1_Dry	1,40,088	122	215.3	5.88	0.96
NJL_1_Wet	1,02,186	211	437.5	5.96	0.95
NJL_2_Dry	1,94,720	346	369.7	6.65	0.97
NJL_2_Wet	1,36,274	207	363.5	4.60	0.79
NJL_3_Wet	1,76,110	373	523.1	5.99	0.95

9.3.2 DNA Analysis

The sequencing of the five samples generated total reads of 7,49,378. These reads were processed and the OTU table was generated. Chao1, Simpson Index and Shannon index of all the samples were calculated (Table 9.2). Chao1 estimate was found higher in NJL3 wetland sediments while Shannon and Simpson estimates were higher in NJL1 wetland sediment. To further examine the sediment bacterial community structure, the taxonomic classification in phylum levels was performed. It was found that the bacterial community in Najafgarh Lake was distributed among 13 known bacterial phyla and 2 belonged to Archaea.

9.3.3 Diversity and Abundance of Bacterial OTUs

The community structure of the Najafgarh wetland shows Firmicutes (29%), Proteobacteria (27%), Bacteroidetes (18%) as the major phyla that are dominating this wetland. Actinobacteria (7%), Aminicenantes (6%), Thermatogae (4%) and Caldiserica (4%) are also part of the top 10 bacterial phyla present in the study site as shown in Fig. 9.1. Many phyla like Thaumarchaeota, Caldiserica, Chloroflexi, Hydrogenedentes and Plantomycetesare also present in the wetland but in very less percentage.

Thaumarchaeota and *Euryarcheota* are the two archaeal phyla present. The archaeal population usually dominates freshwater sediments but in case of Najafgarh, only a few were found, which happens in case of pollution (Boon et al. 1996).

Aminicentes is a poorly categorized bacterial lineage and their ecological role is not known much. They have been found to exhibit the highest abundance in hydrocarbon-impacted wetlands (Farag et al. 2014). The presence of this phyla further collaborates with the fact that our site might be polluted with hydrocarbon residues as well.

Thaumaracheota is an uncultured archaeon and is highly diversified and ancient phylum. These are important ammonia oxidizers in aquatic and terrestrial environments and are the first archaea identified as being involved in nitrification. Thaumarchaeal strain has been shown to be able to use urea as a substrate for nitrification

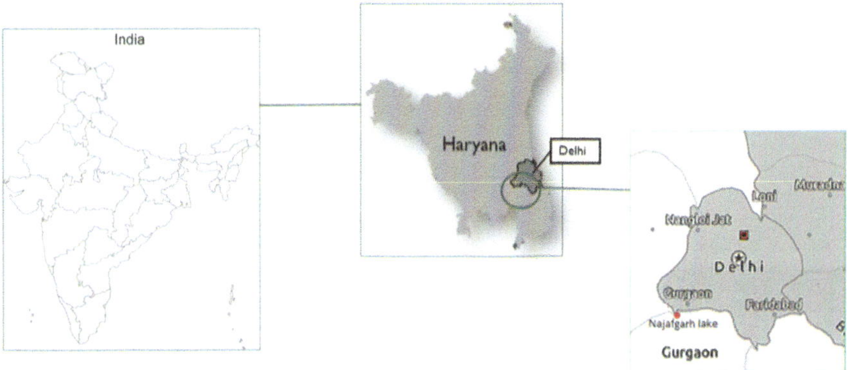

Fig. 9.1 Study area map showing sampling sites at Najafgarh Lake

(Pester et al. 2011; Brochie et al. 2012). As the site contains lots of sanitary waste, these bacteria seem to be fulfilling the role of nitrifiers given the relative absence of nitrifiers in our wetland.

Phylum *Firmicutes* is usually less abundant in freshwater sediments but in case of our site, it is the major phylum. *Sedimentibacter* (35%), *Clostridium* (20%) *Kandleria* (27%) are the major genus of bacteria belonging to phylum *Firmicutes* in Najafgarh Lake Sediments.

Phylum *Proteobacteria* was represented by 27% of total sequences. *Proteobacteria* belonging to all five classes namely *Alpha proteobacteria (20%), Beta proteobacteria (6.7%), Gamma proteobacteria (13.33%), Delta proteobacteria (46.7%) and Epsilon proteobacteria (13.33%)* were found (Fig. 9.2) *Delta*

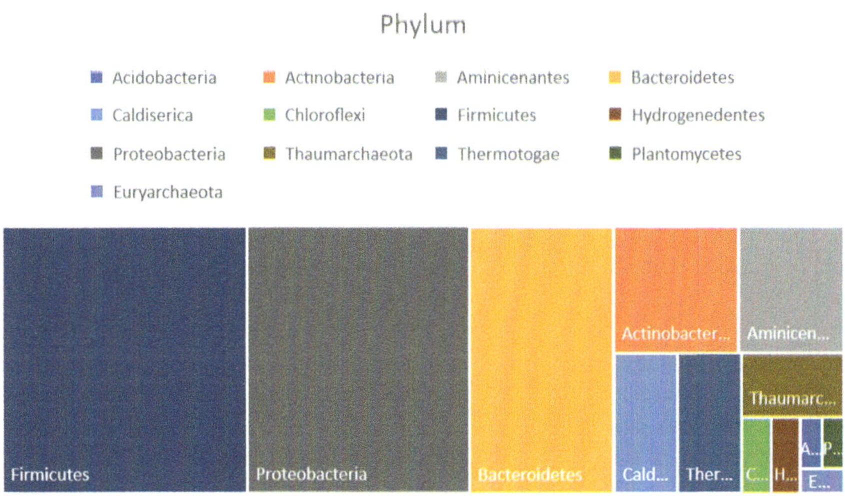

Fig. 9.2 Treemap of observed prokaryotic taxon shown in their hierarchical order

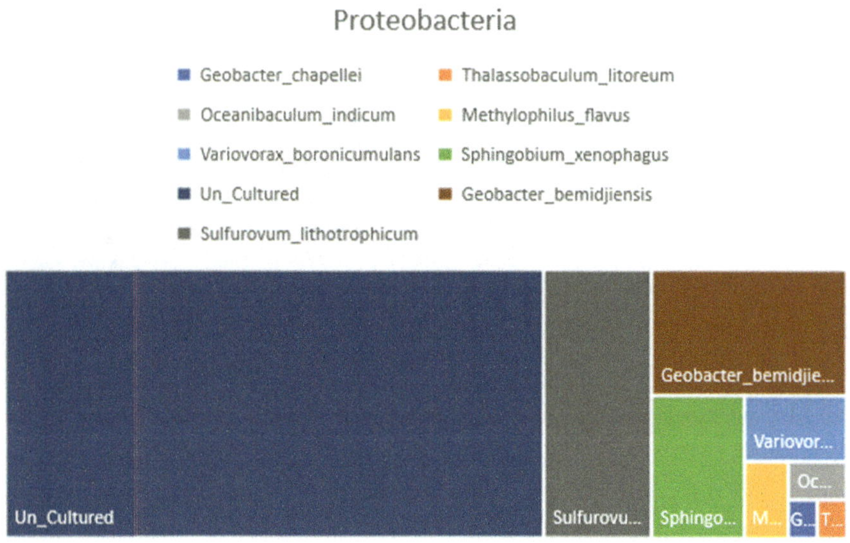

Fig. 9.3 Treemap of observed Proteobacteria taxon shown in their hierarchical order

Proteobacteria represented the most dominant subclass of *Proteobacteria* with *Geobacteraece* and *Syntrophacae* as the major families in this class.

Alpha proteobacteria were the second dominant class with *Sphingonmonad, Rohodospirale* and *Caulobactreace* as the predominant families. *Helicobactraecea* was the dominant family in *Epsilon Proteobacteria* class while *Sulfuricellaceae* and *Pseudumonads* were predominant in beta and gamma proteobacteria class. About 65% of bacteria of this phylum were reported as uncultured (Fig. 9.3).

Bacteroidetes was the third-largest phylum observed in Najafgarh sediments and *Macellibacterium, Lutaonella* and *Mangrovibacterium* were the most abundant genus in the Bacteroidetes phylum. These bacteria seem to play an important role in the degradation of complex biopolymers.

Actinobacteria is usually the dominant phylum in freshwater wetlands but it is not the predominant one in Najafgarh Lake. This points to the fact that our wetland is stressed. *Bifidobacterium* is the only genus of *Actinobacteria* present in our wetland.

The bacterial community structure of our wetlands resembles that of the structure of Anaerobic Digestion Sludge (ADS) from wastewater treatment plants (Yang et al. 2014) showing that our wetland is degraded (Pei et al. 2018). The abundance of saprophytic microbes (putrefying bacteria) in our study positively correlates to the fact that our study area is polluted and the pollution has affected the bacterial community structure by shifting the structure of communities at the phylum level as well as at lower taxonomic levels.

The diversity of the bacterial species within a phylum was also found to be less in accordance with Odum (1985) who stated that species diversity in stressed ecosystems decreases and as our study area is polluted, it has gradually brought about a decline in the microbial diversity in the wetland. Still, this wetland harbors many bacterial species, which have the potential to be used as indicator species like *Geobacter* and *Flavobacterium.*

9.4 Conclusion

This is the first report describing the bacterial diversity using 16S rRNA amplicon library sequencing method in the sediments of Najafgarh Lake. The present study has obtained primary insight into the major bacterial populations that are present in Najafgarh Lake sediments. The analysis has revealed the presence of phylogenetically diverse bacterial populations in this ecological niche and this knowledge at lower taxonomic levels is going to be very useful in designing bioremedial strategies for the restoration of this degraded wetland.

References

Aelion CM, Bradley PM (1991) Aerobic biodegradation potential of subsurface microorganisms from a jet fuel-contaminated aquifer. Appl Environ Microbiol 57(1):57–63

APHA (2005) Standard methods for the examination of water and wastewater, 21st edn. American Public Health Association, Washington, DC, New York

Assessment, M. E (2005) Ecosystems and human well-being, vol 5. Island press, Washington, DC

Baker GC, Smith JJ, Cowan DA (2003) Review and re-analysis of domain-specific 16S primers. J Microbiol Methods 55(3):541–555

Benson AK, Kelly SA, Legge R, Ma F, Low SJ, Kim J, Kachman SD (2010) Individuality in gut microbiota composition is a complex polygenic trait shaped by multiple environmental and host genetic factors. Proc Natl Acad Sci 107(44):18933–18938

Berry D, Mahfoud KB, Wagner M, Loy A (2011a) Barcoded primers used in multiplex amplicon pyrosequencing bias amplification. Appl Environ, Microbiol, p 7778467849

Berry D, Mahfoudh KB, Wagner M, Loy A (2011b) Barcoded primers used in multiplex amplicon pyrosequencing bias amplification. Appl Environ Microbiol 77(21):7846–7849

Blanck H (2002) A critical review of procedures and approaches used for assessing pollution-induced community tolerance (PICT) in biotic communities. Hum Ecol Risk Assess 8(5):1003–1034

Boon PI, Virtue P, Nichols PD (1996) Microbial consortia in wetland sediments: a biomarker analysis of the effect of hydrological regime, vegetation and season on benthic microbes. Marine Freshwater Res 47(1):27–41

Brochier-Armanet C, Gribaldo S, Forterre P (2012) Spotlight on the Thaumarchaeota. The ISME J 6(2):227

Caporaso JG, Lauber CL, Walters WA, Berg-Lyons D, Lozupone CA, Turnbaugh PJ, Knight R (2011) Global patterns of 16S rRNA diversity at a depth of millions of sequences per sample. Proc Natl Acad Sci 108(Supplement 1):4516–4522

Chaineau CH, Rougeux G, Yepremian C, Oudot J (2005) Effects of nutrient concentration on the biodegradation of crude oil and associated microbial populations in the soil. Soil Biol Biochem 37(8):1490–1497

Christen R (2008) Global sequencing: a review of current molecular data and new methods available to assess microbial diversity. Microbes Environ:0809220005-0809220005

DeLaune RD, Reddy KR (2008) Biogeochemistry of wetlands: science and applications. CRC Press
Fang L, Chen L, Liu Y, Tao W, Zhang Z, Liu H, Tang Y (2015) Planktonic and sedimentary bacterial diversity of Lake Sayram in summer. MicrobiologyOpen 4(5):814–825
Farag IF, Davis JP, Youssef NH, Elshahed MS (2014) Global patterns of abundance, diversity and community structure of the Aminicenantes (candidate phylum OP8). PLoS ONE 9(3):e92139
Cochrane G, Akhtar R, Bonfield J, Bower L, Demiralp F, Faruque N, ..., Jang M (2008) Petabyte-scale innovations at the European Nucleotide Archive. Nucleic Acids Res 37(suppl_1):D19–D25
Flood M, Frabutt D, Floyd D, Powers A, Ezegwe U, Devol A, Tiquia-Arashiro SM (2015) Ammonia-oxidizing bacteria and archaea in sediments of the Gulf of Mexico. Environ Technol 36(1):124–135
Freese HM, Karsten U, Schumann R (2006) Bacterial abundance, activity, and viability in the eutrophic river Warnow, Northeast Germany. Microbial Ecol 51(1):117–127
Han Y, Perner M (2015) The globally widespread genus Sulfurimonas: versatile energy metabolisms and adaptations to redox clines. Front Microbiol 6:989
Haukka K, Kolmonen E, Hyder R, Hietala J, Vakkilainen K, Kairesalo T, Sivonen K (2006) Effect of nutrient loading on bacterioplankton community composition in lake mesocosms. Microbial Ecol 51(2):137–146
Huson DH, Beier S, Flade I, Górska A, El-Hadidi M, Mitra S, Tappu R (2016) MEGAN community edition-interactive exploration and analysis of large-scale microbiome sequencing data. PLoS Comput Biol 12(6):e1004957
Kandeler E, Kampichler C, Horak O (1996) Influence of heavy metals on the functional diversity of soil communities. Biol Fertility Soils 23(3):299–306
Konings WN, Albers SV, Koning S, Driessen AJ (2002) The cell membrane plays a crucial role in survival of bacteria and archaea in extreme environments. Antonie Van Leeuwenhoek 81(1–4):61–72
Lauber CL, Hamady M, Knight R, Fierer N (2009) Pyrosequencing-based assessment of soil pH as a predictor of soil bacterial community structure at the continental scale. Appl Environ Microbiol 75(15):5111–5120
Li J, Lin S, Qin S (2016) Characteristics of sediment bacterial community in response to environmental impacts in a sewage polluted river. J Coastal Res 74(sp1):196–206
Liu Z, Liu J (2013) Evaluating bacterial community structures in oil collected from the sea surface and sediment in the northern Gulf of Mexico after the Deepwater Horizon oil spill. MicrobiologyOpen 2(3):492–504
Nealson KH (1997) Sediment bacteria: who's there, what are they doing, and what's new? Ann Rev Earth Planet Sci 25(1):403–434
Purkamo L, Bomberg M, Kietäväinen R, Salavirta H, Nyyssönen M, Nuppunen-Puputti M, ..., Itävaara M (2016) Microbial co-occurrence patterns in deep Precambrian bedrock fracture fluids. Biogeosciences
Oest A, Alsaffar A, Fenner M, Azzopardi D, Tiquia-Arashiro SM (2018) Patterns of change in metabolic capabilities of sediment microbial communities in river and lake ecosystems. Int J Microbiol Ecosyst 2018, Article ID 6234931:15
Parkinson D, Coleman DC (1991) Microbial communities, activity and biomass. Agricult Ecosyst Environ 34(1–4):3–33
Pei Y, Yu Z, Ji J, Khan A, Li X (2018) Microbial community structure and function indicate the severity of chromium contamination of the Yellow River. Front Microbiol 9:38
Pester M, Schleper C, Wagner M (2011) The Thaumarchaeota: an emerging view of their phylogeny and ecophysiology. Curr Opin Microbiol 14(3):300–306
Rothschild LJ, Mancinelli RL (2001) Life in extreme environments. Nature 409(6823):1092
Sandaa RA, Torsvik V, Enger Ø, Daae FL, Castberg T, Hahn D (1999) Analysis of bacterial communities in heavy metal-contaminated soils at different levels of resolution. FEMS Microbiol Ecol 30(3):237–251
Savidge WB, Wilson A, Woodward G (2016) Using a thermal proxy to examine sediment–water exchange in mid-continental shelf sandy sediments. Aquatic Geochem 22(5–6):419–441

Song H, Li Z, Du B, Wang G, Ding Y (2012) Bacterial communities in sediments of the shallow Lake Dongping in China. J Appl Microbiol 112(1):79–89
Tiquia SM (2008) Diversity of sulfate-reducing genes (dsr ab) in sediments from puget sound. Environ Technol 29(10):1095–1108
Tiquia SM (2010) Metabolic diversity of the heterotrophic microorganisms and potential link to pollution of the Rouge River. Environ Pollut 158(5):1435–1443
Tiquia SM, Masson SA, Devol A (2006) Vertical distribution of nitrite reductase genes (nir S) in continental margin sediments of the Gulf of Mexico. FEMS Microbiol Ecol 58(3):464–475
Van der Gucht K, Vandekerckhove T, Vloemans N, Cousin S, Muylaert K, Sabbe K, Vyverman W (2005) Characterization of bacterial communities in four freshwater lakes differing in nutrient load and food web structure. FEMS Microbiol Ecol 53(2):205–220
Wang Y, Qian PY (2009) Conservative fragments in bacterial 16S rRNA genes and primer design for 16S ribosomal DNA amplicons in metagenomic studies. PLoS ONE 4(10):e7401
Woese CR, Fox GE (1977) Phylogenetic structure of the prokaryotic domain: the primary kingdoms. Proc Natl Acad Sci 74(11):5088–5090
Xu M, Zhang Q, Xia C, Zhong Y, Sun G, Guo J, He Z (2014) Elevated nitrate enriches microbial functional genes for potential bioremediation of complexly contaminated sediments. ISME J 8(9):1932
Yang C, Zhang W, Liu R, Li Q, Li B, Wang S, Mulchandani A (2011) Phylogenetic diversity and metabolic potential of activated sludge microbial communities in full-scale wastewater treatment plants. Environ Sci Technol 45(17):7408–7415
Yang Y, Yu K, Xia Y, Lau FT, Tang DT, Fung WC, Zhang T (2014) Metagenomic analysis of sludge from full-scale anaerobic digesters operated in municipal wastewater treatment plants. Appl Microbiol Biotechnol 98(12):5709–5718

Chapter 10
Spatial Distribution of Uranium and Its Associated Water Quality Parameters (Drinking/Ground Water) in Raigad District of Maharashtra

Jyoti G. Koliyar, Priya Patil, S. K. Sahoo, and A. C. Patra

Abstract The present study investigates the concentration of naturally occurring uranium levels and associated water quality parameters of drinking/groundwater samples from Raigad District of Maharashtra. The water samples collected were analyzed for both pre-monsoon and post-monsoon. This study is based on health hazard point of view. Systematic district-wise grid sampling plan was followed in the project and samples were selected on the basis of population density. The estimation of uranium was done by using LED fluorimeter. Along with uranium and other associated physico-chemical parameters such as Temperature, Total Alkalinity, Total Hardness, Magnesium Hardness, Calcium Hardness, Total Dissolved Solids, Chloride, Fluoride, Sulphate, Phosphate, Nitrate, Salinity, ORP, Electrical Conductivity were also analyzed. Statistical tools were applied to analyze the data and its spatial distribution. Finally the variations were crosschecked with respect to recommendation given by BIS/WHO limit. The parameters such as TDS, EC, TH, chlorides, Total Alkalinity are exceeding the permissible limit in some areas whereas other parameters are within the permissible limits. Uranium vary from 0.2 to 0.7 ppb in pre-monsoon and 1–2.6 ppb in post-monsoon which was well below the recommended permissible limits suggested from various organizations. Therefore water in this region is generally fit for drinking purpose and it is recommended that proper treatment methods need to be adopted before the water is used for drinking purpose.

Keywords Uranium · Groundwater · LED fluorimeter · Water quality parameters · BIS/WHO

J. G. Koliyar (✉) · P. Patil
Department of Environmental Sciences, SIES (Nerul) College of Arts, Science and Commerce, Nerul, Navi Mumbai 400706, India
e-mail: jyotijatinderdas20@gmail.com

S. K. Sahoo · A. C. Patra
Health Physics Division, Bhabha Atomic Research Centre, Trombay, Mumbai 400085, India

R. K. Mishra et al. (eds.), *Sustainable Climate Action and Water Management*, Advances in Geographical and Environmental Sciences,
https://doi.org/10.1007/978-981-15-8237-0_10

10.1 Introduction

Water present on the surface of the earth is essential for human existence. Plants and animals need water to survive. If there is no water there would be no life. Apart from drinking to survive, people have many other uses too which include washing, cooking and recreation. Water is also essential for the healthy growth of farm crops etc. With drinking water, people usually check if the water is clear or not and if it looks clean, they guess that it is safe to consume. But this is often not the case. Water looking clean and dirt free may contain some unseen contamination which is not seen through naked eyes and so water quality testing is necessary. Water quality testing is important because it identifies contaminants and prevents water-borne diseases. According to Central Pollution Control Board, 90% of the water supplied in India to towns and cities is polluted, out of which only 1.6% gets treated. Therefore, water quality management is fundamental for human welfare (AERB 2004). About 97.2% of water on earth is salty and only 2.8% is present as fresh water from which about 20% constitutes groundwater (Tripathi et al. 2014). Essentially, water quality testing makes sure that water is safe and meets the local standards. This type of testing can be completed using water sampling techniques and using technology to estimate the amount and level of chemicals. Uranium is a lithophilic, heavy, actinide and last naturally occurring radioactive element in the periodic table. It is the nuclear fuel of the present global and national nuclear power programme. It prevails in the atmosphere, hydrosphere, lithosphere and biosphere with wide variation in concentration naturally since the inception of the earth (UNSCEAR 2016). Uranium is a naturally occuring radioactive element, which is both radiologically and chemically toxic. Uranium is widespread in nature, occuring in granites and various other mineral deposits. Intake of uranium through drinking-water is normally low; however, in circumstances in which uranium is present in a drinking-water source, the majority of intake can be through drinking-water. Uranium gets mixed with groundwater by leaching process and enters into our food chain (Saini et al. 2016). There is a need of assessment of uranium concentration in water and it is considered to be one of the important fuel for nuclear power (Kansal et al. 2011). Under The National Uranium Project (NUP), a study is carried out to establish the spatial distribution of uranium and associated water quality parameters in Raigad district of Maharashtra. This data will be a part of the National database generated by Health physics division (HPD), HSAE Group, BARC, Mumbai. Distribution of Uranium concentration along with associated water quality parameters are studied.

10.2 Materials and Methods

As per the BARC standard protocol a study was carried out for the National Uranium Project. The district was divided into optimized grid size of 6 × 6 km using latitude—longitude as reference coordinates for screening. Minimum one drinking water sample was collected from each grid in pre-acid cleaned polythene bottle in both pre-monsoon and post-monsoon. Gamma radiation was measured at each sampling location. The samples collected were analysed for various physico-chemical parameters such as pH, Electrical Conductivity (EC), Total Dissolved Solids (TDS), Oxidation Reduction Potential (ORP), Salinity, Dissolved Oxygen (DO), Temperatures were measured using portable sensors in the field. Within 24 h of sampling, the total hardness (TH) and total alkalinity (TA) of the water samples were measured by conformist titrimetric method. Uranium was also analysed by standard addition method using LED based fluorimeter. In addition other parameters carbonate, bicarbonate, nitrate, chloride, fluoride, sulphate and phosphate were also analyzed.

10.3 Results and Discussions

Minimum, maximum, average, and median values of gamma radiation 43, 124, 80.5 and 77 nsv/h found in Raigad district during pre-monsoon respectively. Minimum, maximum, average, and median values of gamma radiation found in Raigad district during post-monsoon were 47, 132, 77.4 and 74.5 nsv/h. In the present study, 121 drinking water samples were collected and analyzed for uranium and other water quality parameters for pre-monsoon and post-monsoon. The summary of results is presented in Tables 10.1 and 10.2 for pre- and post-monsoon, respectively. pH, Electrical Conductivity, Total Dissolved Solids, Total Alkalinity, Total Hardness,

Table 10.1 Summary of U and water quality parameters in Raigad district of Maharashtra—Pre-monsoon

Parameter	Pre-monsoon			
	Min	Max	Average	Median
pH	6.2	8.4	7.3	7.4
TDS (ppm)	189	1587	480	439
EC (μS/cm)	295	2365	724	652
Salinity(ppm)	148	1183	362	326
ORP (mV)	116	456	181	167
DO (ppm)	2	6.5	4.4	4.5
F^- (ppm)	0.31	0.99	0.65	0.65
Cl^- (ppm)	74	591	181	163
NO_3^- (ppm)	1.3	6.4	3.5	3.5
SO_4^{2-} (ppm)	11	46	20	19
PO_4^{3-} (ppm)	0.1	0.9	0.5	0.5
TH (ppm)	123	523	257	248
TA (ppm)	52	208	95.4	88
U (ppb)	0.2	0.9	0.7	0.7

Table 10.2 Summary of U and water quality parameters in Raigad district of Maharashtra—Post-monsoon

Parameter	Post-monsoon			
	Min	Max	Average	Median
pH	6.2	8.6	7.4	7.4
TDS (ppm)	76	1187	286	231
EC (μS/cm)	124	1746	430	345
Salinity (ppm)	62	873	215	173
ORP (mV)	124	324	184	172
DO (ppm)	2.3	6.9	4.5	4.5
F^- (ppm)	0.14	0.99	0.64	0.65
Cl^- (ppm)	31	132	107	86
NO_3^- (ppm)	1.2	6.5	2.8	2.7
SO_4^{2-} (ppm)	7.8	32	18	18
PO_4^{3-} (ppm)	0.1	0.9	0.5	0.6
TH (ppm)	60	421	172	156
TA (ppm)	26	683	111	88
U (ppb)	1	1.4	2.6	2.2

Chloride, Fluoride, Sulphate, Phosphate and Nitrate in the water samples were found to vary in the range of 6.2–8.6, 124–2365μcm^{-1}, 76–1587 ppm, 26–683 mg/l, 60–523 mg/l, 31–591 mg/l, 0.14–0.99 mg/l, 7.8–46 mg/l, 0.1–0.9 mg/l, 1.2–6.5 mg/l, respectively. The TDS level was found to be higher than the BIS acceptable limit of 500 ppm. The average TDS level in the district was estimated to be 486 and 286 ppm in pre-monsoon and post-monsoon, respectively. The TDS level was found to be higher than the BIS acceptable limit of 500 ppm. Uranium levels were found to vary from 0.2 to 0.9 ppb and 1–1.4 ppb in pre-monsoon and post-monsoon, with a median value of 0.7 and 2.6 ppb, respectively Graphs 10.1 and 10.2.

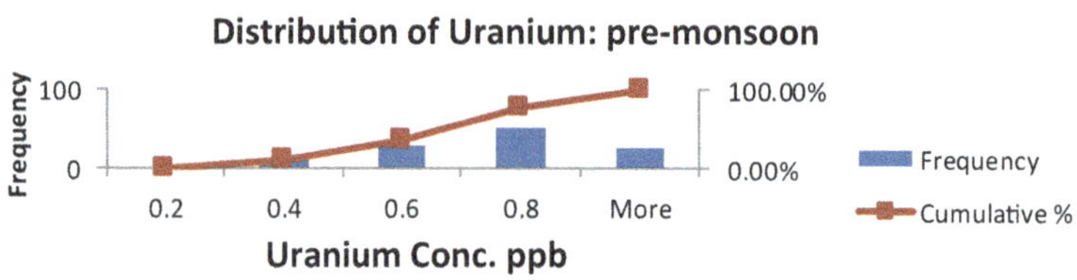

Graph 10.1 Discrepancy in uranium content during pre-monsoon

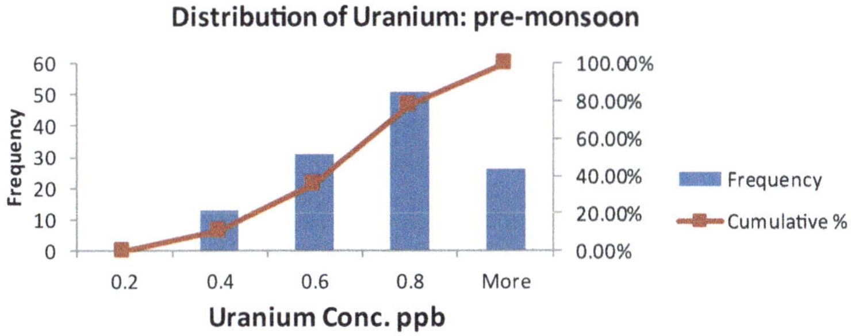

Graph 10.2 Discrepancy in Uranium content during post-monsoon

10.4 Conclusion

The physico-chemical parameters of the collected water samples have been assessed in order to know the overall potable quality of water from Raigad region. Observations revealed that the total number of potable water samples collected and analyzed for various water quality parameters like TDS, EC, Salinity, Alkalinity, Total hardness, ORP, Fluoride, Chloride, DO, Phosphate showed variations in the concentration for both pre-monsoon and post-monsoon. Hence a constant monitoring and water treatment is recommended as parameters like TDS, Total Alkalinity, Hardness, Chloride are exceeding the WHO/BIS recommended values. Measured values of uranium and other water quality parameters showed variation in concentration from place to place. The range of Uranium varies from 0.2 to 0.9 ppb in pre—monsoon and 1–1.4 ppb in post-monsoon respectively. From the study it is observed that Uranium concentration was well below the recommended permissible limits suggested from WHO/AERB limit. Uranium is the naturally occuring radioactive elements present in nature so its main source is natural, due to leaching from natural deposits and also due to human activities like declining ground water table, use of phospate and nitrate fertilizers, etc. Drinking water containing uranium can increase the persons lifetime risk of cancer. Studies of humans exposed to abnormally high levels of uranium show that uranium can be chemically toxic to the kidneys. The potential health effects from uranium in drinking water comes from its heavy metals characteristic and not from its radioactivity, which is very low.

Acknowledgements This study was funded by BRNS, DAE under NUP. The Authors would like to acknowledge members of SIESASCN, TSC-4, NRFCC, BRNS; HPD, HS&E Group, BARC and NUP team members for their continuous support in the execution of the project

References

AERB (2004) Limit on uranium in drinking water, Atomic Energy Regulatory Board

Kansal S, Mehra R, Singh NP (2011) Uranium concentration in ground water samples belonging to some areas of Western Haryana, India using fission track registration technique. J Public Health Epidemiol 3(8):352–357. ISSN 2141-2316

Saini K, Singh Bajwa B (2016) Uranium distribution study in the drinking water samples of SW Punjab, India. Adv Appl Sci Res 7(2):103–108. ISSN:0976-8610

Tripathi AD, Agrawal M, Mahaveer T (2014) Assessment of drinking water quality: A case study of Moradabad Area, Uttar Pradesh, India. Int J Environ Sci 5(2). ISSN 0976-4399

UNSCEAR (2016) Sources and effects of ionizing radiation, United Nations Scientific Committee on the Effect of Atomic Radiation, Annex D—Biological effects of selected internal emitters—Uranium. United Nations, New York

Chapter 11
Water Stress and Water Crisis in Large Cities of India

Priyanka Ghosh

Abstract Large cities of India are facing water shortages, and many cities such as Mumbai, Jaipur, Bhatinda, Lucknow, Nagpur, and Chennai are facing acute water shortages. According to a report of Niti Aayog, Government of India, published in 2018, 21 major cities, including Delhi, Bengaluru, Chennai, and Hyderabad, are expected to deplete their groundwater by 2020, which would directly affect 100 million people. The Composite Water Index Scores published by the Niti Ayog show that only Gujrat, Madhya Pradesh, and Andhra Pradesh have efficient water resource management. A report published by the WaterAid in 2018 also stated that in near future the demand for water will be on the rise due to rapid growth of Indian cities. As the number of people living in urban areas has increased since 2001, many Indian cities will face a water crisis. Cities like Mumbai, Jaipur, Nagpur, Bhatinda, and Lucknow have already started water rationing during the summer months. In November 2018, the Brihanmumbai Municipal Corporation (BMC) declared 10% water supply cut for Mumbai residents. Furthermore, there was a 15% cut in supply time. Similarly, Jaipur also imposed a cut on the daily water supply of the city prior to the onset of monsoon in 2018. Instead of daily cut in water supply, Chennai provides water to its residents every other alternative day. Bhatinda and Lucknow lack any sustainable plan of water conservation and unable to find alternative sources of water. Based on the available literature, this paper examines the problem of water crisis in Indian cities and identifies the key drivers of water shortages. Additionally, the paper aims to provide a comprehensive overview of urban water crisis in India as well as highlights possible solutions.

Keywords Urban water stress · Groundwater · Water crisis · Cities · India

---

P. Ghosh (✉)
Geography and Environmental Science, School of Liberal Arts and Sciences, Era University, Lucknow, UP 226003, India
e-mail: pghos2@gmail.com

R. K. Mishra et al. (eds.), *Sustainable Climate Action and Water Management*, Advances in Geographical and Environmental Sciences,
https://doi.org/10.1007/978-981-15-8237-0_11

11.1 Introduction

Many cities in India including Mumbai, Jaipur, Bhatinda, Lucknow, Nagpur, and Chennai are facing acute water shortages. According to a report published by the Niti Aayog, a think tank of Government of India, 600 million people are facing extreme to high water stress in India. The report says that by 2030, India's water demand is predicted to be twice the supply. Groundwater in India comprises 40% of the country's water supply. Furthermore, 54% of India's groundwater sources are depleting, and 21 major cities including Delhi, Bengaluru, Chennai, and Hyderabad, are expected to deplete their groundwater by 2020, which would directly affect 100 million people (Sengupta 2018). India will lose 6% of its GDP by 2050 if no measures are taken (Koshy 2018).

The Composite Water Index scores published by the Niti Ayog show that only Gujrat, Madhya Pradesh, and Andhra Pradesh have efficient water resource management (Ibid.). The index rank 24 states in total and all states were categorized into two groups: non-Himalayan states and northeastern and Himalayan states. Jharkhand, Bihar, and Haryana ranked last in the non-Himalayan group of states.

A report published by the WaterAid in 2018 also stated that in near future, the demand for water would be on rise due to the rapid growth of Indian cities. As the number of people living in urban areas has increased since 2001, many Indian cities will face a water crisis. Cities like Mumbai, Jaipur, Nagpur, Bhatinda, and Lucknow have already started water rationing during the summer months. In November 2018, the Brihanmumbai Municipal Corporation (BMC) declared 10% water supply cut for Mumbai residents. Additionally, there was a 15% cut in supply time (Sengupta 2018). Since November 2018, Nagpur Municipal Corporation allocated 155 million cubic meters of water instead of its usual 195 million cubic meters (Sengupta 2019). Similarly, Jaipur also imposed a cut on the daily water supply of the city prior to the onset of monsoon in 2018. Instead of providing water daily, residents of Chennai receive water every other alternative day. Bhatinda and Lucknow do not have any sustainable water conservation plan and these cities are unable to find alternative sources of supplying drinking water.

11.2 Materials and Method

Based on available literature on water shortages, this review paper examines the problem of water crisis in Indian cities and identifies the key drivers of water shortages. The reviewed literature comprises national and international peer-reviewed journal articles, published reports and current newspaper articles focusing on the issue of urban water shortages in various parts of the world including India. The paper provides a comprehensive overview of urban water crisis in India as well as possible solutions for urban water shortages. The first section of the paper discusses the water-scarce cities in the world and how those cities have diversified water supply and adopted innovative measures to deal with the problem. In the context of global water crisis, the second section of the paper examines the water shortages in Indian cities. The paper highlights and discusses some possible measures and strategies India can adopt to address urban water shortage and water crisis in the future.

11.2.1 Global Scenario of Water Scarcity

Population growth, affluence, rapid urbanization, in an era of climate change, and unsustainable water resource management have created water-scare cities. In some regions, water insecurity is further aggravated by frequent and sometimes prolonged drought (World Bank 2018). Climate change would affect the world's groundwater reservoirs directly due to change in the recharge pattern of aquifers or indirectly through increased demand, especially through increased demand for water for irrigation (World Bank 2018). According to McDonald and others, 1.9 billion people will be facing seasonal urban water shortages by 2050 (McDonald et al. 2011). Also, by 2050, about 1 billion people will face perennial water shortage within the urban area. Here, the authors have defined water shortages as "the amount of water physically available" and do not include issues related to water quality and distribution (Ibid., 1). Perennial water shortage occurs when less than 100-liter water available for a person per day within the urban extent. In other words, McDonald and others (2011) only considered if water is available nearby to users. It has been estimated that cities in the Middle East and North Africa will face perennial water shortages whereas a large number of cities in India and China will face the problem of seasonal water shortages (Ibid.). Overall, the world may experience a shortage of 2,700 billion cubic meters of water by 2030 (World Bank 2018).

Frequent water scarcity in an urban area leaves an impression of government's failure to provide safe drinking water. It also leads to social conflicts, tensions, and ultimately affects public health (Damania et al. 2017; World Bank 2018). Interrupted water service for a restricted number of hours in a day in many water-scarce cities may result in bacterial growth in the water distribution network. Unsustainable water resource management such as unsustainable management of waste and stormwater leads to depletion of groundwater without adequate recharge. Cities that are rapidly growing create additional pressure on depleting and scarce water resources (World Bank 2018). Cities such as Marrakesh in Morocco, Amman in Jordon, Windhoek, Namibia, and Tucson, Arizona are facing challenges to provide safe potable water due to increase in demand, depletion and deterioration of availability and quality of water resources, climate change, changing priorities in historical sources, and growing competition with other users (Ibid, 6).

Urban water stress in developing countries is linked to deficiencies in water distribution, such as age-old distribution system causing excessive leakages from pipes. It has been seen that leakage of water through pipe breaks is higher in number in developing countries as compared to developed nations (Lee and Schwab 2005). For instance, the city of Bogota, Columbia reported 70 breaks per 100 km per year, whereas the United States experienced 17 breaks on average per 100 km per year (Ibid., 9). Broken and poorly maintained water pipes also increase the chance of contamination as the pipes may open into gutters (Ibid.).

Marrakesh's water demand increased due to its boom in tourism industry. Instead of looking for distant water resources, the city tried to find out an innovative solution for its water stress by treating wastewater. Similarly, Amman, the capital of Jordan, also faces challenges of providing water to its citizens due to rapid population growth and to overcome water deficit, the city is planning to move towards desalinization (World Bank 2018). Certain cities such as Perth in Australia are facing water deficit due to the effects of climate change. Climate change has affected the amount of annual rainfall in Perth. The city had experienced a 20% reduction in its amount of rainfall since 1970s (Ibid.). Water deficits in urban areas may result from change in priorities by the traditional and historic water providers. Examples of change in priorities can be drawn from Singapore, which met its water demand through agreements with Malaysia (Ibid.). Since 1960s, Singapore has imported water from Johor, Malaysia, and Malaysia supplied 40% of its water (Seth 2015). Singapore aspires to reduce and eliminate its dependence on Malaysia by 2060 and to achieve its goal, the city–state aims to tap water from three sources: wastewater, desalinized seawater, and rainwater harvesting (Ibid.). The Public Utility Board of the city–state has already installed a desalinization plant, which can treat 135 million liters of seawater per day (Ibid.). Today, many of the world's water scarce cities are creatively adapting to technical, institutional, and regulatory measures to efficiently manage urban water insecurity. These measures include efficient water use and water conservation practices as well as looking for non-conventional water sources (Ibid.). The following section examines the water scarcity and water stress in urban areas of India and identifies the key drivers of water shortages.

11.2.2 Water Scarcity and Water Pollution in India

Water scarcity is a vital and emerging problem for India and many Indian cities are facing immense challenges to supply safe potable water to their residents. Up to 70% of India's water supply is contaminated (Niti Aayog 2018), based on available government data, Water-Aid, a London-based non-profit organization found that 80% of India's surface water is polluted (Sengupta 2019). According to the Water-Aid, the major drivers of such pollution are domestic sewage, lack of sanitation facilities, and the absence of wastewater and sanitation policies (Ibid.). In India, discharge of sewage through damaged sewers into the rivers is one of the major causes of surface water pollution along with direct discharge of untreated sewage (Dutta, Singh, and Prasad 2010). Also, India's rivers are dying which are major source of country's surface water. According to a report of Central Pollution Control Board (CPCB) published in 2018, out of 29 states of India, 27 states have river stretches, which are critically polluted. As per the CPCB's report, the number of polluted stretches is 351, which are located on 323 rivers and among all the states Maharashtra has the highest number of polluted stretches. Certain river stretches of India are in critically poor condition, which means the Biological Oxygen Demand (BOD) has exceeded 30 mg/liter. The number of critically polluted river stretches in India has also increased from 32 to 45 in the past 3 years. Except pollution, India's rivers are facing problems from biodiversity loss, loss of catchment areas, and sand mining (Ibid.).

India's groundwater resource is under stress, which is a preferred source of water due to its low cost (Wyroll 2012; Sengupta 2019). Groundwater supplies 65% of India's irrigation water (Ibid.) and it supplies 80% of rural water demand. Due to the lack of restriction on volume of groundwater extraction and availability of free or subsidized power in agricultural sector, large-scale farmers in north-western and peninsular India use groundwater inefficiently and unsustainably (Kumar 2019). In urban areas of India, groundwater fulfills 50% of water needs. The annual extraction rate of groundwater is also the highest in the world and 90% of extracted groundwater comes from 21 million operational wells in the country (Saha 2019, 24). In other words, India's dependence on groundwater has constantly increased. In 2004, India used 58% of its net available groundwater, which increased to 62% in 2014 (Ministry of Water Resources, Government of India 2017). Among all Indian states, four states such as Punjab, Rajasthan, Haryana, and Delhi are highly dependent on groundwater consumption where groundwater recharge is less than the consumption (Ibid.). The rapid urbanization in India is characterized by the sprawling of cities over the peri-urban areas (Ruet, Gambiez, and Lacour 2007). Cities like Chennai show such pattern of urban growth. The city has few water bodies around it and, therefore, is highly dependent on groundwater extraction. The peri-urban housing complexes in Chennai simply depend on wells for water supply (Srinivasan et al. 2013).

The major drivers of India's declining groundwater are the utilization of groundwater for irrigated agriculture especially in those places that are far from the surface water sources like river (Wyrwoll 2012). Additionally, water pollution in major rivers in India such as Ganga and Yamuna due to domestic sewage, industrial waste, and toxic chemicals put pressure on country's groundwater resources (Ibid.; Sengupta 2019). Every day, the Ganges receives 3000 million liters of untreated sewage from more than 100 cities and towns located along its banks (Dhillon 2014). It becomes the sixth most polluted river in the world when it reaches Varanasi, which also releases its untreated sewage into the river (Ibid.). In 2018, under the instruction of the National Green Tribunal (NGT), the Central Pollution Control Board (CPCB) of India declared the main stretch of the Ganges River even unfit for bathing. According to the CPCB, the main stretch of the Ganges remains highly polluted even during the post-monsoon season (Sengupta 2019).

The River Yamuna is also considered a holy river, which is completely dead in New Delhi (Sengupta 2006). As the river enters Delhi, the city government extracts 229 million gallons of water to supply drinking water to its residents (Ibid.). When Yamuna lives in the city, it becomes the primary sewer of the city carrying 950 million gallons of waste per day (Ibid.). The river is unsafe for bathing and drinking, and since 2014, the condition has little improved irrespective of Union government's effort to clean the river. The entire stretch of Yamuna River in Delhi remains severely polluted (Sengupta 2019).

Besides the declining groundwater resources, India is also suffering from groundwater pollution. However, unlike surface water pollution, groundwater pollution is not visible (Saha 2019; Sengupta 2019). According to the Central Groundwater Board, groundwater pollution is found in 640 districts of India (Sengupta 2019). The major pollutants that are affecting India's groundwater are chloride, fluoride, nitrate, iron, and arsenic. In 2018, Nicholas School of the Environment at Duke University, USA conducted a study on groundwater pollution and found that aquifers in 16 states are contaminated by uranium, which can lead to chronic kidney disease (The Hindu 2018; Verma 2018). However, the Bureau of Indian Standards' drinking water specifications do not include uranium as a contaminant on their list (The Hindu 2018; Sengupta 2019). Researchers associated with this study found that one-third of the total 324 sampled wells are contaminated by uranium and the level of contamination is above the World Health Organization's (WHO) prescribed limit, which is 30 μg/l (Verma 2018; Sengupta 2019). Among these 324 sampled wells, 226 were in Rajasthan and 98 in Gujarat (The Hindu 2018). According to the lead scientist of this study Avner Vengosh, professor of geochemistry and water quality at the Nicholas School of the Environment at Duke University, nearly one-third of all water wells tested in Rajasthan are contaminated by uranium, which crossed the permissible limit of WHO. Scientists engaged with the study also included 68 previous studies on groundwater contamination, which covered 61 districts of India. Among these 61 districts, 26 districts were spread over in northern India. According to these previous studies, Punjab is the worst affected state and within Punjab Bhatinda, Faridkot, Mansa, and Amritsar reported high levels of uranium in groundwater (Verma 2018).

Water scarcity in large cities of India is also related to loss of water bodies leading to a decline in the underground water table. Cities like Hyderabad lost their water bodies due to urban sprawl. Studies showing the land use and land cover change demonstrate that area under water bodies had declined over time (Ramachandraiah and Prasad 2004). A remote sensing-based study on Hyderabad and its surrounding areas demonstrated that areas under water bodies had shrunk significantly. In 1989, 29.79 sq. km. area was under waterbodies, which reduced to 20.84 sq. km. in 1999 (Ramachandraiah 2002). Like Hyderabad, Chennai also lost water bodies due to population growth and uncontrolled urban sprawl. Between 1971 and 2001, Chennai's population grew by eight times and the population density increased to 80–247 persons per sq. km. (Gupta and Nair 2011). Once the city has 150 water bodies which currently reduced to 27 (Ibid.). Loss of water bodies in Chennai since the 20th century has led to the problem of reduced infiltration, reduced groundwater recharge, and urban flooding (Ibid.). Pollution of water bodies and lakes in urban areas is also an issue that put pressure on groundwater. For instance, the pollution of Hussainsagar in Hyderabad is due to discharge of domestic sewage and toxic industrial chemicals (Ramachandraiah and Prasad 2004). Despite efforts to clean the nearby river and improve the water quality, the government is not able to prevent the release of untreated sewage into the lake (Mahesh 2019). According to the Telangana State Pollution Control Board, the water of the lake is not suitable for drinking and bathing and only can be used for industrial cooling and irrigation (Ibid.).

11.3 Discussion and Conclusion

By 2050, 70% of the world's population will live in urban areas (Srinivasan et al. 2011). The majority of this urban growth would occur in developing countries, especially in two countries—India and China (Ibid.). One of the ways to adapt to the growing water scarcity in urban areas of India is to conserve the water bodies and natural lakes around cities (Dutta, Singh, and Prasad 2010). Wetlands that are located surrounding an urban area are economically more valuable when they are preserved in their natural state. Encroachment of wetlands, water bodies, riverbeds, and paleochannels should be prevented for adequate groundwater recharge (Ibid.). Pumping of groundwater could be monitored by issuing a permit, which is practiced in cities like Tucson, Arizona, United States (World Bank 2018). Also, the municipalities should monitor the quality of groundwater to prevent contamination (Dutta et al. 2010). One of the ways to reduce excessive groundwater usage in the farm sector is to charging electricity based on metered consumption or actual consumption (Kumar 2019). Restricting farmer's electricity consumption through "prepaid energy meters" would limit excessive and unsustainable use of groundwater in agriculture (Ibid., 32).

Instead of depending on conventional water sources, cities in India should diversify their sources of water supply, such as reusing wastewater and use of desalinated seawater. Example of efficient use of desalinated seawater can be drawn from Malta, which started desalination as early as 1880s due to lack of adequate sources of surface water and lack of rainfall during summer months. Today, the country's half of the water needs come from the desalinated seawater. Although the cost of desalination is fairly high, it can be used with a range of other sources to obtain water. Also, desalination works best in those areas where production units and consumers are in proximity to each other. In other words, it works best in coastal areas. Cities in India should build diverse and dynamic water resource portfolio in order to best use the available water resources. The building of diversified and dynamic water resource portfolio for water-scarce cities in India may include a combination of multiple efforts such use of local surface, and groundwater, rainwater harvesting, capturing stormwater, and reusing wastewater. Indian cities should also consider reusing stormwater runoff, which requires less treatment as its quality differs from sewage (World Bank 2018). Overall, a more integrated and holistic approach is needed for water resource management in India.

References

Damania R, Desbureaux S, Hyland M, Islam A, Moore S, Rodella A, Russ J, Zaveri E (2017) Uncharted waters: the new economics of water scarcity and variability. World Bank, Washington, DC

Dutta V, Singh A, Prasad N (2010) Urban sprawl and water stress with respect to changing landscape: Study from Lucknow, India. J Geogr Region Plann 3(5):84–105

Gupta AK, Nair S (2011) Urban floods in Bangalore and Chennai: risk management challenges and lessons for sustainable urban ecology. Curr Sci 100(11):1638–1645

Koshy J (2018, June 14) Demand for potable water will outstrip supply by 2030, says study. The Hindu. https://www.thehindu.com/sci-tech/energy-and-environment/india-faces-worst-water-crisis-niti-aayog/article24165708.ece

Kumar D (2019) A little-known, growing monopoly. A down to earth annual: State of India's environment. Centre for Science and Environment, New Delhi

Mahesh K (2019, June 09) Hussainsaagar's water still not up to standards of cleanliness. The Times of India. https://timesofindia.indiatimes.com/city/hyderabad/hussainsagars-water-still-not-up-to-standards-of-cleanliness/articleshow/69708133.cms

McDonald R, Green P, Balk D, Fekete BM, Revenga C, Todd M, Montgomery M (2011) Urban growth, climate change, and freshwater availability. PNAS 108(15):6312–6317

Ramachandraiah C (2002) Location and Land use Characteristics of Hyderabad, paper prepared for the WHO-NIUA project on Healthy Cities. Hyderabad: Centre for Economic and Social Studies (mimeo)

Ramachandraiah C, Prasad S (2004). Impact of urban growth on water bodies: the case of Hyderabad. Centre for Economic and Social Studies (Working Paper 60), Hyderabad

Saha D (2019) Tainted dream: discourse in India ignores deterioration of groundwater quality, both by natural and anthropogenic causes. A Down To Earth Annual: State of India's Environment. Centre for Science and Environment, New Delhi

Sengupta S (2018, November 20) Cities across India have started water rationing.Down to Earth. https://www.downtoearth.org.in/news/water/cities-across-india-have-started-water-rationing-62187

Sengupta S (2019) Towards day zero: India is fast running out of water due to pollution, over-extraction, and changing climate. A down to Earth Annual: State of India's Environment. Centre for Science and Environment, New Delhi

Srinivasan V, Seto KC, Emerson R, Gorelick SM (2013) The impact of urbanization on water vulnerability: a coupled human–environment system approach for Chennai, India. Global Environ Change 23:229–239

The Hindu (2010) Groundwater in 16 States found to be contaminated by uranium. The Hindu. https://www.thehindubusinessline.com/economy/groundwater-in-16-states-found-to-be-contaminated-by-uranium/article24116833.ece#. Accessed 08 June 2018

Verma R (2018) Uranium in water can cause chronic kidney diseases, but India doesn't monitor it. Down To Earth. https://www.downtoearth.org.in/news/water/uranium-in-water-can-cause-chronic-kidney-diseases-but-india-doesn-t-monitor-it-60833. Accessed 08 June 2018

World Bank (2018) Water scarce cities: thriving in a finite world. World Bank, Washington, DC

Wyrwoll P (2012) 'India's groundwater crisis. Canberra, Australia: Global Water Forum (GWF Discussion Paper 1228)

Chapter 12
Status of River Water Quality in Goa: A Geographical Review

P. R. Morakar and Bala Kishan

Abstract Increasing demand of water for human consumption, irrigation and growing industrial activities has tremendous impact on water quality of rivers. Therefore, an attempt has been made to focus on the causes and consequences of decline in the quality of river water of Goa state. Goa, being the smallest state of India, is located in the west coast region of India. The total geographical area of the state is 3702 km^2. As per 2011 Census, the total population of Goa was 1,458,545. Since it is located in the Western Ghats, it is quite natural that the rivers are active in the study area. Goa is known as tourist paradise, Kashi of the Konkan, Rome of the East, and known for its aesthetic beauty and attract not only the domestic tourists but also foreign tourist. It is also treated as an industrial hub, particularly the pharmaceutical and chemical industries. The mining activities have been responsible for the pollution of river waters. River Mondovi and River Zuari are the major rivers along with the tributaries flowing in the study area and are estuarine waterbodies. Most of the scientists and researchers carried out studies based on the water samples and biochemical oxygen demand, and this resulted in studying the amount of dissolved oxygen used by micro-organisms, which basically help to decompose and mix with water. This biochemical oxygen is basically responsible for increased pollution levels. After liberation of the state, i.e., in 1961, many studies were conducted, especially by the Goa State Pollution Control Board (GSPCB). Accordingly, many rivers of the state are polluted severely and are not fit for human and animal consumptions also. Hence, the data related to water quality of one decade is taken into consideration for the analysis, and the statistical figures of 2015–2016 have witnessed that the pH value of river water is decreasing. Therefore, the proposed study is intended to study the causes for the quality of Goa rivers and proposed to suggest the planning strategies to improve the quality of river water.

Keywords pH value · Mining · Tourism · Industrialization

P. R. Morakar (✉)
Department of Geography, CES College of Arts and Commerce, Cuncolim Salcete, Goa, India
e-mail: prakashmorakar@gmail.com

B. Kishan
Department of Geography, Osmania University, Hyderabad, India

R. K. Mishra et al. (eds.), *Sustainable Climate Action and Water Management*, Advances in Geographical and Environmental Sciences,
https://doi.org/10.1007/978-981-15-8237-0_12

12.1 Introduction

Since time immoral, rivers are essential to human civilization. Rivers provide not only fresh water for drinking purpose but are also used for cleaning, establishment of industries, washing clothes, and many more. Rivers are lifeline to human being. Without potable water there is no future, so rivers basically help man to progress. As everyone knows that most civilizations are noticed along the river beds, religious places are also witnessed along the river basins, so finally one can say that if there is no river, no human life. Significance of the rivers cannot be expressed by words.

Rivers are the backbone of human civilization. They provide us with fresh water, which is helpful for various purposes, such as drinking, cleaning, and washing. Without rivers, life will come to a halt. Rivers just keep on flowing without any stop. We humans exist because of rivers. Importance of rivers cannot be stated in just few words.

Rivers are those who flow naturally, and they are either larger one or small streams. They are seen on every part of the earth crust, and basically fallow the general slope. They are classified into two categories, namely: perennial and non-perennial rivers. Perennial rivers are those which flow throughout the year, and non-perennial rivers flow seasonally, i.e., during rainy season or so. Rivers of the world generally classified into small tributaries, which just 1 km length, and some of them are larger ones, which run thousands of kilometers.

The significance of rivers is known to everyone, and these rivers carry large quantities of water from land to the ocean. The accumulated water evaporates due to insolation and many other reasons. This process is continuous and non-stop also, and leads to formation of clouds. Clouds are nothing but accumulation of moisture. Clouds carry moisture from ocean to land and release this moisture as precipitation. Ultimately rivers of small stream and mighty are formed on the earth surface. This is continuous process and it is known as water cycle. This process continuously supplies fresh water to the earth surface, which is essentially required by almost all living organisms.

12.1.1 Study Area

Goa is the smallest state of India. As per the records of the Town and Country Planning, its geographical area is 3702 km^2. As per 2011 Census record, it has a population of 13,458,546, as against the country's total population of 1,210,569,573 constituting around 0.12% of the country's population. Geographically, it extended by 14°53′54″ North to 15°48′00″ North Latitude and 73°40′33″ East to 74°21′13″ East Longitude. It consists of two districts, namely: South Goa District and North Goa District. Till 2011 the study state was consisting of 11 talukas, but in 2012 another taluka was added by bifurcating Sanguem taluka and one more taluka, i.e., Dharbandora was added. This is 25 state of the Indian Union and often called as "The

Tropical Paradise of Tourists", "Pearl of Rome". It is beautiful and picturesque state among Indian Union and is famous all over the world. Goa State is bounded by the Arabian Sea on the west. It has 105 km of coastline which starts from Goa-Karwar in the south and Terekol in the north, and the distance between east and west measures to 60 km. In the north it is bounded by Sawantwadi taluka of Sindadhurg district of Maharashtra. In the south the territory touches Belgaum and Karwar district of Karnataka. The Goa State, being part of the west coast region of India, has many physical features which are similar to the neighboring states of Maharashtra and Karnataka (see Fig. 12.1).

12.1.2 Objectives of the Study

The specific objectives were formulated to study the status and quality of waters in Goa rivers:

1. To identify the determinants of the river water quality.
2. To discuss the consequences of deteriorated water on impact of biotic organisms.
3. To suggest the remedies to improve the river quality water.

12.1.3 Hypothesis

To strengthen the objectives, the following hypotheses were formulated:

(1) The geographical and non-geographical conditions of the region determine the status of the river water quality in Goa.
(2) The determinants play significant role in changing the structure of agricultural scenario.

12.1.4 Database

The data which were referred to write this article are of secondary modes through which the pH values of the area under major rivers were collected from the published data of Goa State Pollution Control Board (GSPCB) and analyzed by applying suitable statistical techniques and the same were shown with graphs and charts.

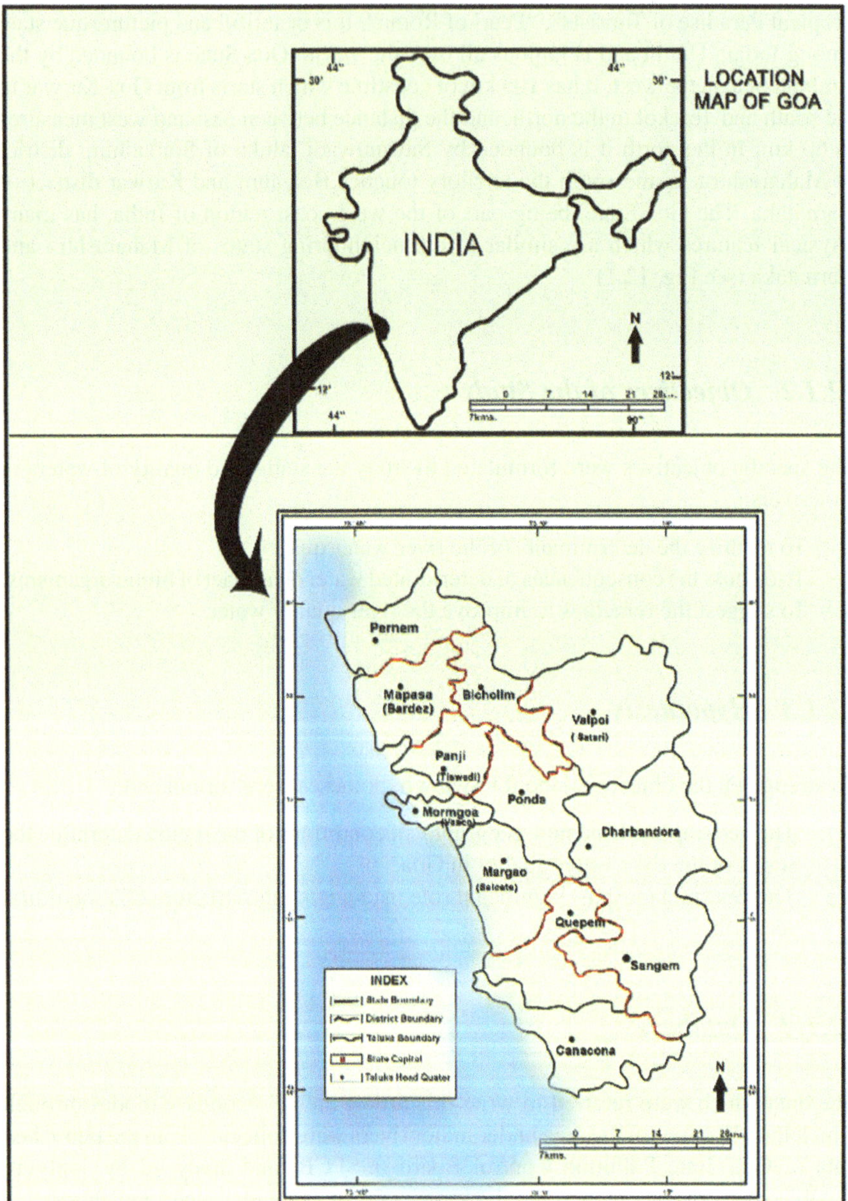

Fig. 12.1 Location map of Goa

12.2 Materials and Methods

For analyzing the status and quality of rivers of Goa, the data collected by Goa State Pollution Control Board were used in this study. The Board collected the information related to temperature, salinity dissolved oxygen, and pH of the selected rivers. They collected the information related to water quality in the specified months of the year, i.e., April to March 2013–2014 and April to March 2017–2018. But the emphasis was given to study the pH values of water quality by assessments.

12.2.1 Significance of Selected Water Quality Parameters

Basically, the parameters which are relevant to study are: Temperature: water has its own temperature; similarly pH value: this involves the acidity and alkalinity of water; dissolved oxygen: this explains the amount of oxygen available to living aquatic organisms; biochemical oxygen demand: this measures the chemical oxidation of inorganic matter, and so on; heavy metals; nutrients: ammonia nitrogen and phosphate phosphorus; and others. These factors play a vital role to determine the status or quality of river water in any study. But the present study is confined to pH values.

Among all the parameters, pH plays an important role to study the quality of water, and many researchers have carried out the study of river water by taking pH into consideration. Accordingly, the pH scale ranges from 0 to 14 (Chapman and Kimstach 1992). If the pH number is lower the water is the more acidic. The higher the number, the more basic it is. A pH of 7 is considered neutral. This is a significant parameter to assess the water quality as it has influence over 'many biological and chemical processes within a water body and all processes associated with water supply and treatment'.

12.2.2 Fluctuation of River Water Quality

Water being a good solvent, varying amounts of dissolved materials are always present in natural waters. The amount of dissolved materials in the water usually differs from the amount present in the body fluids of the organisms living in water, and such difference affects the osmotic exchange between the animals and the environment. The quality of river water fluctuates mainly due to addition of chemicals in the water, which affects the distribution of marine life in particular and other human life in general. So, it is very important to determine the chemical composition of river water, which finally helps to study the living and non-living organisms in the river water.

Goa State Pollution Control Board (GSPCG), which is doing pioneering job to assess the quality of river water in Goa, adopted several methods, especially for collecting information pertaining to quality of river water on monthly basis, and special emphasis was given to collecting pH values, dissolved oxygen, nitrate nitrogen, phosphate, biochemical oxygen demand (BOD), coliform, ammonia, etc.

Figure 12.2 shows the major and minor rives of Goa, and the Goa State Pollution Control Board collected information on pH values of 15 rivers of Goa. Most of the rivers' pH values fluctuate. The reasons behind fluctuations are: mining and tourism activities. In view of the pressure from tourists and mining activities, the pH value stands more acidic, and most of the time the pH value falls below 7 (Tables 12.1, 12.2, 12.3, 12.4 and 12.5).

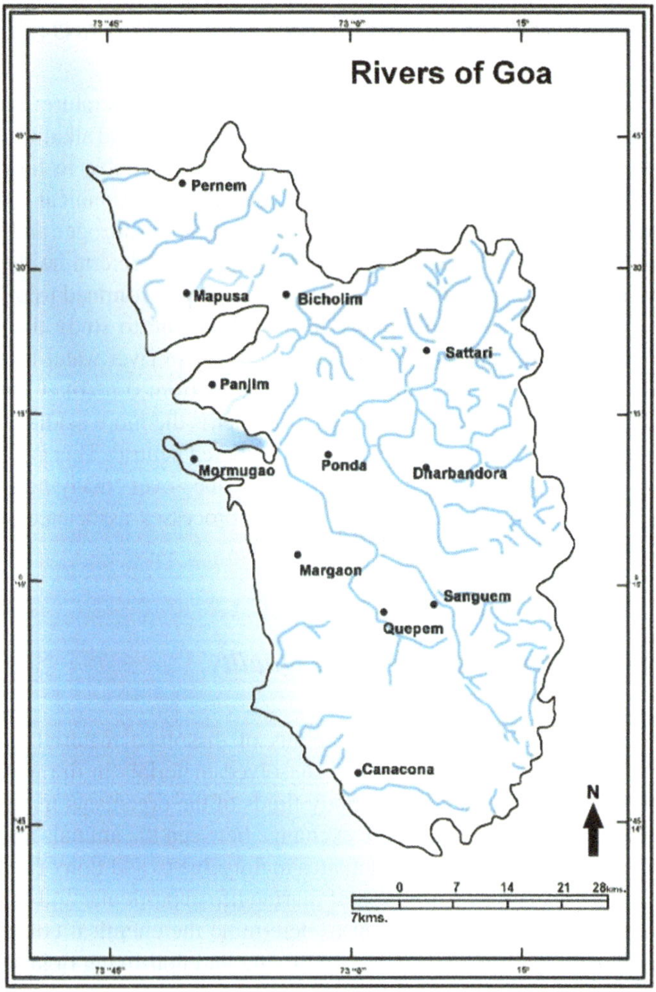

Fig. 12.2 Rivers of Goa

Table 12.1 Water quality assessment of rivers in Goa, pH values (2013–2014)

S. no.	Name of the river	April 2013–2014	May 2013–2014	June 2013–2014	July 2013–2014	August 2013–2014	September 2013–2014	October 2013–2014	November 2013–2014	December 2013–2014	January 2014–2015	February 2014–2015	March 2014–2015
1	Chapora	6.77	7.28	6.9	6.49	6.75	7.06	6.93	7.07	6.46	7.45	6.76	7.06
2	Kalana	6.73	7.18	6.95	6.25	6.76	7.12	6.7	7.06	6.44	7.29	7.03	7.34
3	Mapusa	6.63	7.07	6.87	6.33	6.53	6.92	6.56	6.77	6.71	6.66	6.8	7
4	Assanora	7.15	5.73	5.7	6.25	6.21	6.67	6.19	6.15	6.46	6.06	5.45	6
5	Valvanti	6.74	6.08	6.7	6.61	6.88	7.04	6.22	5.89	6.95	5.88	5.58	6.53
6	Madei	6.37	6.77	5.82	6.98	6.89	7.04	6.42	6.03	6.96	6.17	5.96	6.85
7	Khandepar	6.11	7.4	6.82	5.9	6.58	6.62	5.04	7.79	7.34	6.7	7.11	7.08
8	Kushavati	5.94	7.06	6.96	5.96	6.65	7.14	6.89	6.55	6.64	6.61	6.6	7.3
9	Talpona	6.89	6.93	6.34	5.44	8.02	6.49	6.91	6.03	6.38	6.63	6.95	6.5
10	Sal	6.39	6.83	6.05	5.2	7.71	6.44	6.72	6.09	6.33	6.38	6.5	6.2
11	Biholim	6.9	7.06	6.33	6.54	6.55	6.67	6.48	7.08	6.77	7.15	7.45	7.2
12	Zuari	6.72	7.18	5.98	6.41	5.89	6.97	6.07	7.31	6.86	7.1	6.77	6.72
13	Mandovi	7.8	8.13	7.98	6.64	7.27	7.45	7.42	7.59	7.79	7.3	7.22	7.8
14	Selaulim	7.71	7.51	6.99	6.88	6.9	7.7	7.2	6.76	6.53	6.93	6.95	7.6
15	Tiracol	6.66	6.2	6.85	7.01	7.2	6.96	6.48	6.89	6.32	7	6.59	6.94

Source GSPCB; 2013–2014

Table 12.2 Water quality assessment of rivers in Goa, pH values (2014–2015)

S. no.	Name of the river	April 2014–2015	May 2014–2015	June 2014–2015	July 2014–2015	August 2014–2015	September 2014–2015	October 2014–2015	November 2014–2015	December 2014–2015	January 2015–2016	February 2015–2016	March 2015–2016
1	Chapora	6.88	7.03	6.39	6.53	6.95	6.56	6.83	7.5	7.87	6.62	6.28	7.68
2	Kalana	6.98	6.8	6.27	6.57	6.8	6.63	7.5	7.5	7.95	6.67	7.01	7.03
3	Mapusa	6.9	7.3	6.89	6.61	6.3	6.49	6.79	6.58	7.29	6.35	6.55	6.99
4	Assanora	6.58	6.4	6.88	6.6	6.35	5.69	7.2	6.28	6.9	6.65	6.45	6.54
5	Valvanti	6.81	6.29	6.68	6.4	6.6	5.53	7	5.65	6.88	6.6	6.78	6.96
6	Madei	7.25	6.93	7.2	6.5	7.16	6.06	7.47	6.2	6.84	6.8	7.14	6.82
7	Khandepar	7.43	7.4	7.3	7.08	6.45	6.77	6.97	7.05	7.15	6.07	6.89	7.52
8	Kushavati	7.84	7.25	7.08	6.3	6.11	6.51	7.63	6.9	6.72	6.81	6.91	6.59
9	Talpona	6.51	6.45	6.53	6.67	7.3	6.15	6	7.43	7.24	6.19	6.83	6.9
10	Sal	6.1	6.22	6.09	7.2	6.1	6.18	6.5	6.82	7.05	6.02	6.55	6.12
11	Biholim	6.44	6.3	6.08	6.4	6.3	6.45	7.26	6.45	7.55	7.53	6.9	7.12
12	Zuari	7.16	7.1	6.65	7.6	6.43	6.81	6.9	6.55	6.6	6.65	7.07	6.63
13	Mandovi	7.76	7.64	6.57	7.9	7.08	6.63	7.88	7.19	7.68	7	7.07	7.68
14	Selaulim	7.64	7.5	6.49	6.63	6.51	6.94	7.86	7.12	6.85	7.05	7.27	7.24
15	Tiracol	6.82	6.11	7.02	7.01	6.58	6.66	7.2	7.23	6.58	7.3	7.25	7.45

Source GSPCB; 2014–2015

Table 12.3 Water quality assessment of rivers in Goa, pH values (2015–2016)

S. no.	Name of the river	April 2015–2016	May 2015–2016	June 2015–2016	July 2015–2016	August 2015–2016	September 2015–2016	October 2015–2016	November 2015–2016	December 2015–2016	January 2016–2017	February 2016–2017	March 2016–2017
1	Chapora	6.48	6.83	7.2	7.34	6.93	8.25	7.6	7.08	7.77	7.09	7.12	7.01
2	Kalana	6.81	7.42	6.95	7.35	6.89	8.41	7.7	7.12	8.05	7	7.34	6.8
3	Mapusa	6.82	6.83	6.99	6.85	6.61	7.6	7.4	6.65	7.28	6.85	6.91	7.03
4	Assanora	7.01	6.85	6	6.89	6.56	7.03	6.76	6.85	7.3	7.04	6.72	6.62
5	Valvanti	6.68	6.94	6.02	6.96	6.62	7.17	6.67	6.12	6.9	6.99	6.97	6.63
6	Madei	6.97	6.99	6.17	7.16	6.86	7.4	7.45	5.55	7.12	6.46	6.91	7.26
7	Khandepar	6.57	7.58	7.41	6.86	6.75	6.52	6.63	6.87	7.54	7.54	7.11	7.39
8	Kushavati	6.65	6.74	6.24	6.7	6.64	6.38	6.9	6.38	6.72	6.55	7.33	6.67
9	Talpona	6.61	7.24	7	6.95	7.37	7.7	7.08	6.02	6.61	6.23	9.03	8.81
10	Sal	6.6	7.23	7.15	6.5	6.76	6.5	6.95	6.43	6.48	6.28	8.9	9.3
11	Biholim	7.52	7.12	7.5	6.63	6.36	6.5	6.69	5.84	6.6	7.04	7.2	7.26
12	Zuari	6.31	6.61	6.71	6.76	6.88	6.82	6.1	6.69	6.71	7.06	6.84	7.11
13	Mandovi	7.09	6.18	7.47	7.01	7.09	6.66	6.2	6.1	6.58	7.64	6.12	7.8
14	Selaulim	7	7.4	7.24	7.25	6.98	7.68	7.4	6.24	7.04	7.04	7.17	7.58
15	Tiracol	7.28	6.97	7.16	6.76	6.59	8.05	6.55	5.96	5.99	7.43	5.88	6.77

Source GSPCB; 2015–2016

Table 12.4 Water quality assessment of rivers in Goa, pH values (2016–2017)

S. no.	Name of the river	April 2016–2017	May 2016–2017	June 2016–2017	July 2016–2017	August 2016–2017	September 2016–2017	October 2016–2017	November 2016–2017	December 2016–2017	January 2017–2018	February 2017–2018	March 2017–2018
1	Chapora	6.36	7.1	7.61	6.9	6.93	6.87	6.78	6.85	7.08	6.73	6.41	7.27
2	Kalana	6.73	6.73	7.11	6.8	7.54	6.88	6.98	6.75	7	6.17	6.25	6
3	Mapusa	6.93	6.86	7.29	6.8	6.81	6.97	6.87	6.72	7.1	6.68	7.14	7.26
4	Assanora	6.1	7.81	8.23	5.64	6.07	6.26	6	7.3	7.74	5.98	6.06	6.22
5	Valvanti	6.61	7.69	8.9	5.96	7.15	6.36	6.51	7.09	6.8	6.1	6.37	6.92
6	Madei	7	6.81	8.17	6.48	7.66	6.35	6.83	7.05	7	5.95	6.75	6.96
7	Khandepar	8.05	7.42	7.37	–	7.34	6.79	6.56	6.71	7.62	7.31	7.29	6.75
8	Kushavati	6.57	7.93	6.5	7.62	8	6.49	6.36	7.26	6.78	6.7	6.5	7.2
9	Talpona	7.17	7.42	7.3	7.42	7.05	6.28	6.2	6.82	6.6	6.14	6.4	5.9
10	Sal	6.55	6.63	6.6	7.3	7.24	6.2	6.6	6.57	5.8	5.91	6.41	6.06
11	Biholim	7.51	6.85	7.12	6.64	5.97	6.26	6.45	6.96	7.28	6.18	7.18	7.53
12	Zuari	6.62	6.95	7.14	–	7.36	6.82	6.81	6.96	6.58	6.77	6.82	6.82
13	Mandovi	7.38	7.82	8.07	6.97	6.11	6.94	6.84	7.43	7.63	7.6	7.5	7.97
14	Selaulim	7.56	8.64	7.2	7.61	7.5	6.95	6.55	7.28	6.58	6.9	7.2	6.9
15	Tiracol	6.86	8.05	7.19	6.69	6.08	7.08	6.82	6.43	6.28	6.45	6.44	8.03

Source GSPCB; 2016–2017

Table 12.5 Water quality assessment of rivers in GOA, pH values (2017–2018)

S. no.	Name of the river	April 2017–2018	May 2017–2018	June 2017–2018	July 2017–2018	August 2017–2018	September 2017–2018	October 2017–2018	November 2017–2018	December 2017–2018	January 2018–2019	February 2018–2019	March 2018–2019
1	Chapora	6.58	7.06	6.6	6.49	6.28	7.42	6.59	6.91	7.55	7.4	6.97	7.26
2	Kalana	6.13	6.64	6.38	6.48	6.34	7.29	6.44	7.18	7.29	7.4	7.06	6.78
3	Mapusa	7.16	7.76	6.27	6.23	6.19	7.09	6.4	6.55	6.96	6.9	7.05	6.96
4	Assanora	5.75	6.2	7.03	6.28	5.87	6.85	6.9	6.3	7.09	6.7	6.96	6.77
5	Valvanti	6.23	6.7	6.63	6.55	6.22	7.05	7.23	6.04	7.07	7.24	6.96	7.23
6	Madei	6.68	7.2	6.6	7.11	6.41	7.35	7.65	6.8	7.46	7.38	6.96	7.22
7	Khandepar	7.56	7.13	7.22	6.42	6.1	7.01	6.68	6.62	7.2	7.1	6.95	7.3
8	Kushavati	7.3	6.82	7.21	6.89	6.02	6.99	6.89	6.7	7.23	7.08	6.99	6.75
9	Talpona	9.21	6.78	9.08	6.85	6.55	7.06	5.8	7.06	6.82	7.07	7.14	6.19
10	Sal	6.06	6.99	6.47	6.51	6.34	7.28	6	7.05	6.81	6.79	7.8	6.02
11	Biholim	7	7.34	7.06	6.32	5.6	6.88	6.24	6.8	7.1	7.55	7.4	6.29
12	Zuari	7.13	7.01	7.44	6.01	6.47	7.79	6.91	6.9	6.7	6.9	6.9	7
13	Mandovi	7.6	7.87	7.8	6.93	7.27	7.56	7.58	7.7	7.37	7.7	7.51	7.4
14	Selaulim	6.44	7.61	7.53	6.49	6.25	7.39	7.03	7.14	7.35	7.07	6.99	6.92
15	Tiracol	8.06	7.85	8	7.2	8.16	8.48	8.06	8.1	7.93	8.07	7.6	7.78

Source GSPCB; 2017–2018

The pH value in the above-mentioned river is satisfactory during monsoon and pre-monsoon period, i.e., June to September every year, and then during tourists' peak months, i.e., November to March, the pH value fluctuates. The study region is known for mining activity, and the major ore which is extracted is iron ore. The extraction of iron ore is in full swing after the monsoon season, so the pH value fluctuates, and it is acidic, i.e., mixed with iron and bauxite ore. Similarly, there is pressure on beaches from tourists. All rivers of Goa join the Arabian Sea and estuaries exhibit fluctuation of pH values, which clearly indicates the selected rivers of Goa's pH values. The data shown in the tables are further indicated with line graphs (Figs. 12.3, 12.4, 12.5, 12.6 and 12.7).

GSPCB monitored 52 waterbodies in the state from April 2014 to March 2018 under national water-monitoring program (NWMP). The waterbodies monitored include rivers of Chapora, Kalana, Mapusa, Assnora, Valvanti, Mhadei, Khandepar, Kushavati, Talpona, Sal, Bicholim, Zuari, Mandovi, Salaulim, Tiracol, and Sinquerim,

Very important and significant rivers of Goa are noticed in four talukas, namely Bicholim taluka in North Goa District, Sattari, Sangem, and Quepem talukas of South Goa Districts. Many environmental activists opined that more than 10,000 people of the state lives in the villages of these talukas and they are the ones most affected in

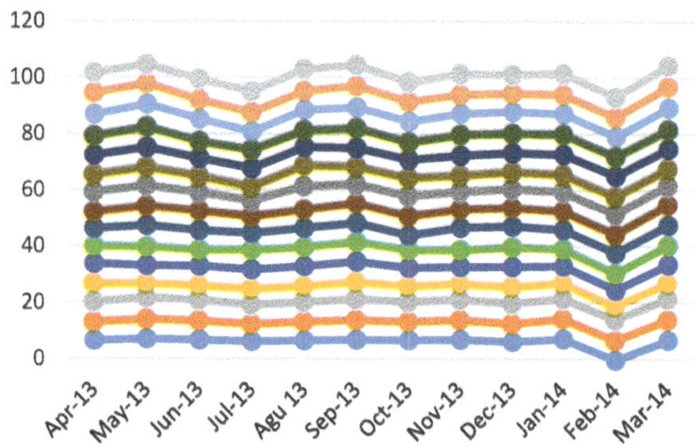

Fig. 12.3 Water quality assement of rivers in 2013–14 (Fig. 1.3)

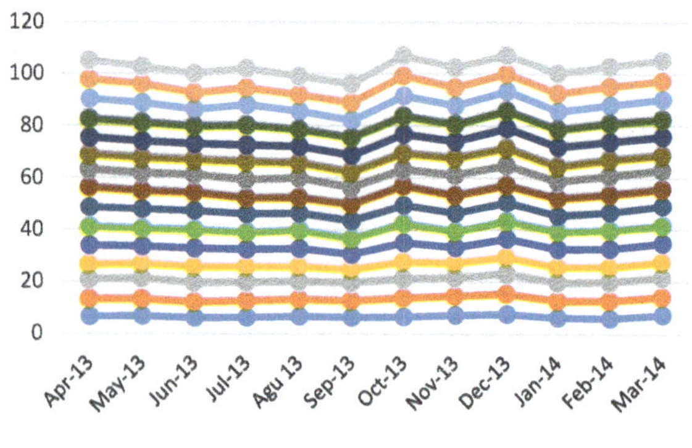

Fig. 12.4 Quality assement of rivers in 2014–15 (Fig. 1.4)

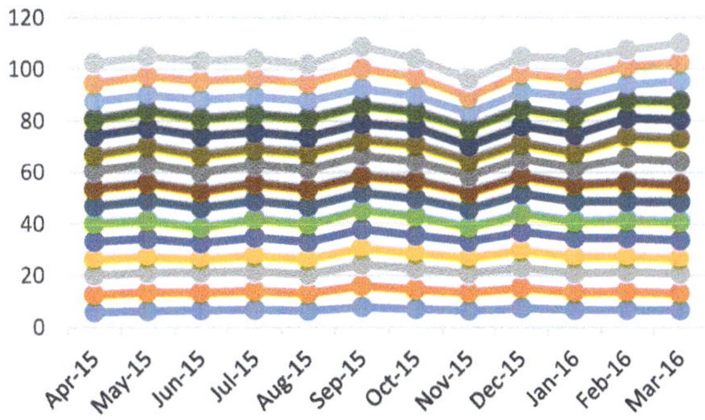

Fig. 12.5 Water Quality Assement of Rivers in 2015–16 (Fig. 1.5)

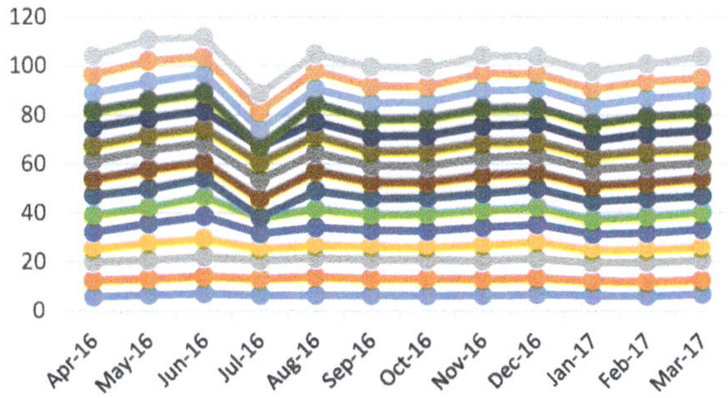

Fig. 12.6 Water qualtiy assement of rivers in 2016–17 (Fig. 1.6)

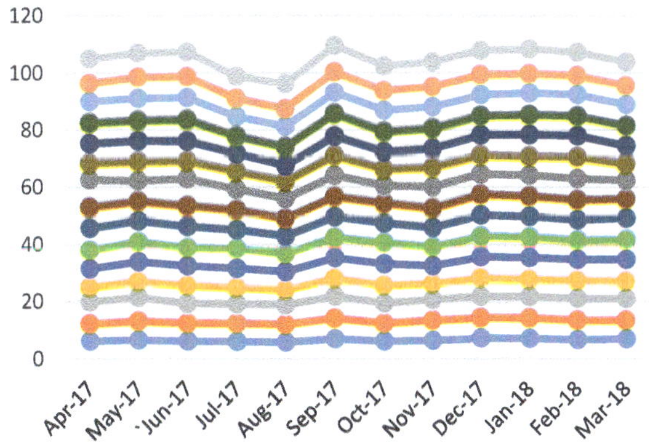

Fig. 12.7 Water Qaultiy Assement of River in 2017–18 (Fig. 1.7)

various ways. These talukas are known for mining activities. Owing to mining activities, people lost their livelihood. Apart from losing their livelihood, they are facing the problems of air, water, and noise pollutions also. According to Ashwini Aghor (2015) "Mining has caused irreversible damage to forests, agriculture, fisheries and water aquifers".

The total production of ore of the state varies from time to time. The actual production of iron ore, manganese, and bauxite is more than 80 million tons annually.

The mines not only extract ore but also soil in the ratio 1:3, which clearly indicates the extraction of 240 million tons of soil every year. Known environmentalist of the State Mr. Alvares (2002) quotes, "Considering that most of the mines are in the Western Ghats, one can well imagine what would happen given the current pace of mining".

12.3 Conclusion

The above facts and figures help to derive conclusion that the impact of mining, tourism, fishing, agriculture, and other activities quantifies the scenario of river waters of Goa, both qualitatively and quantitatively. The pH values which were observed within all four seasons show less variations in the quality of river water, and this change may be due to seasonal changes. The pH value in the states varies from 5 to 11. The highest pH value is noticed in the winter season. This is mainly because of arrival of tourists. In the month of December Christmas is celebrated, whereas in November, IFFI is also celebrated to attract more and more tourists across the world. Monsoon season depicts lower pH value. This is because of influx of South-West Monsoon. An in-depth observation indicates that the anthropogenic activities like mining, fishing, and tourism are highly concentrated in specific areas of the state. The study clearly evidenced that the surface water, i.e., river water of the study area is suitable for drinking and agriculture purposes, except few locations which need adequate treatment before usage.

References

Aghor A (2015) Mining in Goa: damage to the environment greater than loss of revenue. Down to Earth

Alexander Singh TH, Sanamacha Meetei N, Bijen Meitei L (2013) Seasonal variation of some physico-chemical characteristics of three major Riversin Imphal, Manipur: A comparative evaluation. Curr World Environ 8(1):93–102

Chaudhry FN, Malik MF (2017) Factors affecting water pollution: a review. J Ecosyst Ecogr 7(1):1–3

Claude A (2002) Fish curry and rice, a source book on Goa, its ecology and life style. A Goa Foundation Publication

Firmal S (2009) A study on the water quality of NIT Rourkela, an unpublished Thesis, Unpublished PhD Thesis, National Institute of Technology, Rourkela

Gorde SP, Jadhav MV (2013) Assessment of water quality parameters: a review. Int J Eng Res Appl 3(6):2029–2035

Haritah AK, Gaur S, Garg S (2016) Assessment of water quality and suitability analysis of River Ganga in Rishikesh, India. Appl Water Sci. https://doi.org/10.1007/s13201-014-0235-1

Nitya Jacob (2012) Mining—An increasing threat to our rivers, India Water portal, India Water Portal
Singh S, Kant Kamal Rakesh (2014) Assessment of groundwater quality in the mining areas of Goa. Curr World Environ 9(3):94–1000
Tyagi S, Sharma B, Singh P, Dobhal R (2013) water quality assessment in terms of water quality index. Research Gate

Part II
Climate Change: Ethics and Environmental Issues

Chapter 13
Association and Effects of ISMR and El Niño Southern Oscillation on Dengue Outbreaks in India

Netrananda Sahu and Martand Mani Mishra

Abstract Corroboration of El Niño and Southern Oscillation (ENSO) events on the climatic variability in the Indian subcontinent is mentioned in various studies. In the Indian context, if we observe the rising number of dengue cases, it could be treated as a major cause of health concern for the government of this country. Nowadays, an increasing number of dengue incidences are reported from different Indian states which are leading to unfortunate hospitalization and fatal death of the people. Keeping the context in mind, we have examined the relationship between the ENSO events, inter-annual variability in dengue cases in major states of India and Indian summer monsoonal rainfall (ISMR). In this study, we use Pearson's product-moment correlation (hereafter PPMC) to find out the interconnections between '+ winter ONI' (OND, NDJ, DJF, and JFM seasons) events, rainfall index for the JJAS season (June, July, August, September) and annual 10 major state-level data of dengue cases index for the past 14 years. It has been observed through the study that the PPMC result between the 'rainfall index' and 'dengue index' is showing greater variation among different states of the country. Interestingly, we have observed that the correlation result between the '+ winter ONI' and dengue incidence index have (r) value ranging from highly (−) to very high (+) value −0.53 to + 0.83 across the different states.

Keywords ISMR · ENSO · Dengue · Correlation coefficient

N. Sahu (✉) · M. M. Mishra
Department of Geography, Delhi School of Economics, University of Delhi, Delhi 110007, India
e-mail: babunsahu@gmail.com

N. Sahu
Disaster Prevention Research Institute, Kyoto University, Kyoto 6110011, Japan

R. K. Mishra et al. (eds.), *Sustainable Climate Action and Water Management*,
Advances in Geographical and Environmental Sciences,
https://doi.org/10.1007/978-981-15-8237-0_13

13.1 Introduction

The incidences of vector-borne diseases are most sensitive to the prevailing climatic variability and changes (Raheel et al. 2010; Gage et al. 2008; Kovats and Campbell-Lendrum 2000; Gubler et al. 2001a). Climatic factors are mainly responsible for the development, consequent behavior, and survival of the vectors (Gubler et al. 2001b). Changes in the weather pattern determine the range of the incidence whereas variability decides the intensity of impact (Hopp and Foley 2003). Among the vector-borne diseases, dengue is now becoming a serious cause of concern for the worldwide medical community (Gupta et al. 2010). The occurrence of dengue fever in tropical countries has possible links with climatic anomalies (Gagnon et al. 2001; Thai et al. 2010; Corwin et al. 2001). In recent times, dengue is taking the shape of a fiend and treated as an endemic vector-borne disease in India (Guha-Sapir and Schimmer 2005). 'El Niño and Southern Oscillation' (hereafter ENSO) related phenomena have a huge influence on dengue cases. With the passing time, many pioneer studies have been published about the possible impact of ENSO on climate (Saji et al. 1999). ENSO creates disturbances in the local climatic factors like temperature, rainfall, and relative humidity of a region. In this regard, many authors have tried to find out the relationship between ENSO events and dengue cases globally (Arcari and Tapper 2017; Brunkard et al. 2008; Vincenti-Gonzalez et al. 2018)

In the Indian subcontinent, ENSO plays an eminent role in deciding the annual dengue incidences. The occurrence of El- Nino interrupts the normal cycle of monsoonal rainfall which accounts for the maximum rainfall in the country (Kothawale and Kulkarni 2013; Krishnamurthy and Goswami 2000). ENSO events have influenced the amount of rainfall occurring in the country (Sarkar et al. 2004; Kumar 1999; Rajeevan 2004).

Dengue in India is influenced by the rainfall pattern and it is important to note that in many studies it has been observed that dengue infection has a major influence from the pre-monsoon season to the post-monsoon season (Gupta et al. 2006). This study has been taken into account to unearth the relationship between the annual dengue cases in the major 10 states (Mutheneni et al. 2017) (Delhi, Punjab, Haryana, Rajasthan, Gujarat, Maharashtra, Karnataka, Kerala, West Bengal, Tamil Nadu) and 2 union territories (UTs), (Chandigarh and Pondicherry), ISMR and '+ winter Oceanic Niño Index (ONI)' as an indicator of El Niño events in the country. ISMR indicates the four months (JJAS) of heavy monsoonal rainfall.

13.2 Data and Methods

Data for the total number of annual dengue incidences have been collected from the National Vector-Borne Disease Control Board (NVBDCB) for the year 2003 to 2016. Data has been extracted from the year 2003 as dengue infection has nearly taken six–seven years to spread among the Indian states since its major outbreak in Delhi and nearby regions in the year 1996 (Mutheneni et al. 2017). The ONI is the measurement of running three months SST (sea surface temperature) in Niño 3.4 region (5°N—5° S, 120° W—170° W) and the data for the same is collected from Climate Prediction Centre, National Oceanic and Atmospheric Administration (CPC, NOAA). Monthly rainfall data (1970–2016) for the ISMR season of the selected states of the study has been acquired from annual summary reports and monsoon reports downloaded from Indian Metrological Department (IMD).

13.3 Statistical Analysis

ONI value has its ramifications during the winter season in the northern hemisphere as the El Niño events are at the peak (Rajeevan 2004) and therefore, in this study for the calculation of PPMC with the percentage growth of dengue incidences we have taken + ONI years since the year 2002. Positive ONI seasons are taken for October–November–December (hereafter OND), November–December–January (hereafter NDJ), December–January–February (hereafter DJF), January–February–March (hereafter JFM). The years which are selected for the calculation of PPMC are 2002, 04, 06, 09, 14 and 15.

Anomalous percentage deviated is calculated from the year running mean of dengue incidences. The five years moving average for annual dengue cases (2003–2016) reported are calculated to find out the percentage deviated. The percentage growth rate is calculated by applying the formula:

$$\mathit{Dengue\ cases\ Index} = \frac{d - D}{D} * 100$$

where d = Number of cases in a particular year D= Five years moving average.

Cases are taken from the year 2003, 05, 07, 10, 15 and 16 as these years fall in the next year of '+ winter ONI years'. Further, the anomaly for the monsoon season (JJAS) of rainfall has been calculated from the year 1970 to 2016 to correlate with the dengue incidences.

$$\mathit{Calculation\ of\ Rainfall\ Index}(RI) = \frac{r - R}{R} * 100$$

where r = JJAS of a particular year R = mean JJAS from 1970–2016.

PPMC between the rainfall index and the dengue cases index has been calculated in this study to see the impact of rainfall in the major incidence occurring states. For (r) calculation between the rainfall index for the JJAS season and annual dengue index, we have taken similar year (2003, 05, 07, 10, 15 and 16) values.

13.4 Results

13.4.1 PPMC Between Rainfall Index and Dengue Cases Index

It has been observed that during the monsoon season and post-monsoon season there is an outbreak of the incidences in the major states that have been chosen for this study (Chakravarti et al. 2012). As per the result analysis, less amount of rainfall in the monsoon season plays a salient role in surging the number of reported dengue cases in most of the high dengue impacted states and UTs (Fig. 13.1). During the study period (2003–16), the (r) value between ‘ dengue case index’ and ISMR index is showing a value ranging from −0.06 to 0.70 (Fig. 13.1) while during the ‘+ winter ONI’ years (r) value for the states like Haryana, Delhi and Chandigarh, Gujarat, Karnataka, Kerala, Maharastra, Tamilnadu and West Bengal having a negative range from −0.12 to −0.86 shows that the decrease in rainfall in these states led to an outbreak of this disease which is also seen and observed in the past.

During the event years Gujarat, Tamilnadu, and Maharastra are showing a very high (−) correlation with the dengue incidence index (Fig. 13.1). While it is very interesting to note here that in the north, Punjab and Rajasthan are showing the correlation value ranging positively between 0.1 and 0.5. Rajasthan is a rain deficit state and here rainfall provides a positive ground for mosquitoes to breed, therefore depicts a positive indication (Fig. 13.1). It is an indication that in these states, monsoonal rainfall is playing a positive role indicating an increase in dengue cases with its increase.

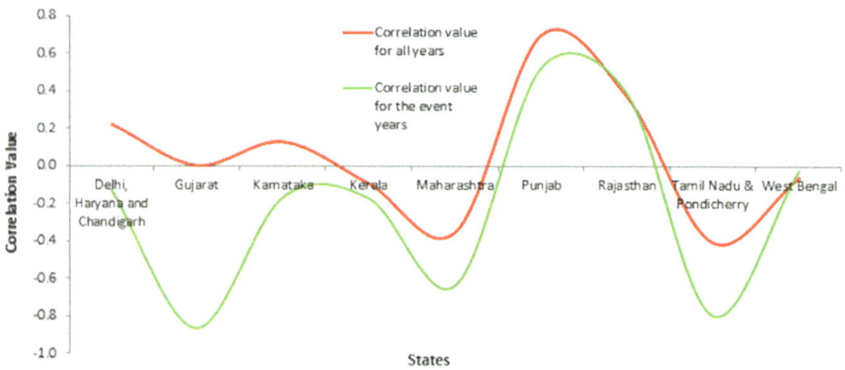

Fig. 13.1 Correlation between rainfall index of JJAS season and dengue incidences index in India from 1970 to 2016

13.4.2 PPMC Between Positive '+ Winter ONI Seasons' and 'Dengue Cases Index'

All four seasons' correlation coefficient has been calculated and it is observed that the states have shown a similar pattern for all the seasons. PPMC value for OND season is ranging from −0.18 to +0.71, for NDJ −0.12 to +0.75, DJF −0.11 to +0.77 and for JFM −0.4 – +0.83 ($p < 0.05$). The states which are showing a positive trend are Gujarat, Karnataka, Kerala, Maharastra, Punjab, Rajasthan, and West Bengal (Fig. 13.2). A positive r value indicates that + winter ONI will lead to an outbreak of cases in an increase of El Niño intensity. Gujarat has shown the highest (+) correlation in the JFM season followed by Rajasthan. Karnataka has shown a very low (r) value signifying no correlation with the dengue incidences except the OND season where the value lies in the moderate zone. In the nearby state of south Kerala and Tamil Nadu, it has shown a moderate to low relationship. Correlation result of Maharastra and Punjab (+) moderate–high to (+) low.

Only two major states are showing a negative value in which Delhi has negligible (−) correlation while Tamil Nadu is showing a moderate negative relationship. Gujarat has the strongest relationship with the '+ winter ONI' followed by Rajasthan and Maharastra (Fig. 13.2). It is clearly evident that ' + winter ONI' during all the seasons have played positively in the maximum number of states.

13.5 Discussion

Climatic variability leading to an outbreak of vector-borne diseases in different parts of the world is now a well-established and documented fact (Semenza 2017; Epstein et al. 1998; Bai et al. 2013; Reiter 2001; Nazareth et al. 2016; Ludwig et al. 2019). Previous pioneer work on ENSO did suggest that there is a strong connection between '+ winter ONI' and ISMR (Dhiman and Sarkar 2017). It has been noticed that '+ winter ONI' has played both negative as well as a positive role in the rainfall in different states of the country (Dhiman and Sarkar 2017).

From the states considered for the study, we have detected that monsoonal rainfall in states like Delhi, Haryana and Chandigarh and Tamil Nadu and Pondicherry has a positive relationship with '+ winter ONI' leading to the occurrence of good rainfall in these states whereas Gujarat and Rajasthan has a highly (−) correlation and in other vulnerable states it has moderate inter-relationship (Dhiman and Sarkar 2017). In Rajasthan, '+ winter ONI' plays both (−) in the west and (+) in the eastern part of the state (Dhiman and Sarkar 2017).

We have noticed that in different states, ISMR has been playing a different role in connection with dengue cases. Excess rainfall will lead to the washing away of larva and eggs of the mosquitoes (Arcari et al. 2007; Hales et al. 1999). It is possible that in most of the states, frequent and good rainfall leads to a decrease in the density of the Aedes mosquitoes during the El Niño event causing fewer dengue cases. In our

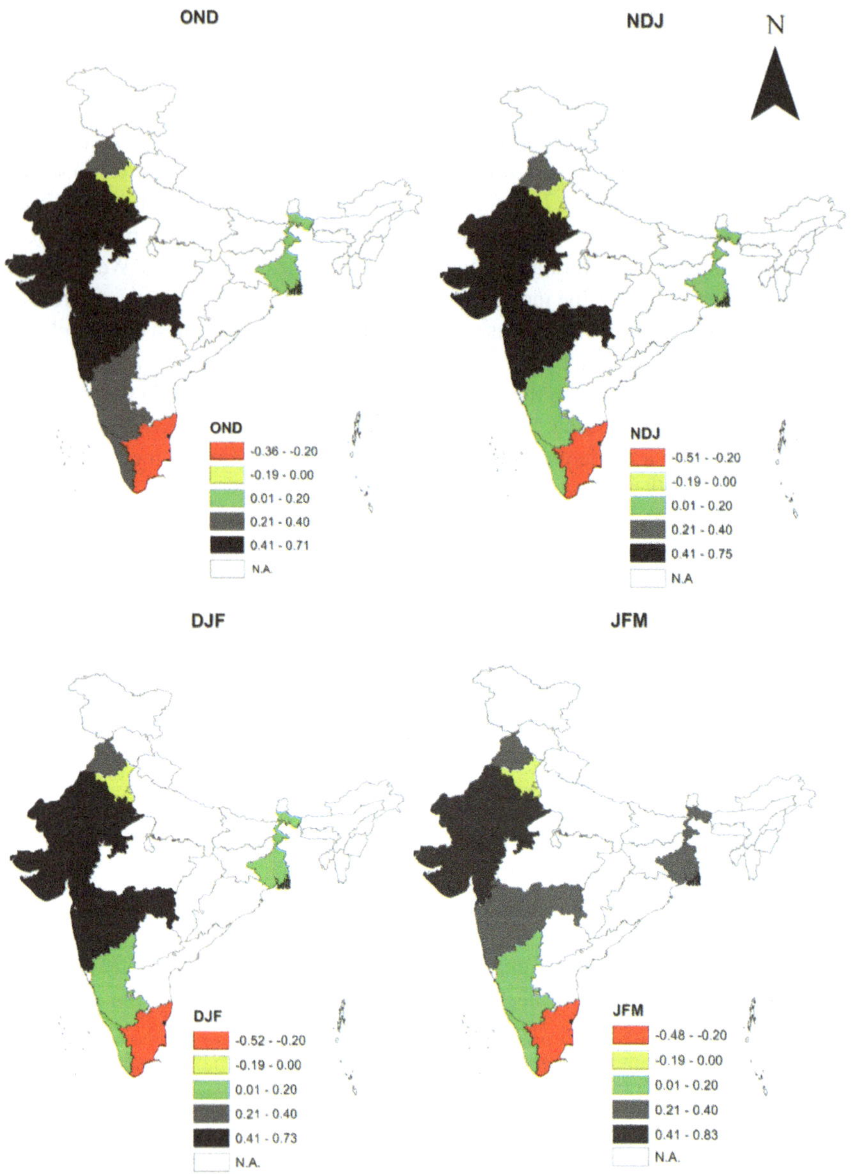

Fig. 13.2 PPMC between '+ winter ONI' and dengue incidences index in major states of India

result section, most of the states are indicating the (+) relationship between '+ winter ONI' and dengue incidences index. Tamil Nadu and Pondicherry are the only major combined state and UT, respectively, which is showing (−) relation as it is also an established fact that during '+ winter ONI' there is a good rainfall in this region.

Several studies have revealed that rainfall plays both a positive as well as a negative role (Gubler et al. 2001a; Kelly-Hope et al. 2004). According to the study of some authors from India, El Niño has generally negative consequences on the monsoonal rainfall in the country (Dhiman and Sarkar 2017). Low to moderate amount of rainfall provides a breeding ground for the mosquitoes that can cause an outbreak of the infection.

In this study, we have been focused to find out the corroboration that how '+ positive ONI' as an indicator of El- Nino event years and dengue cases index are interconnected. Through the map, it is noticed that the most vulnerable region in India lies in the western part of the country. '+ Winter ONI' has shown significantly (+) high to moderate high correlation value in these states. An El Niño condition will surely lead to a rise in dengue cases in this region. In the southwestern state (Karnataka and Kerala) and extreme eastern state (West Bengal) also has (+) moderate to (+) low value. Low rainfall during the El Niño event years and less number of rainy days are the cause for incidences in the states. In Tamil Nadu and Pondicherry, the correlation between '+ winter ONI' and dengue cases are (−) moderately high through all seasons.

Other local climatic factors (temperature, relative humidity, and total number of rainy days in the season) also play an important role in deciding the incidences (Patz 1998; Cazelles et al. 2005; Juliano 1998; Wu et al. 2009) which further needs an investigation into the cases already reported. A combination of these three chief climatic factors, namely, temperature, rainfall, and relative humidity disturb the land use and land cover in a region which has a pivotal role in determining the cases. Increases in temperature help in lengthening the life span of mosquitoes which further leads to another increase.

In India, this disease is increasing at a very rapid pace in other states of the country. Finally, through the study, it can be inferred that a strong El Niño phase ('+ winter ONI') will lead to an outbreak of this disease in most of the vulnerable states.

13.6 Conclusion

In the past two decades, dengue has caused both massive morbidity and mortality in the country (Gupta and Reddy 2013; Hariharan et al. 2019; Kakkar 2012; Gupta and Ballani 2014; Garg et al. 2008) and became a serious cause of concern for the scientists, policymakers and government institutions. In India, there is a need for an additional focus on the prevention and control strategies for this disease. A study related to global and local climatic factors in relation to dengue in India is quite constrained. The study under concern has highlighted how rainfall during the monsoon season and the impact of the El Niño event can be responsible for an outbreak of dengue in the major states of the country. Our result indicates that on the basis of the relationship between dengue, '+ winter Oni' and ISMR's early forecast of the disease can be possible. Limitation in this study is related to the availability of monthly dengue cases as the data is available annually and that too not for all the states, and therefore, we have to take major states for the study purpose. Monthly data availability may have lead to a more precise result. The other factors like social and environmental needs are to be studied as they are also an important cause of the outbreak. These factors could be en route to a further detailed study in the near future.

Acknowledgements Data on the total number of reported dengue cases were obtained from the National Vector Borne Disease Control Board (NVBDCB), Government of India. We are indebted to R. C. Dhiman for providing information about sources of rainfall data, and valuable inputs for writing this paper. His paper related to ENSO and Malaria has been observed as a base paper for our study. We are also thankful to NOAA for proving data on ONI climatic indices. The usage of ArcGIS Version 10.2 has been undertaken to construct a geographical correlation map, SPSS Version 22.0 has been used to calculate the PPMC and R to draw the radar graph. At present, Dr. Netrananda Sahu is working as JSPS Postdoc Fellow at DPRI, Kyoto University. We are thankful to JSPS (ID No. P18360), Japan.

References

Arcari P, Tapper N (2017) The variable impact of ENSO events on regional dengue/DHF in Indonesia. Singap J Trop Geogr 38(1):5–24. https://doi.org/10.1111/sjtg.12179

Arcari P, Tapper N, Pfueller S (2007) Regional variability in relationships between climate and dengue/DHF in Indonesia. Singap J Trop Geogr 28(3):251–272. https://doi.org/10.1111/j.1467-9493.2007.00300.x

Bai L, Morton LC, Liu Q (2013) Climate change and mosquito-borne diseases in China: a review. Global Health 9(1):10. https://doi.org/10.1186/1744-8603-9-10

Brunkard JM, Cifuentes E, Rothenberg SJ (2008) Assessing the roles of temperature, precipitation, and enso in dengue re-emergence on the Texas-Mexico border region. Salud Pública de México 50(3). https://doi.org/10.1590/s0036-36342008000300006

Cazelles B, Chavez M, Mcmichael AJ, Hales S (2005) Nonstationary Influence of El Niño on the Synchronous Dengue Epidemics in Thailand. PLoS Med 2(4). https://doi.org/10.1371/journal.pmed.0020106

Chakravarti A, Arora R, Luxemburger C (2012) Fifty years of dengue in India. Trans R Soc Trop Med Hyg 106(5):273–282. https://doi.org/10.1016/j.trstmh.2011.12.007

Corwin AL, Larasati RP, Bangs MJ, Wuryadi S, Arjoso S, Sukri N et al (2001) Epidemic dengue transmission in southern Sumatra, Indonesia. Trans R Soc Trop Med Hyg 95(3):257–265. https://doi.org/10.1016/s0035-9203(01)90229-9

Cpc. (2001, January 1). NOAA's Climate Prediction Center. Climate Prediction Center. https://www.cpc.ncep.noaa.gov/. Accessed 10 Aug 2019

Dhiman RC, Sarkar S (2017) El Niño Southern Oscillation as an early warning tool for malaria outbreaks in India. Malaria J 16(1). https://doi.org/10.1186/s12936-017-1779-y

Epstein PR, Diaz HF, Elias S, Grabherr G, Graham NE, Martens WJM et al (1998) Biological and physical signs of climate change: focus on mosquito-borne diseases. Bull Am Meteor Soc 79(3):409–417. https://doi.org/10.1175/1520-0477(1998)079%3c0409:bapsoc%3e2.0.co;2

Gage KL, Burkot TR, Eisen RJ, Hayes EB (2008) Climate and vectorborne diseases. Am J Prev Med 35(5):436–450. https://doi.org/10.1016/j.amepre.2008.08.030

Gagnon A, Bush A, Smoyer-Tomic K (2001) Dengue epidemics and the El Niño Southern oscillation. Clim Res 19:35–43. https://doi.org/10.3354/cr019035

Garg P, Nagpal J, Khairnar P, Seneviratne SL (2008) Economic burden of dengue infections in India. Trans R Soc Trop Med Hyg 102(6):570–577. https://doi.org/10.1016/j.trstmh.2008.02.015

Gubler DJ, Reiter P, Ebi KL, Yap W, Nasci R, Patz JA (2001a) Climate variability and change in the United States: potential impacts on vector- and rodent-borne diseases. Environ Health Perspect 109(suppl 2):223–233. https://doi.org/10.1289/ehp.109-1240669

Gubler DJ, Reiter P, Ebi KL, Yap W, Nasci R, Patz JA (2001b) Climate variability and change in the United States: potential impacts on vector- and rodent-borne diseases. Environ Health Perspect 109:223. https://doi.org/10.2307/3435012

Guha-Sapir D, Schimmer B (2005) Dengue fever: new paradigms for a changing epidemiology. Emerg Themes Epidemiol 2(1):1. https://doi.org/10.1186/1742-7622-2-1

Gupta E, Ballani N (2014) Current perspectives on the spread of dengue in India. Infect Drug Resist 337. doi:10.2147/idr.s55376

Gupta B, Reddy BPN (2013) Fight against dengue in India: progresses and challenges. Parasitol Res 112(4):1367–1378. https://doi.org/10.1007/s00436-013-3342-2

Gupta E, Dar L, Kapoor G, Broor S (2006) The changing epidemiology of dengue in Delhi, India. Virol J (3):92–96

Gupta P, Khare V, Tripathi S, Nag VL, Kumar R, Khan MY, Dhole TKNC (2010) Assessment of World Health Organization definition of dengue hemorrhagic fever in North India. J Infect Dev Countries 4(03):150–155. https://doi.org/10.3855/jidc.708

Hales S, Weinstein P, Souares Y, Woodward A (1999) El Nino and the dynamics of vectorborne disease transmission. Environ Health Perspect 107(2):99. https://doi.org/10.2307/3434364

Hariharan D, Das M, Shepard D, Arora N (2019) Pin32 Disease and economic burden cost of dengue illness In India: An Update From 2013 To 2017. Value Health 22. doi:10.1016/j.jval.2019.04.903

Hopp M, Foley J (2003) Worldwide fluctuations in dengue fever cases related to climate variability. Clim Res 25:85–94. https://doi.org/10.3354/cr025085

Juliano SA (1998) Species introduction and replacement among mosquitoes: interspecific resource competition or apparent competition? Ecology 79(1):255–268. https://doi.org/10.1890/0012-9658(1998)079%5b0255:siaram%5d2.0.co;2

Kakkar M (2012) Dengue fever is massively under-reported in India, hampering our response. BMJ 345(dec19 17). https://doi.org/10.1136/bmj.e8574

Kelly-Hope LA, Purdie DM, Kay BH (2004) Differences in climatic factors between ross river virus disease outbreak and nonoutbreak years. J Med Entomol 41(6):1116–1122. https://doi.org/10.1603/0022-2585-41.6.1116

Kothawale DR, Kulkarni JR (2013) Performance of all-India southwest monsoon seasonal rainfall when monthly rainfall reported as deficit/excess. Meteorol Appl 21(3):619–634. https://doi.org/10.1002/met.1385

Kovats RS, Campbell-Lendrum DH (2000) Early effects of climate change: do they include changes in vector-borne disease? Philos Trans: Biol Sci 356(1411):1057–1068. https://doi.org/10.1098/rstb2001.0894

Krishnamurthy V, Goswami BN (2000) Indian Monsoon–ENSO relationship on interdecadal timescale. J Clim 13(3):579–595. https://doi.org/10.1175/1520-0442(2000)013%3c0579:imeroi%3e2.0.co;2

Kumar KK (1999) On the weakening relationship between the Indian Monsoon and ENSO. Science 284(5423):2156–2159. https://doi.org/10.1126/science.284.5423.2156

Ludwig A, Zheng H, Vrbova L, Drebot M, Iranpour M, Lindsay L (2019) Increased risk of endemic mosquito-borne diseases in Canada due to climate change. Can Commun Dis Rep 45(4):91–97. https://doi.org/10.14745/ccdr.v45i04a03

Ministry of Health & Family Welfare-Government of India. National Vector Borne Disease Control Programme (NVBDCP). Home. https://nvbdcp.gov.in/. Accessed 3 Sept 2019

Mutheneni SR, Morse AP, Caminade C, Upadhyayula SM (2017) Dengue burden in India: recent trends and importance of climatic parameters. Emerg Microbes Infect 6(1):1–10. https://doi.org/10.1038/emi.2017.57

Nazareth T, Seixas G, Sousa CA (2016). Climate change and mosquito-borne diseases. Clim Change Manag Clim Change Health:187–199. doi:10.1007/978-3-319-24660-4_12

Official Web Site of India Meteorological Department (IMD), Pune. Official Web Site of India Meteorological Department (IMD), Pune. http://www.imdpune.gov.in/. Accessed 6 Sept 2019

Patz JA (1998) Predicting key malaria transmission factors, biting and entomological inoculation rates, using modelled soil moisture in Kenya. Trop Med Int Health 3(10):818–827. https://doi.org/10.1046/j.1365-3156.1998.00309.x

Raheel U, Faheem M, Riaz MN, Kanwal N, Javed F, Zaidi NUSS, Qadri I (2010) Dengue fever in the Indian subcontinent: an overview. J Infect Dev Countries 5(04):239–247. https://doi.org/10.3855/jidc.1017

Rajeevan M (2004) Tropical Pacific upper ocean heat content variations and Indian summer monsoon rainfall. Geophys Res Lett 31(18). https://doi.org/10.1029/2004gl020631

Reiter P (2001) Climate change and mosquito-borne disease. Environ Health Perspect 109:141. https://doi.org/10.2307/3434853

Saji NH, Goswami BN, Vinayachandran PN, Yamagata T (1999) A dipole mode in the tropical Indian Ocean. Nature 401(6751):360–363. https://doi.org/10.1038/43854

Sarkar S, Singh RP, Kafatos M (2004) Further evidences for the weakening relationship of Indian rainfall and ENSO over India. Geophys Res Lett 31(13). https://doi.org/10.1029/2004gl020259

Semenza J (2017). Climate change and other key drivers of mosquito-borne disease emergence. Eur J Public Health 27(suppl_3). https://doi.org/10.1093/eurpub/ckx187.604

Thai KTD, Cazelles B, Nguyen NV, Vo LT, Boni MF, Farrar J et al (2010) Dengue dynamics in Binh Thuan Province, Southern Vietnam: Periodicity, Synchronicity and Climate Variability. PLoS Neglect Trop Dis 4(7). https://doi.org/10.1371/journal.pntd.0000747

Vincenti-Gonzalez MF, Tami A, Lizarazo EF, Grillet ME (2018) ENSO-driven climate variability promotes periodic major outbreaks of dengue in Venezuela. Sci Rep 8(1). https://doi.org/10.1038/s41598-018-24003-z

Wu P-C, Lay J-G, Guo H-R, Lin C-Y, Lung S-C, Su H-J (2009) Higher temperature and urbanization affect the spatial patterns of dengue fever transmission in subtropical Taiwan. Sci Total Environ 407(7):2224–2233. https://doi.org/10.1016/j.scitotenv.2008.11.034

Chapter 14
Assessment of Carbon Sequestration and Tree Diversity in Gokarna Forest, Kathmandu, Nepal

Sagar Dhakal, Ashutosh Mohanty, and Kedar Prasad Rijal

Abstract Carbon sequestration is one of the most prominent measures for mitigating the impact of climate change. Urban forests depend on the composition of vegetation, species structure, human influence, and management activities as an effective means of balancing carbon in the city areas. The study considered 41 plots, covering an area of 1.025 ha and assessed the role of the urban forest in carbon offsetting. Based on the diameter at breast height (DBH) and height of trees information, the allometric equation is deployed to determine the biomass of the trees. The forest had species richness 39, Shannon–Wiener diversity index (H') 2.4996, Pielou's evenness index 0.6872, and the carbon 720.26tC/ha. The normality of the data was determined using the Shapiro–Wilk test. The relationship of species richness and diversity with carbon stock was determined using Spearman's rank correlation. *Neolitsea cuipala* (119.06tC/ha) had more carbon storage potential followed by *Holarrhena pubescens* (73.57tC/ha) ranging from (0.3–119.06) tC/ha. Long-term monitoring of these tree communities for their conservation management is suggested.

Keywords Urban forest · Diversity · Carbon sequestration · Climate change

14.1 Introduction

Climate change is one major problem of the twenty-first century (Hornsey and Fielding 2020; Hulme 2020). After the 1850s, the industrial revolution has tremendously increased Green House Gas (GHG) emission in the atmosphere (Şahin et al. 2019; Shurpali et al. 2019) increasing the mean temperature (1.5 °C) of the earth (Jana and Roy 2020). Today, the major concern of the world is to stabilize the concentration

S. Dhakal · A. Mohanty (✉)
School of Biological and Environmental Sciences, Faculty of Basic Sciences, Shoolini University, Bajhol, P.O. Sultanpur, Solan 173212, HP, India
e-mail: drashutoship@gmail.com

K. P. Rijal
Central Department of Environmental Science (CDES), Tribhuvan University, Kirtipur, Kathmandu, Nepal

R. K. Mishra et al. (eds.), *Sustainable Climate Action and Water Management*, Advances in Geographical and Environmental Sciences,
https://doi.org/10.1007/978-981-15-8237-0_14

of GHG emissions to a manageable level through different mitigation strategies like adopting renewable energy and promoting more sustainable ways of land and forest (Liu et al. 2020). One such most effective effort is to enhance the natural process to reduce carbon dioxide sink in the forest (Sundarapandian et al. 2013). Carbon gets stored in the forest by the cycling process with the environmental components (Sekhri et al. 2020; Narayana et al. 2020). The storage duration and quantity of carbon depend upon the species, sites, disturbance prevailing in the area as well as on the management practices (Das et al. 2020). The concept of the urban forest is promoted through the conservation of forest patches and conservation programs (Baur et al. 2020). Nowadays, many researchers suggest that urban forests had played a crucial role in carbon sequestration, and highly appreciate those countries that have implemented many programs to conserve the urban forest patch (Meena et al. 2020; Feng and Nursey-Bray 2020). In Nepal, the Forest User Group (FUG), Sacred Forest Conservation Program (SFCP), and Urban Patch Forest Management Programs (UPFMM) promote significant support to carbon storage mitigation (Escobedo et al. 2010; Liu and Li 2012; Strohbach and Haase 2012). Not only carbon sink, but also urban forests are considered to have an important role in biodiversity conservation (Khumbongmayum et al. 2006, Page et al. 2010). Nepal is one of the rapidly urbanizing countries of South Asia and since urban population has increased from 18 to 42% from 2011 to 2015 (Bhusal et al. 2019)) of the total population. The growing population in urban areas has faced the scarcity of potential ecological benefits of forests (Khanal et al. 2019) like urban greenery effects of noise and air pollution from the degradation of urban forest patch (Ebenberger and Arnberger 2019). The biggest threat to human socioeconomic activities resulting from climate change can be reduced from the urban forest conservation offsetting the rate of release of carbon dioxide (Singh et al. 2013).

The urban forest patch is the potentiality to provide a wide range of ecosystem functions, services (Pachauri et al. 2014; Erdős et al. 2019) and is linearly related to the primary productivity of the forest (Forrester and Bauhus 2016) and diversity (Jim and Chen 2009; Saha et al. 2009; Kemeuze et al. 2015). For the whole of Nepal, only a few studies are available on urban forest diversity including carbon sequestration. Therefore, an attempt has been made to (i) assess the diversity of tree species in urban forest patches, (ii) assess the carbon stock in an urban forest patch, and (iii) the relationship between urban forest diversity and carbon sink.

14.2 Methods and Materials

14.2.1 Study Area

The study was carried out in Gokarna forest that is one of the oldest, conserved and protected forests situated at the mid-hill subtropical climatic zone, Kathmandu, Nepal. The forest covers an area of 186 ha and is located in 27°43’09.33”N to

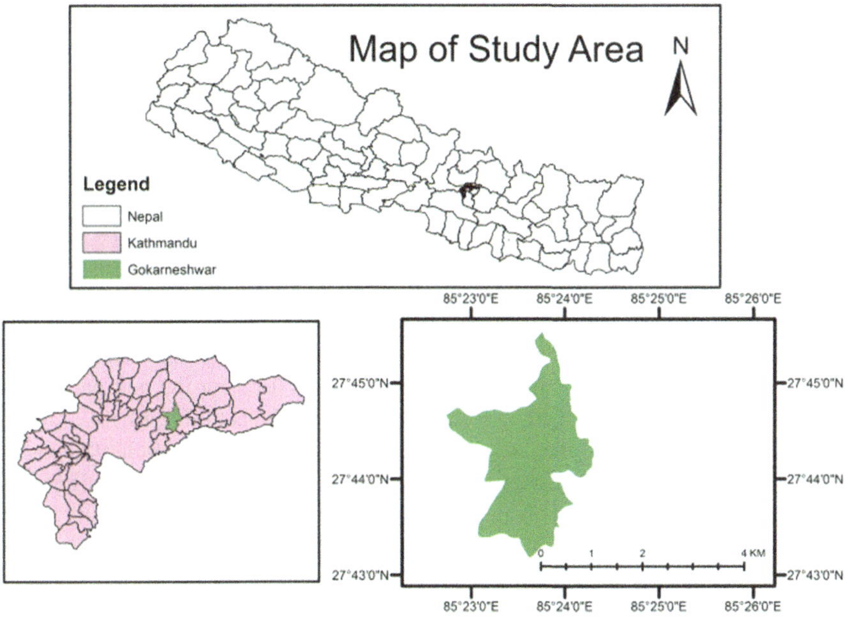

Fig. 14.1 Map of the study area (Scale 1:2500)

27°40'19.05"N latitude and 85°23'03.27"E to 85°24'18.5"E longitude at 1320 m elevation. The average annual temperature of the study area is 19 °C, a maximum of 31.9 °C in June and a minimum of 7 °C in December, and the average annual precipitation is 1500 mm (June to September—80% of rainfall occurs). The forest is composed of several broadleaf evergreen tree species like *Schima wallichii, Neolitsea cuipala, Alnus nepalensis, Pinus roxburghii, Castonopsis indica, Choerospondias axillaris, Walsura trijuga,* etc. (Fig. 14.1).

14.3 Methods

Baseline information about the carbon stock (Ren et al. 2011; Roy et al. 2012) helped to the mitigation of climate change (CC). The amount of carbon stored within the forest was determined using an allometric equation without harvesting vegetation (Chave et al. 2005) with the relationship of wood densities, diameter, and height of the species (Markum et al. 2014).

14.3.1 Sample Design, Field Methods, and Numerical Analysis

Systematic random sampling was taken in 100 m interval with 41 circular plots [area covers more than 0.5% (1.025 ha) out of the total area (186 ha)] at a radius of 8.92 m for the vegetation data collection. All woody plants were greater than 1.5 cm DBH measure at 1.37 m above the ground with the help of a DBH tape (Kinglon diameter tape, Japan). Co-ordinate and altitude of plots measure GPS (eTrex 30). All numerical analysis was carried out using R software (Team 2016) using the vegan package (ter Braak and Šmilauer 2015). The Shapiro–Wilk test was used to determine the normality of the data. Students' t-test was used to test the significance of the different mean and the Wilcoxon signed-rank test was used for the non-parametric test. The following formulae are used for numerical analysis (Table 14.1).

Where D. = Density (no./ha), no. = Number, ha. = hectare, RD = Relative density, F. = Frequency, R.F. = Relative frequency, B.A. = Basal area, R.B.A. = Relative Basal area, I.V.I. = Importance Value Index, H' = Shannon–Wiener diversity index, n_i = number of individuals of species in a sampling unit of concern, N = Total number of species in the sampling unit of concern, ni/N = Ratio of individuals to total number, ln = natural log, Pi = Proportion of individuals in the *i*th species, i.e. (*ni/N*), C = Dominance index, e = Pielou's evenness index, S = Number of species in the sampling unit of concern, AGTB = Above-ground tree biomass [kg], 0.0509 = Constant value, ∂ = Wood density (g/cm^3), D = Tree diameter at breast height [m], H = Tree height [m], AGTC = Above-ground trees carbon 0.47 = Conversion factor or default carbon fraction, BGTB = Below-ground trees biomass, and BGTC = Below-ground trees carbon.

14.4 Results

The studied urban forest was a total of 844 trees with 38 different species in all developmental/growth stages (i.e. seedlings, saplings, and trees) which were recorded with 562 trees/ha. They were found to be significantly different ($P < 0.05$) plot-wise. Species richness was found to be normally distributed ($W = 0.95117$, p-value = 0.07695) with significant differences among the plots ($t = 19.247$, $df = 40$, and p = 2.2e-16). The average species in each plot ranged from 4.08 to 5.03 at a 95% confidence level. The height of trees ranged from 4 to 31 m (average 14.68 m) while DBH ranges from 10 to 115 cm (average 40 cm). The height of trees was found with the increment of DBH and the relation is significant (intercept = 8.34, slope = 0.10, and $p < 0.05$) as shown in Fig. 14.3a. The highest height class was found to be 10-15 m with density 279 trees/ha whereas the lowest height class was 30-35 m with a density of 3.01 trees/ha. The distribution of trees grouped in different diameter classes had shown that the greater number of trees size class 10-20 cm (195 trees/ha) and occupied 23.15% of the total tree density. The highest density was of *Neolitsea*

Table 14.1 List of formulas that are applied in numerical analysis for the study

S. No.	Formula	Analysis parameter	References
1	$D = \frac{\text{Total no.of individuals of a species}}{\text{Total no.of plants studied X area of each quadrat}} \text{ x } 10000$	The numerical strength of a species in a community	Sharma et al. (2018)
2	$RD = \frac{\text{Total no.of individuals of a particular species in all quadrats}}{\text{Total no.of individuals of all species in all quadrats}} \text{ x } 100$	The numerical strength of a species to the total number of individuals of all species	Sharma et al. (2018)
3	$F = \frac{\text{No.of quadrats in which the species occured}}{\text{Total no.of quadrates studied}} \text{ x } 100$	Dispersion of species in a community	Sharma et al. (2018)
4	$RF = \frac{\text{No.of quadrats in which the species occurs}}{\text{Total no.of all species in the quadrats}} \text{ x } 100$	Particular species to the total frequency of all the species	Sharma et al. (2018)
5	$BA = \frac{\pi(DBH^2)}{4}$	The ground penetrated by the stems	Sharma et al. (2018)
6	$RBA = \frac{\text{Basal area of a particular relative species in quadrats}}{\text{Total basal area of all species in quadrats}} \text{ x } 100$	The amount of an area occupied by stems to the total occupied by the stems	Sharma et al. (2018)
7	$\mathbf{IVI = RD + RF + RBA}$	Dominance and ecological success of any species	Sharma et al. (2018)
8	$\mathbf{H' = -\sum (ni/N) \ln (ni/N) = -\sum pi \ln pi}$	Quantitative description of species	Smith and Grassle (1977)
9	$\mathbf{C = \Sigma (pi)^2}$	Reflects the dominance	(Simpson, 1949)
10	$e = \frac{H'}{\log S}$	Quantifies how equal the community is	Smith and Wilson (1996)
11	$\mathbf{AGTB = 0.0509 * \partial * D^2 H}$	Assessment of above-ground biomass of Trees	MacDicken (1997), Subedi et al. (2010)
12	$\mathbf{AGTC = 0.47 \times AGTB}$	Assessment of above-ground carbon stock of trees	MacDicken (1997), Subedi et al. (2010)

(continued)

Table 14.1 (continued)

S. No.	Formula	Analysis parameter	References
13	**BGTB = root-to-shoot ratio 1:5 (20% of AGTB)**	Assessment of below-ground biomass of trees	MacDicken (1997), Subedi et al. (2010)
14	**BGTC = BGTC = 0.47 × BGTB**	Assessment of below-ground carbon stock of Trees	MacDicken (1997), Subedi et al. (2010)
15	**TC = AGTC + BGTC**	Estimation of total carbon stock of trees	MacDicken (1997), Subedi et al. (2010)

cuipala (300.67 trees/ha), with a frequency of occurrence 88.33%, and *Dodecadenia grandiflora* was the rarest with a density of 0.667 tree/ha. The total basal area of the trees was 83.38 m^2/ha. The basal area variation among plots was significantly different ($p < 0.05$). The Shannon–Wiener diversity index of the forest was 2.4996 with plot-wise variation from 0.2573 to 1.8095, while Pielou's evenness index was 0.6872. Albeit, a normal distribution pattern was found significantly in both, the Shannon diversity index (W = 0.94552, p-value = 0.04877) and Pielou's evenness index (W = 0.92366, p-value = 0.008944). From the Wilcoxon signed-rank test a significant difference was found in the diversity (V = 861, p-value = 2.516e-08) and evenness index (V = 861, p-value = 2.515e-08) plot-wise, respectively (Fig. 14.2).

14.5 Carbon Stock

The carbon stock of trees was 720.26 tC/ha within the study area. Species-wise carbon stock ranged from 0.30 to 720.26 tC/ha forest while *Spondias pinnata* (0.30tC/ha) and the highest carbon stock sequestrated in *Neolitsea cuipala* (119.06 tC/ha). Plot-wise carbon stock was observed to be normally distributed (W = 0.89118, p-value = 0.0009199) and found to be significantly different (V = 861, p-value = 9.095e-13). The relationship between height (m) of trees and DBH (cm) is shown in Fig. 14.3a with value Rho = 0.4537, and relationship between species diversity and carbon stock of the forest is shown in Fig. 14.3b. with value Rho = 0.328.

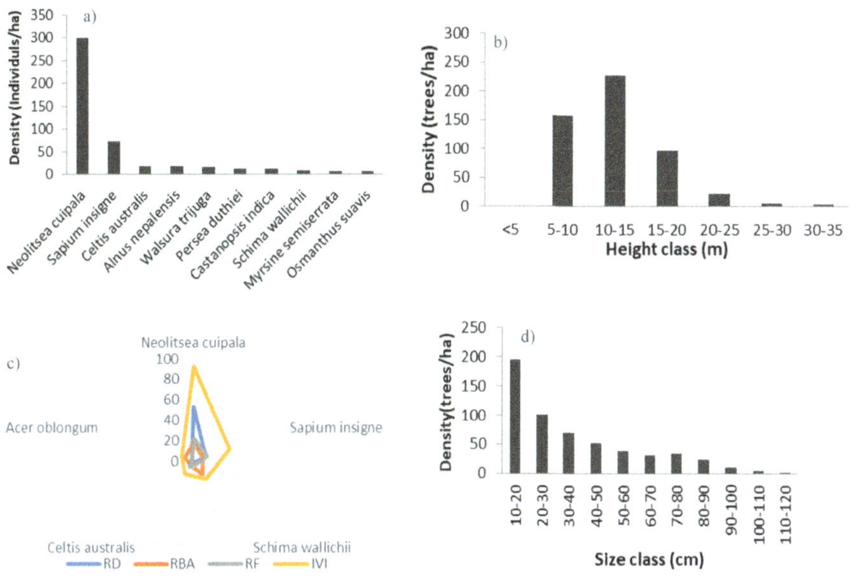

Fig. 14.2 Community attributes. **a** Density of dominant tree species, **b** Density of trees in different height classes, **c** Photograph of main dominant trees in the forest, **d** Density of trees in different diameter classes

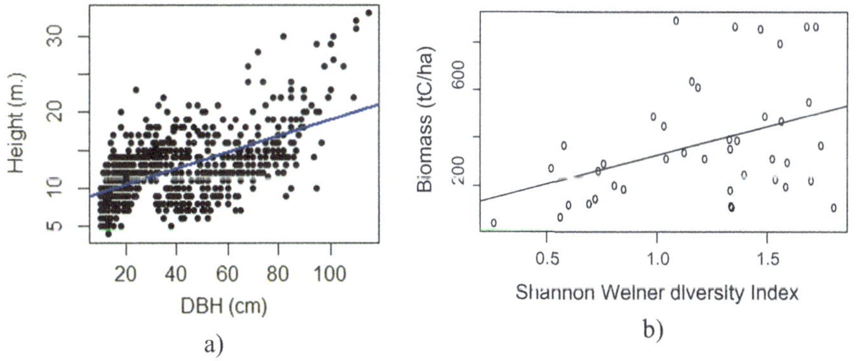

Fig. 14.3 **a** Relationship between Height (m) and DBH (cm), **b** Relationship between Species Diversity and Biomass

14.6 Discussions

The result demonstrated that the forest was dense in falling within 390–1460 trees/ha as reported by (Khadka and Schmidt-Vogt 2008). It was proved as similarly reported by (Sobuj and Rahman 2011) ranging from 140 to 750 trees/ha in Pindari forest, Kathmandu. The high species richness found in the forest means there was no grazing and good conservation. The presence of high diversity with high-class girth trees indicated that the forest was matured and well preserved (Jim and Chen 2009). *Spondias pinnata* and *Litsea polyantha* densities were found to be low because of only a few occurrences in the forest (Bhattarai and Kindlmann 2012). The basal area is considered an important criterion for evaluating timber production in the forest ecosystem (Agrawal 1992). The basal area also indicated that the forest found natural fertility (Krebs 1972). A study of forest basal area of trees has reported higher than Sikre forest 76.15 m^2/ha, Nagarjun hill forest 48.65 m^2/ha, Bhabar forest 34.30 m^2/ha, and 70.00 m^2/ha Budanilkantha community forest reported by (Sapkota 2009) (all forest are adjoining Shivapuri Nagarjun National Park, Kathmandu) indicates this urban forest is matured and high contain biomass. The diameter distribution class was found in a small class size maximum means the forest considered in good regeneration. From another aspect, the Dominance Diversity (DD) curve found a gentle slope in the study of species that interpreted the forest is in sustainable growth (Roy et al. 2012). Shannon Weiner index and evenness index was obtained higher but Simpson's index of dominance was lower which suggested species diversity increases as the study forest ecosystem (Odum 1967). The diversity indices is found to be higher as compared to the results reported by authors, which is illustrated in Table 14.2.

Species diversity is a controlled habitat condition as well as the interactions of the component species (Simpson 1949). The evenness index is found to be greater than 0.5 in most of the plots indicating that the tree community in the study area represents the mature community to the later stage of secondary succession; the resources are shared evenly by the species and there is a lack in the dominance of single species (Odum and Barrett 2005). It is said that the species with the greatest importance value are leading dominants of the forest (Feng and Nursey-Bray 2020) but the study area shows less dominance in each species which means high ecological success (Krebs 1972) and less competence (Laakkonen et al. 2019). As per the recent study it reported 79.02 as the Importance Value Index (IVI) for *Alnus nepalensis* in Sikre forest while in Hill Sal forest IVI = 206.9, in the central region of Nepal.

Biomass varied in different species and plots of the forest due to variation in age, size of the tree, forest composition, and tree density (Bluffstone et al. 2020). Moreover, (Thapa-Magar and Shrestha 2015) noticed that micro-climatic variables

Table 14.2 H' in different forests

S. No.	H' index	Study forest	References
1.	0.63	Sub-alpine forest(Langtang National Park)	Sayers and Norconk (2008)
2.	2.33	Sikre forest (Kathmandu)	Sapkota (2009)
3.	1.38	Bhabar lowland forest (Nawalparasi)	Sapkota (2009)
4.	1.74–2.78	Siwalik Hills (Ilam)	Sapkota (2009)
5.	2.08	Siwalik Hills (Palpa)	Bhusal et al. (2019)
6.	2.499	Gokarna forest (Kathmandu)	In this work

could play a significant role in the amount of carbon stock in the forest. The carbon stock was found to be more than 270.46 t/ha in the urban forest which means good sequestration of carbon (Zhao et al. 2010). Similarly, (Dangal et al. 2017) have estimated the above-ground tree biomass of *Pinus roxburghii* forest (269.205 t/ha) in the mixed broadleaf forest of Palpa district and (76.65 t/ha) in Gadhanta-Bardibash Community forest management (CFM) in Sarlahi district. Total carbon stock (720.26 t/ha) estimated in the study forest was found to be higher compared with (Bradshaw and Warkentin 2015) reported value in the global boreal forest and in the tropical seasonal forest (192.87t/ha) in Southwestern China (Liu and Li 2012). This indicated that the study forest has no harvesting activities taking place because of the strict protection mechanism in carbon sequestration (Nave et al. 2010). The higher percentage of biomass in the present study indicated that urban forest patches had found good protection (Van Kooten et al. 1995).

In this study, an attempt was made to establish the relationship of carbon stock with species diversity with no emphasis given to functional diversity (Petchey and Gaston 2006). The moderate positive relationship between species diversity with carbon stock (Chave et al. 2005) and species richness with carbon stock was drawn to be similar to other studies (Markum et al. 2014). The degree of variation in carbon stock with the species richness and species diversity was explained with the relationship of carbon stock with productivity (Mattsson et al. 2015).

14.7 Conclusions

This paper showed that the urban forest has significantly played a role of acarbon sink and supported the mitigation of climate change. In addition, two important issues were identified, firstly, a great dependency of carbon mitigation from the urban forest, and secondly ecosystem services. Moreover, this kind of study is essential for the conservation of urban forests. Behind the climate change mitigation, the urban forest should be promoted for their ability to provide other environmental services such as air and noise pollution amendment potential, and resource of drinking water which could be the topic of further research.

Acknowledgements We are thankful to the Gokarna forest committee for helping us by giving permission for research. Thanks go to the forest guide. We are also grateful to Ramesh Sapkota for his immense help. Finally, we thank and acknowledge the anonymous reviewers for their constructive comments.

Appendix

See Table 14.3.

Table 14.3 Carbon Stock (tC/ha) of all species in the study area

Name of Species	Biomass (kg)	Biomass (Tons)	Biomass (tC/ha)	Carbon Stock (tC/ha)
Neolitsea cuipala	63328.4	63.33	253.31	119.06
Holarrhena pubescens	39134.29	39.13	156.54	73.57
Celtis australis	37874.35	37.88	152.5	71.21
Alnus nepalensis	27293.48	27.29	109.17	51.31
Schima wallichii	25641.64	25.64	102.57	48.21
Osmanthus suavis	20213.83	20.21	80.86	38
Fraxinus ornus	17923.52	17.92	71.69	33.7
Engelhardia spicata	16360.02	16.36	65.44	30.76
Persea odoratissima	15982.33	15.98	63.93	30.05
Castonopsis indica	14116.61	14.12	56.47	26.54
Walsura trijuga	11024.29	11.02	44.1	20.73
Acer oblongum	10804.64	10.8	43.22	20.31
Quercus semecarpifolia	8651.76	8.65	34.61	16.27
Choerospondias axillaris	7774.96	7.77	31.1	14.62
Euonymus sp.	7403.37	7.4	29.61	13.92
Sepium insigne	7331.39	7.33	29.33	13.78
Acer japonicum	6500.17	6.5	26	12.22
Syzygium cumini	6057.67	6.06	24.23	11.39
Albizia sp.	4392.44	4.39	17.57	8.26
Eurya acuminata	4340.77	4.34	17.36	8.16
Populus nigra	4063	4.06	16.25	7.64
Betula alnoides	3626.6	3.63	14.51	6.82
Mallotus philippensis	3315.25	3.32	13.26	6.23
Dodecadenia grandiflora	3087.59	3.09	12.35	5.8
Ficus semicordata	2998.3	3	11.99	5.64
Myrsine semiserrata	2311.35	2.31	9.25	4.35

(continued)

Table 14.3 (continued)

Name of Species	Biomass (kg)	Biomass (Tons)	Biomass (tC/ha)	Carbon Stock (tC/ha)
Ziziphus mauritiana	2107.99	2.11	8.43	3.96
Michelia champaca	1629.15	1.63	6.52	3.06
Schleichera trijuga	1583.19	1.58	6.33	2.98
Magnolia campbellii	1323.4	1.32	5.29	2.49
Cleistocalyx operculatus	1114.1	1.11	4.46	2.09
Ficus benghalensis	801.46	0.8	3.21	1.51
Fraximus floribunda	699.09	0.7	2.8	1.31
Litsea polyantha	664.44	0.66	2.66	1.25
Siziya sp.	587.64	0.59	2.35	1.1
Pyrus pashia	545.14	0.55	2.18	1.02
Persea duthiei	342.03	0.34	1.37	0.64
Spondias pinnata	157.79	0.16	0.63	0.3
Total	**383107.44**	**383.08**	**1533.45**	**720.26**

References

Agrawal S (1992) Fundamental of ecology. Ashish publishing House, New Delhi

Baur JW, Ries P, Rosenberger RS (2020) A relationship between emotional connection to nature and attitudes about urban forest management. Urban Ecosystems 23:187–197

Bhattarai BP, Kindlmann P (2012) Impact of livestock grazing on the vegetation and wild ungulates in the Barandabhar Corridor Forest, Nepal. Himalayan biodiversity in the changing world. Springer

Bhusal DR, Ghimire KC, Dulal S, Baniya P, Thakuri S (2019) Spatial relation of bumblebees (Hymenoptera-Apidae) with host-plant and their conservation issues: an outlook from urban ecosystem of Kathmandu Valley, Nepal. Eur J Ecol 5:1–7

Bluffstone R, Dannenberg A, Martinsson P, Jha P, Bista R (2020) Cooperative behavior and common pool resources: Experimental evidence from community forest user groups in Nepal. World Dev 129:104889

Bradshaw CJ, Warkentin IG (2015) Global estimates of boreal forest carbon stocks and flux. Global Planet Change 128:24–30

Chave JR, Andalo C, Brown S, Cairns MA, Chambers JQ, Eamus D, Fölster H, Fromard F, Higuchi N, Kira T (2005) Tree allometry and improved estimation of carbon stocks and balance in tropical forests. Oecologia 145:87–99

Dangal SP, Das AK, Paudel SK (2017) Effectiveness of management interventions on forest carbon stock in planted forests in Nepal. J Environ Manage 196:511–517

Das A, Yadav G, Layek J, Lal R, Meena R, Babu S, Ghosh P (2020) Carbon management in diverse land-use systems of Eastern Himalayan Subtropics. Carbon management in tropical and sub-tropical terrestrial systems. Springer

Ebenberger M, Arnberger A (2019) Exploring visual preferences for structural attributes of urban forest stands for restoration and heat relief. Urban Forest Urban Green 41:272–282

Erdős L, Krstonošić D, Kiss PJ, Bátori Z, Tölgyesi C, Škvorc Ž (2019) Plant composition and diversity at edges in a semi-natural forest–grassland mosaic. Plant Ecol 220:279–292

Escobedo F, Varela S, Zhao M, Wagner JE, Zipperer W (2010) Analyzing the efficacy of subtropical urban forests in offsetting carbon emissions from cities. Environ Sci Policy 13:362–372

Feng H, Nursey-Bray M (2020) Adaptation by herders on the Qinghai-Tibetan Plateau in response to climate change and policy reforms: the implications for carbon sequestration and livelihoods. Carbon management for promoting local livelihood in the Hindu Kush Himalayan (HKH) Region. Springer

Forrester D, Bauhus J (2016) A review of processes behind diversity—productivity relationships in forests. Curr For Rep 2:45–61

Hornsey MJ, Fielding KS (2020) Understanding (and reducing) inaction on climate change. Soc Issues Policy Rev

Hulme M (2020) Is it too late (to stop dangerous climate change)? An editorial. Wiley Interdiscipl Rev: Clim Change 11:e619

Jana SK, Roy J (2020) Biodiversity and impacts of climate change in home gardens: evidence from a study in West Bengal, India. Current State and Future Impacts of Climate Change on Biodiversity, IGI Global

Jim CY, Chen WY (2009) Ecosystem services and valuation of urban forests in China. Cities 26:187–194

Kemeuze V, Mapongmetsem P, Sonwa D, Fongnzossie E, Nkongmeneck B (2015) Plant diversity and carbon stock in sacred groves of semi-arid areas of Cameroon: case study of Mandara mountains. Int J Environ 4:308–318

Khadka SR, Schmidt-Vogt D (2008) Integrating biodiversity conservation and addressing economic needs: an experience with Nepal's community forestry. Local Environ 13:1–13

Khanal N, Uddin K, Matin MA, Tenneson K (2019) Automatic detection of spatiotemporal urban expansion patterns by fusing OSM and landsat data in Kathmandu. Remote Sens 11:2296

Khumbongmayum AD, Khan M, Tripathi R (2006) Biodiversity conservation in sacred groves of Manipur, northeast India: population structure and regeneration status of woody species. Human exploitation and biodiversity conservation. Springer, New York

Krebs CJ (1972) Ecology the experimental analysis of distribution and abundance

Laakkonen A, Hujala T, Pykäläinen J (2019) Integrating intangible resources enables creating new types of forest services-developing forest leasing value network in Finland. Forest Policy Econ 99:157–168

Liu C, Li X (2012) Carbon storage and sequestration by urban forests in Shenyang, China. Urban for urban green. Elsevier GmbH, New York

Liu L, Huang G, Baetz B, Guan Y, Zhang K (2020) Multi-dimensional hypothetical fuzzy risk simulation model for greenhouse gas mitigation policy development. Appl Energy 261:114348

Macdicken KG (1997) A guide to monitoring carbon storage in forestry and agro-forestry projects. Winrock National Institute for Agricultural Development, Arlington, USA

Markum M, Soesilaningsih EA, Suprayogo D, Hairiah K (2014) Plant species diversity in relation to carbon stocks at Jangkok Watershed, Lombok Island. AGRIVITA, J Agric Sci 35:207–217

Mattsson E, Ostwald M, Nissanka S, Pushpakumara D (2015) Quantification of carbon stock and tree diversity of homegardens in a dry zone area of Moneragala district, Sri Lanka. Agrofor Syst 89:435–445

Meena RS, Kumar S, Yadav GS (2020) Soil carbon sequestration in crop production. Springer, Nutrient dynamics for sustainable crop production

Narayana J, Nanda A, Savinaya MS (2020) Carbon sequestration potential of trees in Kuvempu University Campus Forest Area, Western Ghats, Karnataka. Socio-economic and eco-biological dimensions in resource use and conservation. Springer

Nave LE, Vance ED, Swanston CW, Curtis PS (2010) Harvest impacts on soil carbon storage in temperate forests. For Ecol Manage 259:857–866

Odum EP, Barrett GW (2005) Fundamentals of ecology. 5th edn. Thomson Brooks/Cole, Belmont, CA

Odum HT (1967) Biological circuits and the marine systems of Texas

Pachauri RK, Allen MR, Barros VR, Broome J, Cramer W, Christ R, Church JA, Clarke L, Dahe Q, Dasgupta P (2014) Climate change 2014: synthesis report. Contribution of Working Groups I, II and III to the fifth assessment report of the Intergovernmental Panel on Climate Change, Ipcc

Page NV, Qureshi Q, Rawat GS, Kushalappa CG (2010) Plant diversity in sacred forest fragments of Western Ghats: A comparative study of four life forms. Plant Ecol 206:237–250

Petchey OL, Gaston KJ (2006) Functional diversity: back to basics and looking forward. Ecol Lett 9:741–758

Ren Y, Wei X, Wei X, Pan J, Xie P, Song X, Peng D, Zhao J (2011) Relationship between vegetation carbon storage and urbanization: a case study of Xiamen, China. For Ecol Manage 261:1214–1223

Roy S, Byrne J, Pickering C (2012) A systematic quantitative review of urban tree benefits, costs, and assessment methods across cities in different climatic zones. Urban Forest Urban Green 11:351–363

Saha SK, Nair PR, Nair VD, Kumar BM (2009) Soil carbon stock in relation to plant diversity of homegardens in Kerala, India. Agrofor Syst 76:53–65

Şahin ÜA, Onat B, Ayvaz C (2019) Climate Change and Greenhouse Gases in Turkey. Springer, Recycling and Reuse Approaches for Better Sustainability

Sapkota IP, Tigabu M, Odén PC (2009) Tree Tree diversity and regeneration of community-managed Bhabar Lowland and Hill Sal forests in central region of Nepal 300:57–68

Sayers K, Norconk MA (2008) Himalayan Semnopithecus entellus at Langtang National Park, Nepal: diet, activity patterns, and resources. Int J Primatol 29:509

Sekhri S, Kumar P, Fürst C, Pandey R (2020) Mountain specific multi-hazard risk management framework (MSMRMF): Assessment and mitigation of multi-hazard and climate change risk in the Indian Himalayan Region. Ecol Ind 118:106700

Sharma CM, Tiwari OP, Rana YS, Krishan R, Mishra AK (2018) Elevational behaviour on dominance–diversity, regeneration, biomass and carbon storage in ridge forests of Garhwal Himalaya, India. For Ecol Manage 424:105–120

Shurpali N, Agarwal A, Srivastava V (2019) Introduction to greenhouse gas emissions. Springer, Greenhouse gas emissions

Simpson EH (1949) Measurement of diversity. Nature 163:688–688

Singh A, Unnikrishnan S, Naik N, Duvvuri K (2013) Role of India's forests in climate change mitigation through the CDM and REDD+. J Environ Plann Manage 56:61–87

Smith B, Wilson JB (1996) A consumer's guide to evenness indices. Oikos:70–82

Smith W, Grassle JF (1977) Sampling properties of a family of diversity measures. Biometrics:283–292

Sobuj NA, Rahman M (2011) Assessment of plant diversity in Khadimnagar National Park of Bangladesh. Int J Environ Sci 2

Strohbach MW, Haase D (2012) Above-ground carbon storage by urban trees in Leipzig, Germany: Analysis of patterns in a European city. Landsc Urban Plann 104:95–104

Subedi BP, Pandey SS, Pandey A, Rana EB, Bhattarai S, Banskota TR, Charmakar S, Tamrakar R (2010) Forest carbon stock measurement: guidelines for measuring carbon stocks in community-managed forests. Asia Netw Sustain Agric Bioresour Fed Community For Users, Int Cent Integr Mt Dev Katmandú, Nepal 69

Sundarapandian S, Amritha S, Gowsalya L, Kayathri P, Thamizharasi M, Dar J, Srinivas K, Gandhi S (2013) Estimation of biomass and carbon stock of woody plants in different land-uses. Forest Res 3:2

Team RC (2016) R: A language and environment for statistical computing [computer software](version 3.1. 3.). R Foundation for Statistical Computing, Vienna, Austria. https://www.r-project.org (Last viewed June 30, 2016)

ter Braak CJ, Šmilauer P (2015) Topics in constrained and unconstrained ordination. Plant Ecol 216:683–696

Thapa-Magar KB, Shrestha BB (2015) Carbon stock in community managed hill sal (Shorea robusta) forests of central Nepal. J Sustain For 34:483–501

van Kooten GC, Binkley CS, Delcourt G (1995) Effect of carbon taxes and subsidies on optimal forest rotation age and supply of carbon services. Am J Agr Econ 77:365–374

Zhao M, Kong Z-H, Escobedo FJ, Gao J (2010) Impacts of urban forests on offsetting carbon emissions from industrial energy use in Hangzhou, China. J Environ Manage 91:807–813

Chapter 15
Geospatial Analysis to Understand the Linkage Between Urban Sprawl and Temperature of a Region: Micro- and Meso-Scale Study of Mumbai City

Hemant Bherwani, Suman Kumar, Narendra Kumar, Anju Singh, and Rakesh Kumar

Abstract Micro climate of any city has its unique features depending upon location and topographical features, including presence of waterbodies and forest areas. In the present study, micro/meso scale climate of four selected regions of Mumbai city and its suburbs is assessed in terms of land use land cover (LULC) and temperature profile of the study area. The study is aimed at understanding the effect of urbanisation on micro-climate of an area, conisdering aspect of temperature. The outcome of the study can provide information to policy makers and developers dealing with town and country planning to consider the contributing factors and adopt proper mitigation measures to avoid undesirable shift in micro-climate of an area. Remote sensing data of Landsat-5 and Landsat-8 are used for LULC analysis and land surface temperature (LST) analysis, and change over a time period of 20 years is analysed. The study revealed that a rapid increase in population in the city has led to considerable changes in land use pattern and surface temperature of the city. In the current analysis of LULC data, it is found that the built-up area increased by about 140%, whereas all other topographical features like waterbody, mangrove forest, deciduous forest, barren land

H. Bherwani (✉) · S. Kumar · N. Kumar · R. Kumar
CSIR—National Environmental Engineering Research Institute (CSIR-NEERI), Nagpur-440017, Maharashtra, India
e-mail: h.bherwani@neeri.res.in

S. Kumar
e-mail: kumar.suman.93087@gmail.com

N. Kumar
e-mail: narendra121tit@gmail.com

R. Kumar
e-mail: rkumar@neeri.res.in

H. Bherwani
Academy of Scientific and Innovative Research (AcSIR), Ghaziabad-201002, Uttar Pradesh, India

A. Singh
National Institute of Industrial Engineering (NITIE), Mumbai-400087, Maharashtra, India
e-mail: danjusingh@gmail.com

R. K. Mishra et al. (eds.), *Sustainable Climate Action and Water Management*, Advances in Geographical and Environmental Sciences,
https://doi.org/10.1007/978-981-15-8237-0_15

and agricultural land have shown a percentage reduction as 19, 61, 18, 27 and 65%, respectively, over a period of 20 years (1998–2018). Due to this change in LULC pattern and overall other global climatic changes, an increase in snapshot temperature in the range of 3.4–8.8 °C has been observed at the regions where analysis has been done. The protected areas are the least affected, while a major effect is seen in densely populated and commercial areas. This snapshot analysis is done to understand the effects of urbanisation on local climatic factors.

Keywords Micro climate · Remote sensing · Land use land cover (LULC) · Land surface temperature (LST)

15.1 Introduction

Any locality, urban area or a city, has its local climate pattern referred to as micro-climate of that particular city or area. It depends on various features, like geographical features, land use pattern, location of the area, presence of waterbodies, forest areas etc., and as soon as any change occurs in these features it results in some changes in the micro-climate of that particular area (Bherwani et. al., 2020a). Geographical features do not change significantly in a short span of time but land use pattern of all cities is changing in a fast pace due to urban agglomeration. As more and more people are migrating towards cities, large number of buildings and other infrastructure are being constructed to suit the changing lifestyle of the urban population. The temperature of large urban areas is generally higher as compared to those of the surrounding rural or suburban areas (Santamouris et al. 2018). Use of materials like concrete, asphalt etc. in urban area that effectively retain heat for a longer time is one of the major causes of urban heat island (UHI) (Shahmohamadi et al. 2011). The energy consumption for cooling purpose is also more in an urban area (Yang and Santamouris 2018). Mumbai city, in the Indian state of Maharashtra, is the largest city in India in terms of population as well as economy. People from all parts of country are moving towards Mumbai for jobs, business, education and other purposes. Owing to rapid increase in population of Mumbai city, its land use pattern is getting changed significantly, and because of these changes, micro-climatic features, particularly local temperature, is changing which is assessed in the present study with the help of remote sensing techniques.

15.1.1 Study Area

The present study of LULC and LST change analysis is done for areas of Mumbai city and its suburbs. Four locations, namely Kurla region, Sanjay Gandhi National Park, outskirts of Mira Bhayandar and Kolegaon, were selected for the study. Figure 15.1 shows geographical location of the study area. The selected regions consist of one

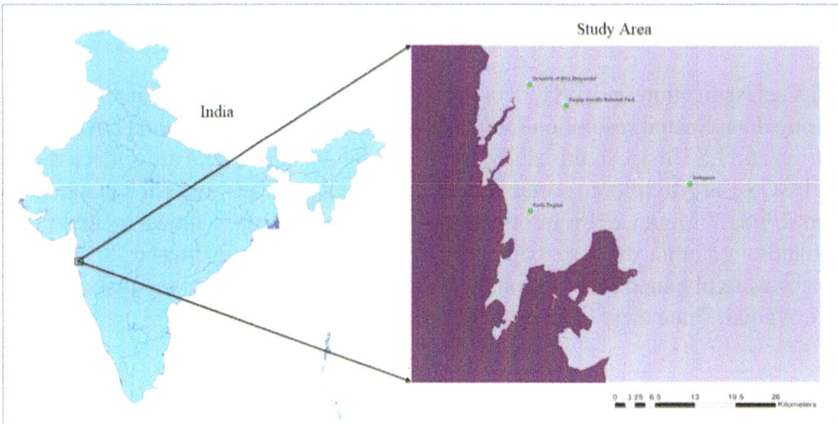

Fig. 15.1 Location of the study area in India

forest area—like Sanjay Gandhi National Park, and one central area of Mumbai—like Kurla region and the remaining two are suburban areas of Mumbai.

15.1.2 Objectives of the Study

Mumbai (latitude 19.1089° N, longitude 72.8813° E) is considered for the study of micro-climatic parameter (temperature) and land use land cover pattern change. The major objectives of the present study are as follows:

(i) To assess the change in LULC pattern of the study area during the period November 1998 to November 2018.
(ii) To assess the change in temperature of the study area for the period November 1998 to November 2018 by using Landsat remote sensing data.

15.2 Materials and Methods

Remote sensing tools and techniques are useful for the study of urban heat island for small and large cities (Weng et al. 2004). In the present study, Landsat satellite data are used for assessing LULC changes and LST changes. The methodology adopted for satellite data acquisition and preparation of LULC and LST maps is described in the following sections.

15.3 Satellite Data Acquisition

LULC classification and LST data are extracted using Landsat (5 and 8) remote sensing data. Spatial resolution (30 m) data is used for generating land cover classification and LST map. In this study, three cloudless satellite data are collected from the USGS Earth Explorer portal [USGS] to develop LULC and LST maps of the years 1998, 2011 and 2018. Information of the satellite data is shown in Table 15.1.

Landsat satellite data are obtained for the years 1998, 2011 and 2018. These images were obtained for the month of November during which the quality of data is very good, since there is no cloud.

15.3.1 Developing LULC Map and LST Map from Landsat Data

LULC: In order to prepare LULC map for the study area, Landsat 5 and 8 satellite data are used, taken from United States Geological Survey (USGS). False colour composite (FCC) is generated for identifying land types in the study area. Supervised classification is done and maximum likelihood method is adopted to get LULC map (Rahman et al. 2011). The cross validations are done using Google Earth Imageries (Naqvi et al. 2014).

LST: In order to prepare LST map for the study area, Landsat 5 and 8 satellite data are used, same as LULC, taken from USGS. The LST map is generated for identifying land surface temperature of selected regions in the study area. To extract LST from Landsat 5 and Landsat 8, Band 6 and Band 10 are used, respectively, which are thermal bands of each satellite. In Landsat 8, Band 10 provides better results than using Bands 10 and 11 in combination (Yu et al. 2014). To retrieve LST, RS and GIS tool is used in QGIS. The tool performs well-defined algorithms on raw satellite data to produce popularly used outputs like, LST, satellite brightness temperature, normalized difference vegetation index (NDVI). To get LST from the respective bands as highlighted above, digital numbers (DN) are converted into radiance followed by computation of brightness temperature, assessment of land surface emissivity (LSE), assessment of near surface air temperature and brightness temperature emissivity/atmospheric correction (Isaya and Avdan 2016).

Table 15.1 Landsat data used in the study

S. no.	Imagery date	Spacecraft identifier	Path/row	Resolution (m)	Coordinate system	UTM zone	Source
1	1998/11/26	LANDSAT_5	148/047	30	WGS84	43	USGS
2	2011/11/04	LANDSAT_5	148/047	30	WGS84	43	USGS
3	2018/11/17	LANDSAT_8	148/047	30	WGS84	43	USGS

15.4 Results and Discussion

It is evident from the analysis of LULC and LST data, obtained from Landsat 8 and Landsat 5, that over the years both LULC pattern and LST of study area have changed significantly. The comparative results of land use land cover pattern in the years 1998, 2011 and 2018 are given in Table 15.2 and graphical representation of land use land cover changes is shown in Fig. 15.2.

Remote sensing and GIS-based map showing LULC pattern of study area for the years 1998, 2011 and 2018 are shown in Fig. 15.3. Change in LULC pattern is the outcome of urbanization where natural features of an area are altered by impermeable surfaces like asphalt and concrete (Grover and Singh 2015). This change is visible

Table 15.2 Changes in LULC pattern in the study area during 1998, 2011 and 2018

Area changes (km^2)							
S. no.	Area	1998	2011	2018		% Change (2011 to 2018)	% Change (1998 to 2018)
1	Waterbody	112	101	91	−10	−10	−19
2	Mangrove forest	262	177	101	−32	−43	−61
3	Deciduous forest + scrubs	1369	1140	1118	−17	−2	−18
4	Barren land	228	225	166	−1	−26	−27
5	Built-up	529	1024	1270	94	24	140
6	Agriculture	378	211	132	−44	−37	−65
	Total area (km^2)	2878	2878	2878			

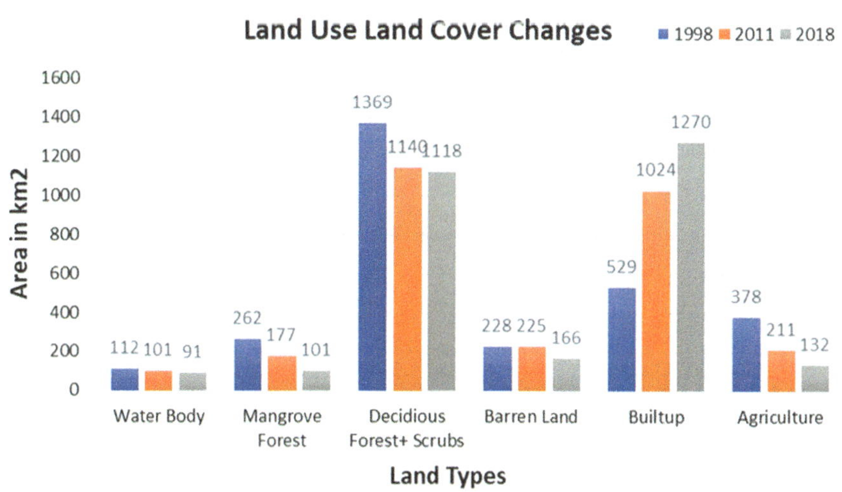

Fig. 15.2 LULC changes in the study area in November 1998, 2011 and 2018

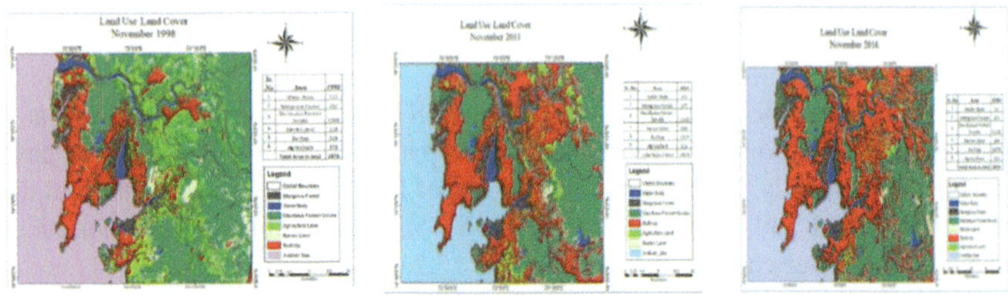

November 1998 November 2011 November 2018

Fig. 15.3 LULC map of the study area in November 1998, 2011 and 2018

in the present study as built-up area increased by 140% over a period of 20 years. All other physical features of the study area, including waterbody, mangrove forest, deciduous forest, barren land and agricultural land have shown reduction in the range of 19, 61, 18, 27 and 65%, respectively, for the years 1998 to 2018. Most of the parts of these reduced areas have been converted into built-up area which is visible from Fig. 15.3. The reduction in mangrove forest is a concern as it works as a carbon sink and can play a vital role to maintain air quality and biodiversity along with controlling temperature rise in a city like Mumbai.

Similarly, comparative results of land surface temperature for November 1998, 2011 and 2018 are given in Table 15.3 and graphical representation of land surface temperature changes is shown in Fig. 15.4. Temperature of the regions under study is found to have increased over a period of 20 years (1998–2018). Remote sensing and GIS-based map showing land surface temperature of the study area in November 2011 and November 2018 is shown in Fig. 15.5. There is an increase in temperature in all the regions under study. A minimum shift in temperature is found at Sanjay Gandhi National Park and maximum increase is found at outskirts of Mira Bhayandar as 3.4 and 8.8 °C, respectively, for the year 1998–2018. One of the major reasons for increase in temperature of Mira Bhayandar is the growth of transportation hub in this area which was evident from ground truthing. The other regions, namely Kurla region which is a major commercial place and Kolegaon, have also shown an increase in

Table 15.3 Changes in LST in the study area over the period November 1998, 2011 and 2018

Temperature changes (°C)							
S. no.	Name of the Area	1998	2011	2018	1998–2011	2011–2018	1998–2018
1	Kurla region	31.2	32.9	39.3	1.6	6.4	8.1
2	Sanjay Gandhi National Park	23.7	25.0	27.1	1.3	2.1	3.4
3	Outskirts of Mira Bhayandar	27.1	31.2	35.9	4.1	4.6	8.8
4	Kolegaon	25.8	28.8	34.0	2.9	5.2	8.2

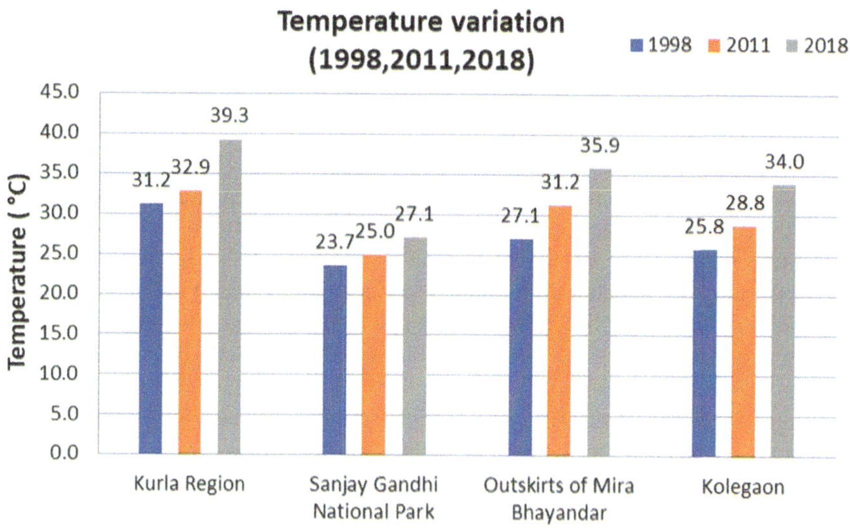

Fig. 15.4 Changes in land surface temperature of selected regions of the study area over the period November 1998, 2011 and 2018

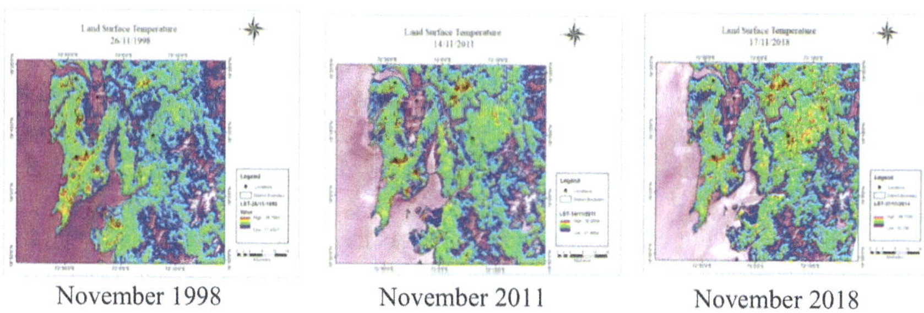

Fig. 15.5 LST map of the study area during the period November 1998, 2011 and 2018

temperature as 8.1 and 8.2 °C, respectively. The increase in land surface temperature causes thermal discomfort to the dwellers which in turn affects ecological footprint due to the excess energy consumption to improve the quality of living (Bherwani et. al, 2020b; Patz et al. 2005). There are many factors which affect the intensity of LST and UHI. These factors may include greenery, waterbodies, use of metals, impervious concrete and asphalt, LULC pattern and surface roughness (Kuang et al. 2015).

15.5 Conclusion

The present study reflects the change or shift in temperature with the continuous change in land use land cover (LULC) pattern of Mumbai city during the period 1998, 2011 and 2018. It is found that over the period of 1998 to 2018, the built-up area increased by 140%, whereas all other topographical features including waterbody, mangrove forest, deciduous forest, barren land and agricultural land have shown percentage reduction as 19, 61, 18, 27 and 65%, respectively. Temperature, being one of the most significant micro-climatic parameters, increased over the years and a difference of minimum 3.4 °C to maximum 8.8 °C has been observed in the study area. A minimum change is found in the temperature of Sanjay Gandhi National Park, which is a forest area. At the same time, change in land use pattern is also minimum in this forest region. Increase in temperature in Sanjay Gandhi National Park area may be due to an increase in the built-up area in the surrounding localities and overall global climatic changes due to other factors. A maximum change in temperature is found at the outskirts of Mira Bhayandar. This change in snapshot temperature is largely due to land use land cover changes. The rise in temperature of the above areas includes rise due to local changes as well as the effect of overall global climatic changes, particularly temperature. It can be said that vegetation and presence of waterbodies help in reducing the chances of increase in temperature.

To maintain a favourable micro-climatic pattern for a city like Mumbai, it is necessary to retain the existing physical features, like waterbodies, mangrove forest, deciduous forest, agricultural land etc. Uncontrolled growth of built-up area in a limited area can lead to a significant rise in temperature of the area, hence this development should be distributed in different parts of the city by providing sufficient open space.

Wind is also a major climatic parameter and its free flow can bring a considerable change in the climate of an urban area. A highly urbanised city like Mumbai is becoming more and more compact, hence wind profile is getting changed. The effect of wind may also be considered for analysis of micro-climate of an area as well as its role with rise in temperature of a region in subsequent studies.

Conflict of Interest On behalf of all other authors, the corresponding author states that there exists no conflict of interest.

References

Bherwani H, Singh A, Kumar R (2020a) Assessment methods of urban microclimate and its parameters: a critical review to take the research from lab to land. Urban Clim 34:100690

Bherwani H, Nair M, Kapley A, Kumar R (2020b) Valuation of ecosystem services and environmental damages: an imperative tool for decision making and sustainability. Eur J Sustain Dev Res 4(4):em0133. https://doi.org/10.29333/ejosdr/8321

Grover A, Singh RB (2015) Analysis of urban heat island (UHI) in relation to normalized difference vegetation index (NDVI): A comparative study of Delhi and Mumbai. Environments 2(2):125–138
Isaya Ndossi M, Avdan U (2016) Application of open source coding technologies in the production of land surface temperature (LST) maps from Landsat: a PyQGIS plugin. Remote Sens 8(5):413
Kuang W, Liu Y, Dou Y, Chi W, Chen G, Gao C, Yang T, Liu J, Zhang R (2015) What are hot and what are not in an urban landscape: Quantifying and explaining the land surface temperature pattern in Beijing, China. Landsc Ecol 30(2):357–373
Naqvi HR, Siddiqui L, Devi LM, Siddiqui MA (2014) Landscape transformation analysis employing compound interest formula in the Nun Nadi Watershed, India. Egyptian J Remote Sens Space Sci 17(2):149–157
Patz JA, Campbell-Lendrum D, Holloway T, Foley JA (2005) Impact of regional climate change on human health. Nature 438(7066):310–317
Rahman A, Aggarwal SP, Netzband M, Fazal S (2010) Monitoring urban sprawl using remote sensing and GIS techniques of a fast growing urban centre, India. IEEE J Sel Top Appl Earth Obs Remote Sens 4(1):56–64
Santamouris M, Haddad S, Saliari M, Vasilakopoulou K, Synnefa A, Paolini R, Fiorito F (2018) On the energy impact of urban heat island in Sydney: Climate and energy potential of mitigation technologies. Energy Build 166:154–164
Shahmohamadi P, Che-Ani AI, Maulud KNA, Tawil NM, Abdullah NAG (2011) The impact of anthropogenic heat on formation of urban heat island and energy consumption balance. Urban Stud Res
United States Geological Survey (USGS) Available online: www.earthexplorer.usgs.gov. Accessed 06 Aug 2019
Weng Q, Dengshang L, Jacquelyn S (2004) Estimation of land surface temperature—Vegetation abundance relationship for urban heat island studies. Remote Sens Environ 89(4):467–483
Yang J, Santamouris M (2018) Urban heat island and mitigation technologies in asian and australian cities—impact and mitigation. Urban Sci 2(3):74
Yu X, Guo X, Wu Z (2014) Land surface temperature retrieval from Landsat 8 TIRS—Comparison between radiative transfer equation-based method, split window algorithm and single channel method. Remote Sensing 6(10):9829–9852

Chapter 16
Climate Change Mitigation Measures in Telangana State, India

Ashok Kumar Lonavath and Karunakar Virugu

Abstract Climate change is the result of long-term fluctuations in temperature, precipitation, wind and other atmospheric phenomena and such changes can occur from both natural and man-made processes. The changing climatic elements lead to risks and hazards for mankind and cause problems for livelihood. The mitigation measures are needed to adapt climatic variations. Telangana state is geographically located in the semi-arid area and has a predominantly hot and dry climate. The mean monthly maximum temperature recorded is 41 °C during May. December is the coldest month in Telangana and the mean minimum temperature recorded is 15.83 °C. The monsoon which enters into the State in the June month and lasts until September records 713 mm of precipitation. The state government adapted many climate change mitigation measures such as Mission Kakatiya, Mission Bhagiradha and Telanganaku Harithaharam, to tackle climate change impacts. In Telangana state the net area under tank irrigation is 9.1% and historically these tanks act as the important source for minor irrigation and drinking water purpose. Most of the tanks in the state were filled with erosion material and in order to restore these tanks 'Mission Kakatiya' program was launched by the Telangana government with an objective to fill up all minor irrigation tanks and lakes in the state. Similarly, 'Harithaharam' program is planned for large-scale tree plantation in the state and was launched on 5 July, 2015 with an objective to increase forest cover in the state from 22.66% to 33%. 'Mission Bhagiradha' is another flagship program of the state government, introduced to ensure safe and sustainable piped drinking water supply from surface water source to the entire state.

Keywords Climate change · Livelihood issues · Mitigation measures · Environment management

---

A. K. Lonavath (✉) · K. Virugu
Department of Geography, University College of Science, Osmania University, Hyderabad 500 007, India
e-mail: ashokouprof@gmail.com

R. K. Mishra et al. (eds.), *Sustainable Climate Action and Water Management*, Advances in Geographical and Environmental Sciences,
https://doi.org/10.1007/978-981-15-8237-0_16

16.1 Introduction

Climate change poses a challenge to sustainability of social and economic development, livelihood of communities and environmental management etc. The vulnerability of climate change started all around the world during the industrial period with the emission of greenhouse gases. The international and national level committees on climate change have been highlighting about its impact since 1980s. It is because the world has noticed that the effect of climate change is leading to melting of polar glaciers due to consistent increase in temperature, which further causes rise in sea level and submergence of coastlands and islands, desertification, droughts and heat waves, uneven rainfall, stronger and more intense cyclones, etc.

Wilby, R. L. highlighted about the climate change impacts on built-up environment in four areas; specifically, urban ventilation and cooling, urban drainage and flood risk, water resources, and outdoor spaces (including air quality and biodiversity). Watkins et al. pointed out that the management of urban heat by the way of urban design can improve thermal comfort. He stated that changing albedo of the building can reduce the impact of solar radiation by reducing the radiant temperature of the environment.

The resultant effects of climate change lead to action by various governments in the world. The Intergovernmental Panel on Climate Change (IPCC) was established in 1988 to assess the availability of scientific information on climate change, to assess environmental and socio-economic impacts of climate change and to formulate response strategies (*Source* Climate Change 1992, The supplementary report to the IPCC Scientific Assessment).

In December 2015, the 21st session of the Conference of the Parties to the United Nations Framework Convention on Climate Change, which also called as COP 21, targeted on reduction of greenhouse gases. In compliance to that, the Government of India released a National Action Plan on Climate Change (NAPCC) in 2008 as part of an ambitious domestic action plan to address climate change. The NAPCC focuses on adaption of climate change (CC) and protecting the vulnerable sections of the society through inclusive and sustainable development strategy that would enhance the ecological sustainability and innovation.

The National Forest Policy of India envisages a minimum of 33% of the total geographical area under forest cover in order to maintain environmental stability and ecological balance that are vital for the sustenance of all lifeforms. India has a long history of protection of trees. Historically, it is evident that Gautama Buddha was preaching everyone to plant trees, and his follower, the great king Ashoka planted trees along the roadside in his entire kingdom. Considering the sustainable climate as the top-most priority, the Government of Telangana introduced 'Harithaharam' program which is meant for plantation of trees in the entire state of Telangana. It was launched on 5 July, 2015, and the objective of the program is to increase the forest cover from 22.66% to 33%. The plantation of trees is done in both notified and non-notified forest regions.

The 'Harithaharam' program is designed to be achieved by a multi-pronged approach by rejuvenating the degraded forest and soil conservation in the Telangana state. The program is aimed to improve the green cover and biodiversity, maintaining ecological balance, ensuring sustainable livelihoods, good rainfall and reducing carbon dioxide emissions.

16.1.1 Objectives

1. To study the spatial and temporal distribution of forest cover in Telangana state.
2. To map the spatial distribution of Harithaharam program in Telangana state.
3. To study the percent increase of forest cover due to Harithaharam program.
4. To correlate natural vegetative cover with respect to Harithaharam program.

16.2 Methodology

The study is based on both primary and secondary data. The secondary data is collected from the State Forest Department and Forest Survey of India. The digital data source for the study is LISS-IV Imageries. For the analysis of data and mapping, statistical methods like mean is applied. Geographical information system (GIS) and image processing techniques are used for spatial and temporal mapping and analysis.

16.2.1 Different Plantation Models Initiated by the Government of Telangana

1. **Avenue plantation**: Under this model, tree plantation is done on the roadside of national and state highways and also along the roads which are connecting the towns and villages.
2. **Block plantation**: Under this model, tree plantation is done at the vicinity of the villages mostly in waste lands, Panchayat lands and common lands.
3. **Institutional plantation**: Under this type, tree plantation is done in the premises of schools, colleges, universities, all government institutions, hospitals, graveyards and private institutions, industries.
4. **Barren hill plantation**: Under this model, tree plantation is done in the barren lands and hillocks.
5. **Homestead plantation**: Under this model, tree plantation is done in and around the residential houses and colonies.
6. **Agro-forestry plantation**: Under this model, tree plantation is done in the farm lands which are taken care of by farmers.

The target of program is to plant 230 crore saplings in three phases during 2015–2016, 2016–2017 and 2017–2018. The target year is mentioned in the vision of the program where different phases were formulated. A total of 82 crore saplings is the target of the government to plant trees in all the three phases. In the first phase, 15.861 crore saplings were planted; in the second phase, 31.674 crore saplings were planted; and in the third phase, 34 crore saplings were planted till 30 July, 2017 and the program is still ongoing. The Government of Telangana has so far spent Rs. 1645.00 crore for the development of nurseries in various districts (*Source* State action plan on Climate Change, 2015).

The total number of nurseries developed district-wise under Harithaharam program in Telangana state varies with respect to its natural vegetative cover and availability of land. A total of 3,549 nurseries were developed in the state under Harithaharam program with a target to grow 4,218.6 lakh saplings. A highest number of 590 nurseries were developed in erstwhile Karimnagar district with a target to grow 521 lakh saplings. The least number of 73 nurseries were developed in the concrete jungle of Hyderabad city with a target to grow 231 lakh saplings. District-wise number of nurseries and target given to them is mentioned in Table 16.1.

According to agricultural statistics at a glance published by the Department of Economics and Statistics, Telangana state has 22.66% of forest cover during 2015–2016. The highest 42% of forest cover is found in Adilabad district, followed by Khammam 41.9%, Warangal 29%, and Nizamabad and Karimnagar 21%. The least 5.8% of forest cover is found in Nalgonda. No forest cover is found in Hyderabad district.

When district-wise total number of nurseries and the target given to them to grow number of saplings as mentioned in Table 16.1 and the percentage of forest cover district-wise as mentioned in Table 16.2 are compared, it is evident that except in Adilabad district, number of nurseries developed was more in those districts

Table 16.1 District-wise number of nurseries and saplings grown in lakhs during 2016–2017

S. no.	District name	No. of nurseries	Saplings target in lakhs
1	Adilabad	499	404.8
2	Hyderabad	73	231
3	Karimnagar	590	521.4
4	Khammam	208	321.2
5	Mahabubnagar	482	505.4
6	Medak	372	557.7
7	Nalgonda	400	488.4
8	Nizamabad	331	347.6
9	Ranga Reddy	265	321.9
10	Warangal	329	519.2
	Total	3,549	4,218.6

Source State Harithaharam Program, Aranyabhavan, Hyderabad

Table 16.2 Percent forest cover in Telangana state, 2015–2016

S. no.	District	Geographical area (sq. km)	Forest cover (sq. km)	% Forest area
1	Adilabad	16105	6895.17	42.81
2	Nizamabad	7956	1693.43	21.28
3	Karimnagar	11823	2504.1	21.18
4	Medak	9700	913.9	9.42
5	Hyderabad	217	0	0.00
6	Ranga Reddy	7493	730.75	9.75
7	Mahabubnagar	18432	2555.96	13.87
8	Nalgonda	14240	830.73	5.83
9	Warangal	12846	3710.14	28.88
10	Khammam	13266.1	5566.83	41.96
	Total	112078.1	25401.01	22.66

Source Agricultural Statistics at a Glance, Telangana, 2015–2016, DES, Hyderabad

which are having less forest cover and the target fixed for them to grow number of saplings is high. The aim of the program is to increase the forest cover more or less uniformly in all the districts. An increase of forest cover in the entire region will reduce environmental imbalance in the state.

According to the Forest Survey of India, the forest cover in Telangana state is 18.22% of its total geographical area during 2016 2017. The highest percentage of forest cover was recorded as 35.32% in Adilabad district, followed by Khammam 33.42%, Warangal 22.72% and Mahabubnagar 12.47%. The least percentage of forest cover was recorded as 2.88% in Nalgonda district. The other districts recorded least percentage of forest cover: Hyderabad 7.83% and Medak 7.69% (Table 16.3). Overall there was a decline of total forest cover from 2015–2016 to 2016–2017. Interestingly, highly urbanized districts like Hyderabad, Medak and Ranga Reddy have noticed an increasing pattern of forest cover during this period, whereas all the remaining districts have noticed a decreasing pattern of forest cover. The possible reason is the Telangana state was formed in the year 2014, and from then the urbanization process was accelerated in the entire state; as a result, the forest cover in GHMC/HMDA area covering three districts, namely Hyderabad, Ranga Reddy and Medak, has increased. The forest cover in remaining eight districts has decreased. However, in October 2016, the existing 10 districts were divided and the number of districts increased to 33, adding 23 new districts in the state. This effect has given further scope for the growth of small and medium size towns, so the loss of forest cover was noticed in all the districts except Hyderabad, Ranga Reddy and Medak.

Further, the growth of forest area in Hyderabad (GHMC/HMDA), Ranga Reddy and Medak districts is also due to the introduction of new policies and programs by the state government from time to time, and in turn this has helped the protection of forest area (Figs. 16.1).

Table 16.3 Percent forest cover in Telangana state, 2016–2017

S. no.	District	Geographical area (sq. km)	Forest cover (sq. km)	% Forest area
1	Adilabad	16105	5688	35.32
2	Nizamabad	7956	1187	14.92
3	Karimnagar	11823	1962	16.59
4	Medak	9700	746	7.69
5	Hyderabad	217	17	7.83
6	Ranga Reddy	7493	760	10.14
7	Mahabubnagar	18432	2298	12.47
8	Nalgonda	14240	410	2.88
9	Warangal	12846	2918	22.72
10	Khammam	13266.1	4433	33.42
	Total	112078.1	20419	18.22

Source Forest Survey of India, India

Harithaharam program which is meant for large-scale tree plantation in Telangana state was started in 2015–2016 and a total of 1586.07 lakh saplings were planted covering 2,440 sq. km area under first phase. The percentage of saplings planted outside and inside the forest area was recorded as 81.51% and 18.5%, respectively. The highest percentage of saplings planted outside forest area was recorded as 100% in Hyderabad district followed by Medak district 90%, Khammam district 89.79%, and the lowest percentage of saplings planted was recorded as 72.55% in Nizamabad district followed by Warangal district 74.59% and Mahabubnagar district 79.55%. The highest percentage of saplings planted inside the forest area was recorded as 27.45% in Nizamabad district followed by Warangal 25.41%, Adilabad 20.59%, and the lowest percentage of saplings planted was recorded as 0% in Hyderabad district followed by Medak 9.29% and Khammam 10.21% (Fig. 16.2, Table 16.4).

Under second stage of Harithaharam program during 2016–2017, a total of 3,168 lakh saplings were planted in 4873 sq. km area. The percentage of saplings planted outside and inside the forest area was recorded as 87.39% and 12.61%, respectively. The highest percentage of saplings planted outside forest area was recorded as 100% in Hyderabad district followed by Nalgonda 96.28% and Mahabubnagar with 95.52% and lowest percentage of saplings planted was recorded as 67.86% in Adilabad district followed by Khammam 83.86% and Karimnagar 87.49%. The highest percentage of saplings planted inside the forest area was recorded as 32.14% in Adilabad followed by Khammam 16.14% and Karimnagar 12.51%, and the lowest percentage of saplings planted was recorded as 0% in Hyderabad followed by Nalgonda 3.72% and Mahabubnagar 4.48% (Fig. 16.3, Table 16.5).

Under third phase of Harithaharam program in 2017–2018, a total of 3,408 saplings were planted in 5,243 sq. km area in Telangana state. The percentage of saplings planted outside and inside forest was recorded as 88.44% and 11.56%, respectively. The highest percentage of saplings planted outside forest area was

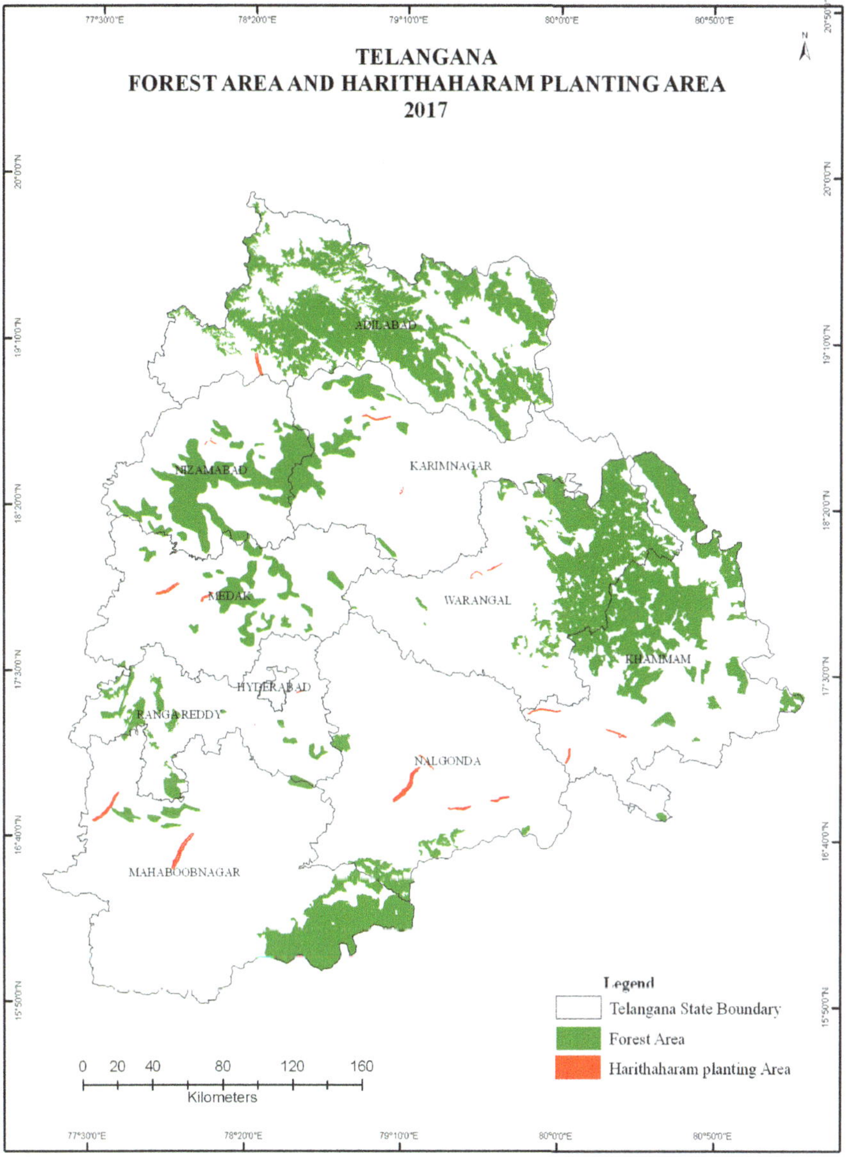

Fig. 16.1 Telangana forest area and Harithaharam planting area 2017

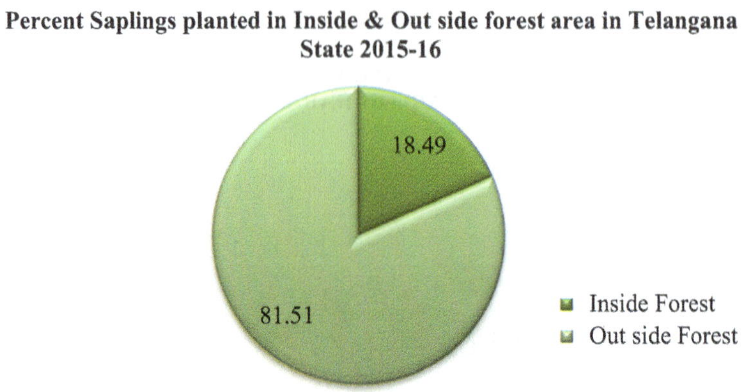

Fig. 16.2 Percent saplings planted inside and outside forest area in Telangana State 2015–2016

Table 16.4 Number of saplings planted in Telangana state during 2015–2016 (in lakhs)

S. no.	District name	Area under HH in sq. km (2015–2016)	Total no. of saplings planted in lakhs (2015–2016)	% Saplings planted inside forest 2015–2016	% Saplings planted outside forest 2015–2016
1	Adilabad	461.22	299.79	20.59	79.41
2	Hyderabad	37.91	24.64	0	100
3	Karimnagar	336.09	218.46	17.89	82.11
4	Khammam	335.43	218.03	10.21	89.79
5	Mahabubnagar	211.72	137.62	20.45	79.55
6	Medak	178.66	116.13	9.29	90.71
7	Nalgonda	207.54	134.9	18.95	81.05
8	Nizamabad	227.25	147.71	27.45	72.55
9	Ranga Reddy	126.72	82.37	15.42	84.58
10	Warangal	317.57	206.42	25.41	74.59
	Total	**2440.11**	**1586.07**	**18.49**	**81.51**

Source Haritha Haram Program, State Forest Department, Telangana state [Note: Per hectare 650 saplings planted or per one seedling 3 × 3 sq. m]

recorded as 100% in Hyderabad district followed by Nizamabad 97.44% and Mahabubnagar with 97.26%, and the lowest percentage of saplings planted was recorded as 62.57% in Adilabad district followed by Khammam 82.16% and Karimnagar 85.55%. The highest percentage of saplings planted inside forest area was recorded as 37.43% in Adilabad district followed by Khammam 17.84% and Karimnagar 14.45%, and the lowest percentage of forest cover was recorded as 0% in Hyderabad followed by Nizamabad 2.56% and Mahabubnagar 2.74% (Fig. 16.4, Table 16.6).

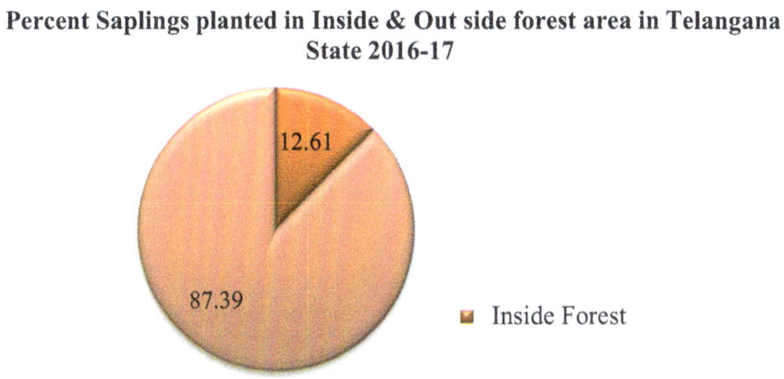

Fig. 16.3 Percent saplings planted inside and outside forest area in Telangana State 2016–2017

Table 16.5 Number of saplings planted in Telangana state during 2016–2017 (in lakhs)

S. no.	District name	Area under HH Harithaharam (sq. km) 2016–2017	No. of saplings planted in lakhs 2016–2017	% Saplings planted inside the forest 2016–2017	% Saplings planted outside the forest 2016–2017
1	Adilabad	655.06	425.79	32.14	67.86
2	Hyderabad	279.90	181.94	0.00	100.00
3	Karimnagar	665.2308	432.4	12.51	87.49
4	Khammam	597.6462	388.47	16.14	83.86
5	Mahabubnagar	333.8615	217.01	4.48	95.52
6	Medak	438.5385	285.05	12.04	87.96
7	Nalgonda	434.1846	282.22	3.72	96.28
8	Nizamabad	555.8615	361.31	12.34	87.66
9	Ranga Reddy	214.2308	139.24	6.63	93.37
10	Warangal	698.4769	454.01	8.22	91.78
	Total	**4873**	**3167.45**	**12.61**	**87.39**

Source Harithaharam Program, State Forest Department, Telangana state. [Note: Per hectare 650 saplings planted or per one sapling 3 × 3 sq. m]

16.3 Conclusion

The Harithaharam program which is designed by Telangana state government in order to increase the green belt in the state for balancing the climatic stability has achieved good results. The program has been implemented in three phases. During the first phase in 2015–2016, 1586.07 lakh saplings were planted; similarly, during the second and third phases in 2016–2017 and 2017–2018, a total of 6575.39 lakh saplings were planted under two categories, i.e., inside the forest area and outside the forest area. The program was very much successful. New nurseries were developed under this

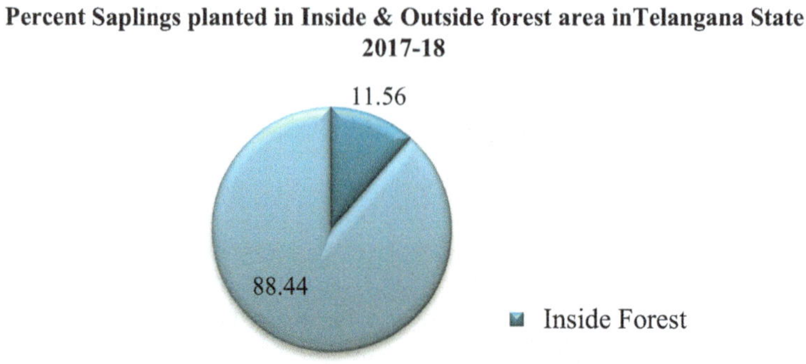

Fig. 16.4 Percent saplings planted inside and outside forest area in Telangana State 2017–2018

Table 16.6 Number of saplings planted in Telangana state during 2017–2018 (in lakhs)

S. no.	District name	Area under HH Harithaharam (sq. km) 2017–2018	Number of saplings planted in lakhs (2017–2018)	Number of saplings planted in lakhs (inside the forest 2017–2018)	Number of saplings planted in lakhs (outside the forest 2017–2018)
1	Adilabad	544.30	353.79	37.43	62.57
2	Hyderabad	257.22	167.19	0.00	100.00
3	Karimnagar	600.98	390.635	14.45	85.55
4	Khammam	557.56	362.42	17.84	82.16
5	Mahabubnagar	2,133.98	1,387.09	2.74	97.26
6	Medak	2,829.23	1,839.00	9.24	90.76
7	Nalgonda	610.60	396.89	5.77	94.23
8	Nizamabad	496.23	322.55	2.56	97.44
9	Ranga Reddy	292.94	190.41	7.39	92.61
10	Warangal	677.73	440.53	6.29	93.71
	Total	5,242.98	3407.94	11.56	88.44

Source Harithaharam Program, State Forest Department, Telangana state. [Note: Per hectare 650 saplings planted or per one sapling 3 × 3 sq. m]

program in all the districts and each nursery is given a target of growing number of saplings. Maximum number of nurseries was developed in those districts having highest forest cover. This program has increased the green belts in the state by planting trees mostly along the national and state highways and in the premises of educational institutions, government offices and villages etc. Inside forest cover category is aimed to improve the environmental balance in urban and rural areas which has less forest cover. Forestry was also developed in residential zones and agriculture belts so as to regain the regular monsoon and to reduce heavy pollution, greenhouse gases emission, heat waves etc. The program was implemented successfully under three phases whereby the percentage of total forest cover has increased in the state.

References

Agricultural Statistics at a Glance, Telangana, 2015–16, DES, Hyderabad

Forest Survey of India, India 2016–17 report

State action plan on climate change in Telangana State, 2015

Watkins R, Palmer J, Kolokotroni M (2007) Increased temperature and intensification of the urban heat island: implications for human comfort and urban design. Alexandrine Press. doi:https://doi.org/10.2148/benv.33.1.85

Wilby RL (2007) A review of climate change impacts on the built environment. Alexandrine Press. doi:https://doi.org/10.2148/benv.33.1.31

Chapter 17
Examining Land Surface Temperature from Agglomerative Spectra Using Hyperspectral Dataset

K. Nivedita Priyadarshini, V. Sivashankari, Sulochana Shekhar, and K. Balasubramani

Abstract Rapid urbanization yields a bleak impact by bolstering the climate change, subsequently escalating the urban heat island (UHI) effect. In order to identify a specific urban material that aggravates more heat, a methodical approach is incorporated by assessing spectral signatures acquired using field-based and image-based analysis. The study area considered is Kalaburagi, Karnataka and the dataset used is AVIRIS NG reflectance corrected level 2 data ranging from 376 to 2500 nm with a ground sampling distance of 4–8 m and sampling interval of 5 nm. Comparison of these libraries provides accurate information about the reflectance spectra signifying unique urban construction materials present in the study area. This paper portrays about the distinct endmembers from an amalgamation of various material composites that are used for urban construction. The corresponding land surface temperature (LST) for the material is retrieved from multispectral sensor datasets with which radiative skin temperature can be estimated. Further, from the identification of thermal characteristics of urban materials widely present in the study area, it is found that construction elements like cement concrete, asbestos sheets and barren land black soil aggravate more heat. Proper analysis on proportions of urban construction material that contributes less heat liberation can be replaced in near future, thus paving the way for environmental sustainability.

Keywords AVIRIS NG · Kalaburagi · Spectral signatures · LST · UHI

17.1 Introduction

Thermal behavior of urban materials has a significant impact on the increase in temperature. The construction materials used in cities and towns to build blocks of buildings use an amalgamation of heat absorption elements which subsequently triggers the urban heat island (UHI) phenomena (Chow et al. 2006; Md Din et al. 2012). The heat retaining properties of urban materials are higher during the day due

K. N. Priyadarshini (✉) · V. Sivashankari · S. Shekhar · K. Balasubramani
Department of Geography, Central University of Tamil Nadu, Neelakudy, India
e-mail: nivi.darshini@yahoo.com

R. K. Mishra et al. (eds.), *Sustainable Climate Action and Water Management*, Advances in Geographical and Environmental Sciences,
https://doi.org/10.1007/978-981-15-8237-0_17

to solar radiation, whereas areas with abundant foliage cover represent significantly lower land surface temperature (LST). In recent years, research has been carried out in discovering materials that are eco-friendly and liberates low heat as a way to tackle environmental issues (Doulos et al. 2004). In this study, an attempt has been made to identify urban construction materials using pure spectra extracted from samples collected from field survey. High-resolution hyperspectral datasets yield more information about the existing land cover features that appear obvious, thus easing the process of classification. The reflectance properties of the earth's surface are distributed well among the spectrum, thus exposing the terrain features explicitly (Priyadarshini et al. 2019a).

The objectives used to fulfill the study include:

- To classify the hyperspectral dataset using the extracted spectral libraries that represent distinct endmembers
- To examine the LST retrieved from multispectral sensor dataset for the classified output and identify the material that aggravates more heat.

17.2 Materials and Methods

17.2.1 Datasets Used

Airborne Visible/Infrared Imaging Spectrometer—Next Generation (AVIRIS NG) hyperspectral dataset for a portion of Kalaburagi region, Karnataka is used in this study. AVIRIS NG provides a spectrum of 432 bands ranging from a wavelength of 376–2500 nm. The dataset is atmospherically corrected to obtain reflectance spectrum of totally 425 bands which serves for further processing (Priyadarshini et al. 2019b).

Landsat 8 dataset is robust and widely used for a variety of remote sensing applications. The thermal infrared sensor (TIRS) collects the images for two thermal bands of 100 m (resampled to 30 m) spatial resolution. The TIRS bands 10 and 11 (10.895 and 12.005 μm) are used for retrieval of LST. The data was collected in WRS path 145 and row 48, respectively. Dataset of summer is used for retrieval of LST.

The major software used for manipulation and mapping were ENVI and Arc Map.

17.2.2 Methodology

The methodical procedure followed for this paper is explained in the below mentioned subsections (Fig. 17.1).

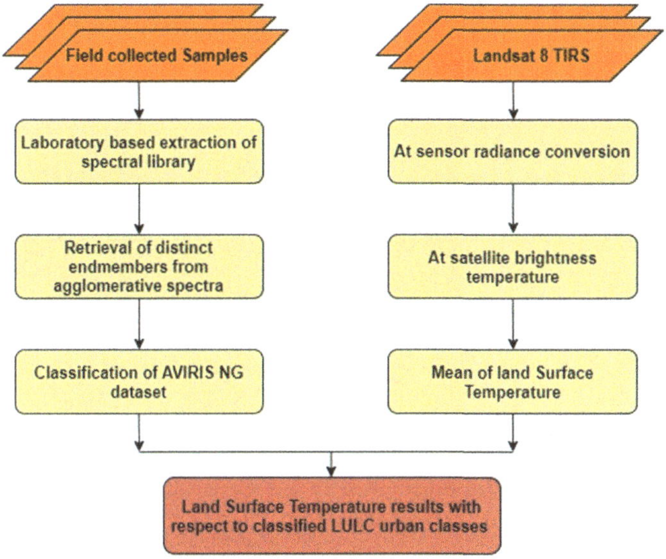

Fig. 17.1 Formulated workflow

17.2.2.1 Spectral Library Extraction from Urban Materials

Urban construction materials are complex due to mixture of constituent elements in varied chemical proportions. Identification of distinct substance is difficult as it is coalescence. In order to overcome this, in situ sample collection was carried out and numerous urban material composites with several proportions were collected from study area extent. These materials were processed under lab conditions for obtaining exclusive spectral libraries. A total of 40 different materials were collected where each material was placed on a dark room setup which nullifies external noise under the illumination of halogen lamp that supplies constant energy source. A pistol grip is used to record the spectrum by setting up the white reference every 10–15 min. The measurements are recorded by using the ASD Fieldspec 3 sensor and viewed using ViewSpec Pro (Kumar et al. 2015).

17.2.2.2 Classification of Endmembers

The endmembers represent pure spectra of materials that are present in Fig. 17.2. The lab spectra are collected from smaller sample size to clearly define the urban material. Likewise, 40 different materials are grouped into 10 classes on the whole based upon the spectral library's reflectance properties. These aggregated spectral libraries are applied to the benchmark hyperspectral dataset. Spectral angle mapper (SAM), a classification algorithm that matches image pixels shown in Table 17.1

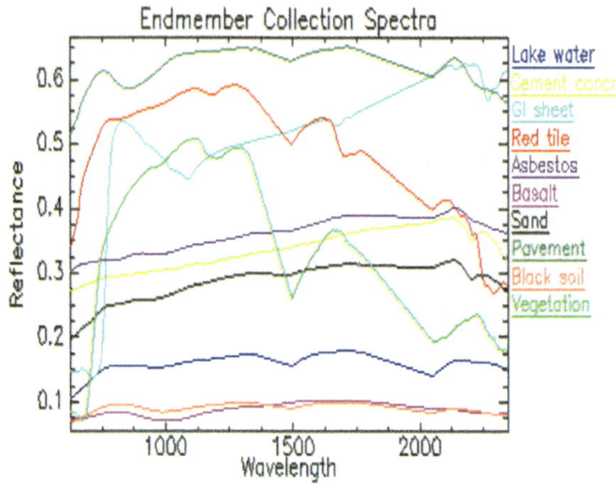

Fig. 17.2 Endmember spectra of urban materials

Table 17.1 Associated pixels for each class

LULC classes	No. of pixels
Lake water	267
Cement concrete	96
GI sheets	38
Red tile	28
Asbestos	14
Basalt	37
Sand	37
Pavement blocks	71
Black soil	245
Vegetation	234
Unclassified	196

with the corresponding reflectance spectra, is incorporated to obtain distribution of urban material from the classified results.

17.2.2.3 Retrieval of LST from Landsat 8 TIRS

A split window (SW) algorithm approach is used for calculating LST. Fetching LST from thermal infrared bands includes various steps. The raw multispectral data collected from Landsat 8 sensors, OLI and TIRS contains information as scaled digital numbers (DN). The DN values must be rescaled to radiance and reflectance using information provided in the metadata file and rescaling coefficients. Surface emissivity is a key parameter for LST retrieval. Emissivity for every pixel having maximum temperature is taken into account. Temperature is estimated using SW algorithm for Landsat 8 proposed first by McMillin in 1975. LST for Landsat is calculated using

$$LST = TB_{10} + C_1(TB_{10} - TB_{11}) + C2(TB_{10} - TB_{11})^2 + C_0 + (C_3 + C_4W)(1 - \varepsilon) + (C_5 + C_6W)\Delta\varepsilon \quad (1)$$

where W is atmospheric water vapor content, TB_{10} and TB_{11} are brightness temperature of bands 10 and 11, ε and $\Delta\varepsilon$ are the mean and difference of surface emissivity for bands 10 and 11, respectively.

17.3 Results and Discussion

17.3.1 Analysis of Classified Urban Materials

The classification result in Fig. 17.3 shows distribution of unique urban materials over the benchmark AVIRIS NG dataset sample of Kalaburagi. In SAM classification, as the angle mentioned is smaller, features matches well with that of the pixels in the imagery. The remaining pixel that overreaches the specified angular limit remains unclassified in the result. Urban materials such as cement concrete, barren

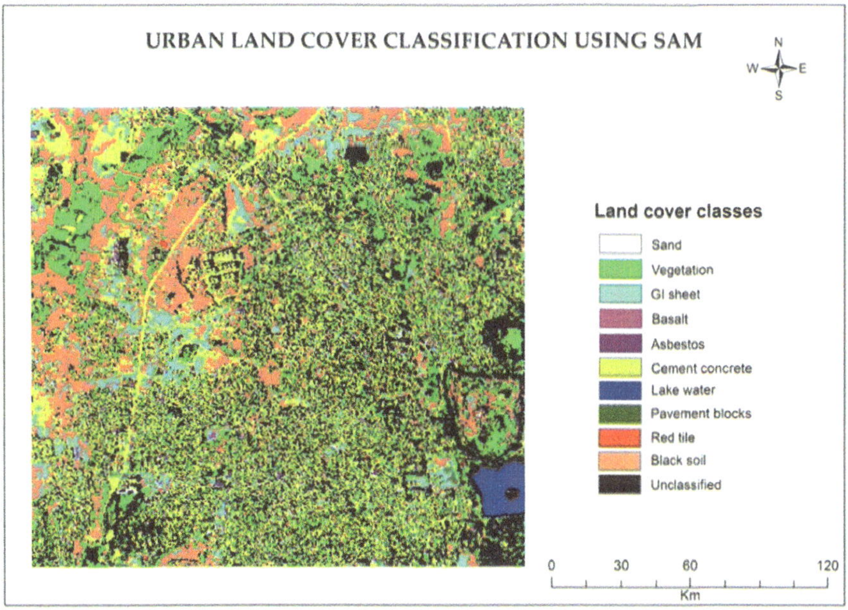

Fig. 17.3 Urban land cover classes representing endmember spectra

area black soil, vegetation etc. are predominant classes observed in the sample dataset considered and are intensively classified. From the classified output, it is evident that most of the endmember collection spectra have outperformed well for the applied SAM classification algorithm. The overall accuracy observed is 82.79% and the kappa coefficient is 0.7819.

17.3.2 LST with Respect to Urban Land Cover

The LST is a value emphasizing the external skin temperature of any urban material or ground that exists in the study area. Mean LST is estimated for every endmember spectra used for classification. The results exhibit temperature of distributed land covers which is of course resolution when compared to that of the classified results. The values of the retrieved LST (30 m resolution) is matched with the classified hyperspectral results where the GPS coordinated for every endmember is verified with Google Earth imagery and the temperature of each material is retrieved. It is observed that there is a fluctuation in temperature with the presence of greenery inside the urban sphere. On the other hand, materials like basalt, pavement blocks, cement concrete exhibit more heat, thus aggravate the UHI phenomena. The Fig. 17.4 and Table 17.2 represent the LST value for each endmember spectra collected.

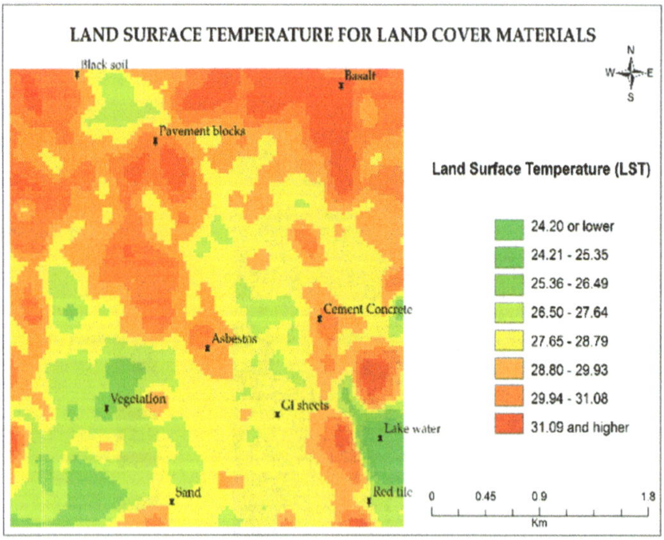

Fig. 17.4 LST for urban land cover

Table 17.2 LST for endmembers

LULC	LST (in °C)
Lake water	23.14
Cement concrete	29.92
GI sheets	27.50
Red tile	26.62
Asbestos	28.01
Basalt	31.16
Sand	26.95
Pavement blocks	29.98
Black soil	28.94
Vegetation	25.74

17.4 Conclusion

Urban land cover consisting of heat-inducing construction materials can be replaced by changing the proportions and chemical compositions of mixture which may have a negligible impact on the environment. Suggestion on the usage of lower heat liberating materials for construction is considered as a future scope of this study. Measures for minimal usage of materials fostering UHI effect have to be inculcated to the society in order to preserve environment sustainability.

Acknowledgements This research was carried out under the support of DST funded Network Project on Imaging Spectroscopy & Applications (NISA).

References

Arun Prasad K, Gnanappazham L, Selvam V, Ramasubramanian R, SekarKar C (2015) Developing a spectral library ofmangrove species of Indian east coast using field spectroscopy. Geocarto Int 30(5):580–599. https://doi.org/10.1080/10106049.2014.985743

Chow WT, Roth M (2006) Temporal dynamics of the urban heat island of Singapore. Int J Climatol. https://doi.org/10.1002/joc.1364

Doulos L, Santamouris M, Livada I (2004) Passive cooling of outdoor urban spaces.The role of materials. Solar Energy 77(2):231–249. https://doi.org/10.1016/j.solener.2004.04.005

Md Din MF, Dzinun H, Ponraj M, Chelliapan S, Noor ZZ et al (2012) Investigation of thermal effect on exterior wall surface of building material at urban city area. J Civil Environment Engg 2:110. https://doi.org/10.4172/2165-784X.1000110

Priyadarshini KN, Sivashankari V, Shekhar S, Balasubramani K (2019a) Assessment on the potential of multispectral and hyperspectral datasets for land use/ land cover classification. In: 2nd International electronic conference on geosciences, pp. 1–6. MDPI. https://doi.org/10.3390/iecg2019-06211

Priyadarshini KN, Sivashankari V, Shekhar S, Balasubramani K (2019b) Comparison and evaluation of dimensionality reduction techniques for hyperspectral data analysis. In: 2nd International electronic conference on geosciences, pp. 1–6. MDPI. (https://doi.org/10.3390/iecg2019-06209)

Chapter 18
Assessment of Agricultural Drought Vulnerability in Tamil Nadu Using MODIS NDVI, NDWI and VSDI

S. Latha

Abstract Agriculture continues to be sensitive to drought conditions despite the technological developments in weather analysis and crop management in Tamil Nadu. Agricultural drought vulnerability is defined as exposure, sensitivity and adaptive capacity of an agricultural area to the situation of inadequate soil moisture availability. Geographic information systems-based assessment of agricultural drought vulnerability map that was created through the development of a numerical weighting scheme helps to evaluate the drought potential of the classes within each factor that can lead to mitigation-oriented drought management. The severity and extent of drought in Tamil Nadu state is believed to be the result of aberrations in rainfall, overexploitation of ground water, lower reservoir levels and crop stress conditions. Hence in the present study, the agricultural drought is assessed for Tamil Nadu from 2006 to 2018, using MODIS's Normalized Difference Vegetation Index (NDVI), Normalized Difference Water Index (NDWI) and Visible and Shortwave Infrared Drought Index (VSDI). From these qualitative approaches of assessing the agricultural drought vulnerability, the values are calculated, mapped and categorized into five classes: Very slightly vulnerable, Slightly vulnerable, Moderately vulnerable, Highly vulnerable and Very highly vulnerable, for all the above-mentioned indices. From the results, top five acute drought-prone districts in Tamil Nadu, particularly the area under highly vulnerable and very highly vulnerable classes are identified for all three indices. In the study area, different areas are differentially exposed to drought and have different levels of vulnerability. This assessment of agricultural drought vulnerability would help in planning and preparedness in the study area. In addition, recommendations are also put forward for consideration.

Keywords Agricultural drought vulnerability · NDVI · NDWI · VSDI

S. Latha (✉)
Department of Geography, Bharathidasan University, Tiruchirappalli, India
e-mail: lathapala2017@gmail.com

R. K. Mishra et al. (eds.), *Sustainable Climate Action and Water Management*, Advances in Geographical and Environmental Sciences,
https://doi.org/10.1007/978-981-15-8237-0_18

18.1 Introduction

Agriculture continues to be sensitive to drought conditions despite the technological developments in weather analysis and crop management (Murthy 2014). Agricultural drought is defined as the decline in the productivity of crops due to irregularities in rainfall, increase in temperature etc., which cause a decrease in soil moisture (Sruthi and Mohammed Aslam 2015). It is described as a relative short-term shortage of precipitation or large increase in evapotranspiration which can cause loss of water within the soil, making water unavailable for use for both natural and cultivated vegetation (Chandrasekar et al. 2009). The impact of droughts on agriculture in China was analysed by different scholars at different year periods (Fu, 1991; Fang et al. 1997; Wang et al. 2002; He et al. 2013).

The term vulnerability refers to the frangible nature of a system threatened by different disasters (Lui 2013). The vulnerability changes over time, incorporating social responses as well as new rounds of hazardous events (Downing and Bakker 2000). The vulnerability to agricultural drought means the extent of agriculture potential sensibility to the drought threat (Liu et al. 2002). Agricultural drought vulnerability is defined as exposure, sensitivity and adaptive capacity of an agricultural area to the situation of inadequate soil moisture availability (Murthy 2014). Research on the vulnerability to agricultural drought is of great significance to mitigate drought losses and guarantee the national food security (Jiang 2012). Measuring agricultural drought vulnerability is a prerequisite for targeting interventions to improve and sustain the agricultural performance of both irrigated and rain-fed agriculture (Murthy et al. 2015).

The use of remote sensing and GIS for integrating data from different sources was found essential in many studies regarding drought. Monitoring and delineation of drought-affected area from space-borne platform will be very useful for early warning of drought disaster and better preparedness by developing contingency plans by policy makers (Chakraborty, Sehgal 2010). Geographic information systems-based assessment of agricultural drought vulnerability map that was created through the development of a numerical weighting scheme helps to evaluate the drought potential of the classes within each factor that can lead to mitigation-oriented drought management (Matthews et al. 1994). A GIS-based methodology can be used to assess the possible impacts of climate change on the susceptibility of soil drought (Olga et al. 2002). Remote sensing studies of vegetation normally use specific wavelengths selected to provide information about the vegetation present in the area from which the radiance data emanate (Tucker and Seller 1986).

The long-term agricultural drought for a period of 20 years from 1984 to 2003 is assessed for the state of Tamil Nadu, India (Vaani, Porchelva 2017). The severity and extent of drought in Tamil Nadu state is believed to be the result of aberrations in rainfall, overexploitation of ground water, lower reservoir levels and crop stress conditions (Nathan 1998). Hence in the present study, the agricultural drought is assessed for Tamil Nadu from 2006 to 2018, using MODIS's Normalized Difference

Vegetation Index (NDVI), Normalized Difference Water Index (NDWI) and Visible and Shortwave Infrared Drought Index (VSDI).

18.2 Objectives

- To assess the agricultural drought for Tamil Nadu from 2006 to 2018, using
  - MODIS's Normalized Difference Vegetation Index (NDVI),
  - Normalized Difference Water Index (NDWI) and
  - Visible and Shortwave Infrared Drought Index (VSDI).
- To calculate the mean for these year periods.
- To identify the acute (highly vulnerable and very highly vulnerable classes) drought-prone districts in Tamil Nadu.
- To suggest recommendations to overcome the agricultural drought.

18.3 Study Area

Tamil Nadu is located in the southern part of India, which extends from 8°5′N to 13°35′N latitudes and from 78°18′E to 80°20′E longitudes (Chandrasekar et al. 2009) (Fig. 18.1). At present, there are 38 districts, covering an area of 130,058 sq. km, with a total population of 7.21 crores as per 2011 census.

18.4 Study Area—Tamil Nadu

18.4.1 Methodology

Step 1: Data Products (MOD13Q1)-Pre-Processing (Mosaicking, Re-Projecting & Cloud Screening)

Step 2: Season's Maximum NDVI (Mean) Generation

Step 3: Season's Maximum NDWI (Mean) Generation

Step 4: Season's Maximum VSDI (Mean) Generation

Step 5: Acute Agricultural Drought-Prone Districts (Highly and Very Highly Vulnerable).

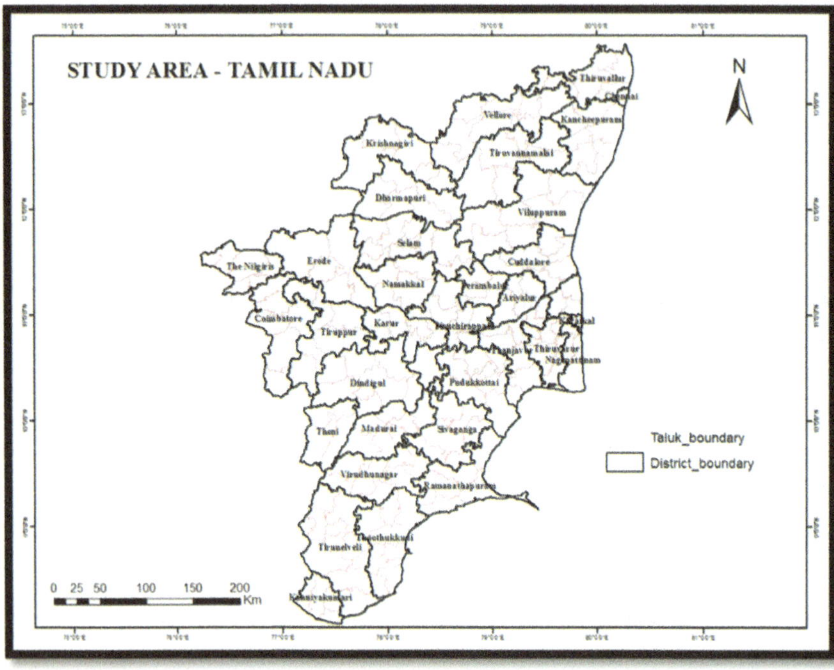

Fig. 18.1 Study area—Tamil Nadu

18.5 Data Used

18.5.1 MODIS

Moderate Resolution Imaging Spectroradiometer (MODIS) data will improve our understanding of global dynamics and processes occurring on the land, in the oceans and in the lower atmosphere, and it plays a vital role in the development of validated, global, interactive earth system models and is able to predict global change accurately enough to assist policy makers in making sound decisions concerning the protection of our environment (MODIS). The high temporal resolution of MODIS images played a major role in monitoring the floods (Akhtar Ali Memon 2015). The satellite-derived indices from the Moderate Resolution Imaging Spectroradiometer (MODIS) provided an opportunity to study correlations between soil moisture and vegetation indices (Gu 2008).

To monitor the progress of agricultural drought and for its relative estimation, 16 days composite image of surface reflectance product of MODIS called MOD13Q1 can be downloaded from website http://edcimswww.cr.usgs.gov./pub/imswelcome/ (Murthy et al. 2015). Surface reflectance data of the MODIS satellite with TERRA sensor and MOD09A1 during 8-day period was used to analyse the agricultural drought severity (Das Prabal 2017). The 250-m spatial resolution is the finest available from the MODIS NDVI dataset (MOD13Q1, collection 5), and the 16-day composite was selected in order to ensure a high probability of having the best quality pixel (reduced cloud effects) representing the NDVI within each 16-day period and this product is used to analyse the influence of drought (Zribi et al. 2016). Hence in the present study, the above-mentioned products are used, from October to December, for the years 2006 to 2018 and Terra MODIS 250 m 16-day composite is used. The software Arc GIS 10.1 ERDAS IMAGINE 2014 Microsoft Office Suite 2007 is used to process the data.

18.5.2 NDVI

NDVI index uses radiances or reflectance from a red channel around 0.66 μm and a near-IR channel around 0.86 μm, where the red channel is located in the strong chlorophyll absorption region, while the near-IR channel is located in the high reflectance plateau of vegetation canopies (Gao, Goetz 1995). The NDVI has become the primary tool for description of vegetation changes and interpretation of the impact of environmental phenomena (Kogan 1990). The Normalized Difference Vegetation Index (NDVI) is used to study the changes in vegetation and trends in occurrence of agricultural drought in any area (Sruthi and Mohammed Aslam 2015).

The NDVI is used to assess the agricultural drought, where NDVI was calculated based on Landsat band RED and near-infrared (NIR) by the following equation:

$$\text{NDVI} = \text{NIR}(5)\text{-RED}(4)/(\text{NIR}(5) + \text{RED}(4)$$

where NIR and RED are the reflectance in the near-infrared band (0.76–0.9 μm) and Red band (0.63–0.69 μm), for Landsat 5 TM (Legesse, and Suryabhagavan, 2014). However, for Landsat 8 OLI differed slightly: near-infrared band (0.85–0.88 μm) and Red band (0.64–0.67 μm). The values ranged from −1 to 1 (Legesse, and Suryabhagavan, 2014). The higher value of NDVI infers the presence of dense vegetation in the area, while its lower value is the indicator of sparse vegetation (Legesse, and Suryabhagavan, 2014). The same equation is used to assess NDVI in Assam (Krishna Das 2017). A high NDVI indicates a strong level of photosynthetic activity (Myneni 2009). The variation of the NDVI with a positive value indicates normal conditions while a negative value indicates severe drought conditions (Vaani, Porchelvan 2017).

As agricultural drought occurs due to lack of soil moisture and the consequent water stress in the vegetation, a water index should also be used along with the NDVI to obtain a clear picture of the early season agricultural drought (Chandrasekar et al. 2010).

18.5.3 NDWI

The Normalized Difference Water Index (NDWI) is a widely used remote sensing indices which involves spectral bands of multi-spectral and multi-temporal satellite datasets for vegetation, water resources and wetland studies (Ashraf et al. 2015; Ji et al. 2009; Rokni et al. 2014). NDWI uses both the bands in near-infrared region and demonstrates its potential applicability for canopy-level water content estimation based on the liquid water absorption band centred at 1240 nm enhanced by canopy scattering (Gao 1996).

The NDWI uses two near IR channels and is calculated using the following equation:

NDWI = NIR(5)-SWIR(7)/NIR(5) + SWIR (7)

where NIR range with wavelengths 0.841–0.876 nm and SWIR is a part of range with wavelengths 1.628–1.652 nm (Gao 1996). The values ranged from −1 to 1 (Gao 1996). The higher values correspond to high plant water content and cover with low vegetation. During the period of water stress, the NDWI rate will decrease (Gao 1996). Rogers and Kearney introduced NDWI using red and short-wave infrared (SWIR) channels of Landsat TM (Rogers et al. 2004). Normalized Difference Water Index (NDWI) and multi-spectral LANDSAT satellite data has been used to delineate the water surface area (Mohit Singh 2016).

One of the ways to extract water information using the multi-band method is through analysing signature features of each ground target among different spectral bands and finding out the signature differences between water and other targets based on the analysis, and then using an if–then–else logic tree to delineate land from open water (Yu 2001). And the other way is a band-ratio approach using two multi-spectral bands (Xu 2002). NDWI is widely used as it can suppress non-water features but not remove them (Mcfeeters 1996). The modified NDWI (MNDWI) can enhance open water features while efficiently suppressing and even removing built-up land noise as well as vegetation and soil noise (Hanqiu 2006). NDWI calculated from the 500-m SWIR band of MODIS is used to detect and monitor the moisture condition of vegetation canopies over large areas (Xiao et al. 2002; Jackson et al. 2004; Maki et al. 2004; Chen et al. 2005; Delbart et al. 2005). A five-year comparative study of NDVI and NDWI using MODIS data is used to analyse the grassland drought assessment (Gu 2007).

18.5.4 VSDI

Visible and Shortwave Infrared Drought Index (VSDI) is robust and reliable in the estimation of surface dryness in China, at the regional and national scales, as this index has the highest correlation with soil moisture (Zhang et al. 2013). The VSDI is a newly developed drought index for drought monitoring of both soil and vegetation surfaces and this index is a combination of MODIS Blue (band 3), Red (band 1), and SWIR (band 6) bands, and is defined as VSDI = RED(1)-SWIR(7)/RED(1) + SWIR(7), where $0 < \text{VSDI} \leq 1$. The smaller the value, the drier the condition it indicates (for the surface of farmland or any surface that can be simply classified as soil, vegetation and the combination of the two) and $1 < \text{VSDI} \leq 2$, Water or Snow Water Equivalent (including water bodies, snow, and ice cover) (Zhang et al. 2013).

By analysing the spectral response to water stress of plants and soils, SWIR and Red bands are found sensitive to moisture variation for both types of surfaces, thus they are used as the moisture measuring bands, where the Blue band is less sensitive to water changes and can serve as the moisture benchmark (Zhang et al. 2013). In this way, VSDI is constructed based on the difference between moisture-sensitive bands (SWIR and Red) and reference band (Blue). The combination of (RSWIR – RBlue) and (RRed – RBlue) may maximize the moisture variation and give the potential to estimate surface water independent of land cover types and finally, "(RSWIR – RBlue) + (RRed – RBlue)" is subtracted from 1 to make VSDI positively correlated to moisture variation (Zhang et al. 2013).

18.6 Results and Discussion

18.6.1 Season's Maximum NDVI Mean

The NDVI mean values are calculated, district-wise, in the study area for the years from 2006 to 2018, and the same is shown in Fig. 18.2.

Some of the previous studies done to assess the agricultural vulnerability have grouped their results as classes like very high, high, medium, low and very low (Chandrasekar et al. 2009; Johanna et al. 2020). Similarly, in the present study, the resulted values are grouped into five vulnerability classes as very highly vulnerable (< 0.3), highly vulnerable (0.31–0.6), moderately vulnerable (0.61–0.7), slightly vulnerable (0.71–0.8) and very slightly vulnerable (> 0.81).

The areas covered by each vulnerability classes for each district in the study area are calculated and listed in Table 18.1.

From Fig. 18.2 and Table 18.1, the top five districts with area covered under the class very highly vulnerable and highly vulnerable are Tirunelveli, Ramanathapuram, Pudukkottai, Sivagangai and Kancheepuram. Hence these districts are classed as acute agricultural drought-prone districts in the study area, between the years 2006 and 2018.

18.6.2 Season's Maximum NDWI Mean

The NDWI mean values are calculated, district-wise, in the study area for the years from 2006 to 2018, and the same is shown in Fig. 18.3.

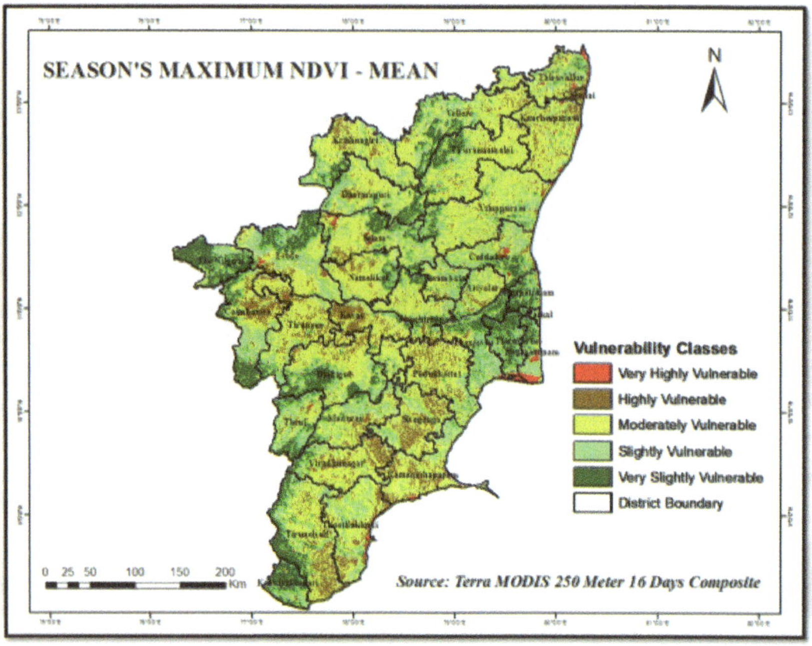

Fig. 18.2 Season's maximum NDVI Mean 2006–2018

The resulted values are grouped into five vulnerability classes as very highly vulnerable (< 0.35), highly vulnerable (0.36–0.45), moderately vulnerable (0.46–0.55), slightly vulnerable (0.56–0.65) and very slightly vulnerable (>0.66).

The areas covered by each vulnerability classes for each district in the study area are calculated and listed in Table 18.2.

From Fig. 18.3 and Table 18.2, the top five districts with area covered under the class very highly vulnerable and highly vulnerable are Tiruppur, Karur, Krishnagiri, Dindigul and Coimbatore. Hence these districts are classed as acute agricultural drought-prone districts in the study area, between the years 2006 and 2018.

18.6.3 Season's Maximum VSDI Mean

The VSDI mean values are calculated, district-wise, in the study area for the years from 2006 to 2018, and the same is shown in Fig. 18.4.

The resulted values are grouped into five vulnerability classes as very highly vulnerable (< 0.20), highly vulnerable (0.21–0.40), moderately vulnerable (0.41–0.60), slightly vulnerable (0.61–0.80) and very slightly vulnerable (>0.81).

The areas covered by each vulnerability classes for each district in the study area are calculated and listed in Table 18.3.

Table 18.1 Area covered by each vulnerability classes (sq. km) for each district—NDVI Mean

District	Very highly vulnerable	Highly vulnerable	Moderately highly vulnerable	Slightly highly vulnerable	Very slightly vulnerable
Ariyalur	7.35	266.02	981.53	604.80	90.64
Chennai	6.23	143.07	13.58	5.63	1.13
Coimbatore	8.53	1024.30	1299.44	1243.73	1149.01
Cuddalore	59.30	411.29	1135.60	1423.19	582.91
Dharmapuri	16.58	344.90	2066.09	1347.52	729.14
Dindigul	1.13	739.28	2413.14	1437.78	1465.64
Erode	32.52	540.13	2115.41	1996.43	1298.04
Kancheepuram	59.25	1210.51	2527.61	655.46	37.57
Kanniyakumari	18.84	37.78	97.78	487.70	1031.60
Karikal	0.54	21.14	35.20	88.92	14.49
Karur	10.04	1166.13	1348.16	265.35	119.30
Krishnagiri	3.22	707.41	2576.49	1361.74	473.70
Madurai	4.88	604.69	1481.52	1243.20	380.00
Nagapattinam	231.51	204.68	271.06	1036.75	849.46
Namakkal	2.52	475.58	1935.15	637.97	373.24
Perambalur	1.45	179.99	667.48	785.49	127.35
Pudukkottai	2.09	1370.17	2427.31	782.86	72.72
Ramanathapuram	63.32	1354.66	1972.66	689.16	34.08
Salem	54.95	543.03	2329.91	1380.58	939.61
Sivagangai	3.92	1357.34	1991.28	843.61	32.09
Thanjavur	31.34	289.09	624.76	1494.83	1005.30
Nilgiris	0.00	7.94	116.56	678.86	1737.55
Theni	8.26	252.92	790.48	1113.38	768.69
Thiruvallur	83.50	692.92	1498.91	898.24	232.26
Thiruvarur	109.05	45.02	96.65	851.44	1168.71
Thoothukkudi	74.86	1149.87	2163.01	118.35	65.31
Thiruchirappalli	15.83	581.46	2285.20	1061.97	565.46
Tirunelveli	6.71	1503.95	2592.59	1358.31	1346.12
Tiruppur	8.16	1205.04	2743.93	780.12	218.79
Thiruvannamalai	3.01	538.26	3751.70	1096.15	827.46
Vellore	4.56	593.05	2796.68	1600.17	1091.54
Villupuram	25.81	685.14	4392.24	2042.64	490.66
Virudhunagar	0.80	1230.42	2094.53	685.73	239.88

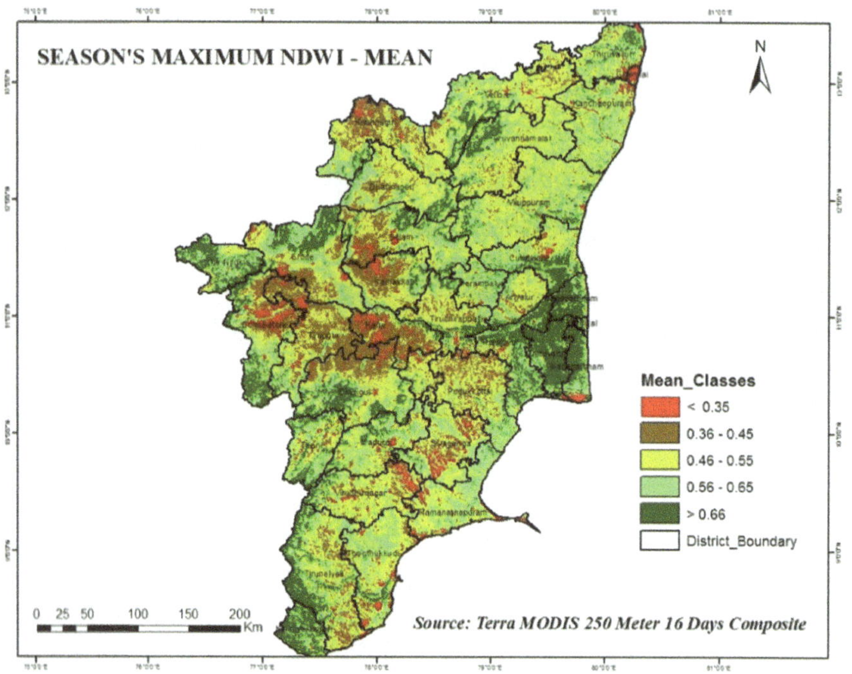

Fig. 18.3 Season's maximum NDWI Mean 2006–2018

From Fig. 18.3 and Table 18.2, the top five districts with area covered under the class very highly vulnerable and highly vulnerable are Tiruppur, Krishnagiri, Tirunelveli, Dindigul and Karur. Hence these districts are classed as acute agricultural drought-prone districts in the study area, between the years 2006 and 2018.

18.7 Recommendations

About 68% of cropped area in India is vulnerable to drought, of which 33% receives less than 750 mm of mean annual rainfall and is classified as "chronically 14 drought-prone" while 35% which receive mean annual rainfall of 750–1125 mm is classified as "drought-prone", that are confined primarily to the arid, semi-arid, and sub-humid regions of peninsular and western India (Drought Management Plan 2017).

The assessment of agricultural drought vulnerability in Tamil Nadu proves that the values of NDVI, NDWI and VSDI have shown changes, district-wise, over these 13 years, in the study area. From the results, acute agricultural drought-prone areas (the area under highly and very highly vulnerable classes) are identified. In the study area, the levels of agricultural drought vulnerability differ among the districts. The

Table 18.2 Area covered by each vulnerability classes (sq. km) for each district—NDWI mean

District	Very highly vulnerable	Highly vulnerable	Moderately highly vulnerable	Slightly highly vulnerable	Very slightly vulnerable
Coimbatore	493.12	1124.27	899.74	1023.71	1197.31
Karur	427.55	1671.28	408.84	209.02	193.09
Sivagangai	388.80	902.00	1798.03	1062.24	77.87
Virudhunagar	300.74	857.35	2033.46	774.22	284.37
Ramanathapuram	278.41	518.94	1653.46	1547.21	120.21
Tiruppur	242.89	2376.70	1481.41	660.67	196.41
Thoothukkudi	230.87	724.20	2010.92	1399.84	268.81
Namakkal	189.17	1208.10	1111.40	595.79	320.00
Erode	184.28	1107.96	1725.69	1690.22	1282.91
Krishnagiri	177.84	1628.99	2066.09	1004.07	259.95
Salem	177.58	1152.29	1746.41	1347.84	823.32
Kancheepuram	155.04	416.44	1886.64	1803.29	230.76
Tirunelveli	153.05	1159.85	2323.00	1585.95	1599.37
Thiruvallur	145.43	440.26	1306.41	1080.70	448.74
Pudukkottai	145.11	1365.87	2114.76	873.77	15.70
Dindigul	144.30	1553.27	1669.02	1357.45	1333.30
Nagapattinam	129.92	61.07	182.03	496.80	1623.89
Tiruchirappalli	128.80	844.41	2058.63	813.50	664.53
Chennai	127.19	26.94	9.02	5.90	1.13
Cuddalore	98.37	271.38	1137.36	1164.25	940.63
Madurai	96.54	461.14	1320.42	1310.38	526.24
Vellore	91.93	763.70	2866.98	1485.97	89.17
Villupuram	62.20	518.99	378.37	2604.35	627.63
Thanjavur	58.55	218.52	435.27	977.88	1754.83
Theni	43.79	362.88	750.66	962.91	751.57
Ariyalur	41.27	248.47	971.33	501.28	188.52
Thiruvarur	33.92	13.47	61.12	291.08	1871.61
Kanniyakumari	24.79	23.99	100.19	402.49	1126.80
Thiruvannamalai	18.03	534.07	3271.88	1610.15	782.65
Dharmapuri	13.47	692.60	2093.08	1141.07	566.54
Perambalur	12.72	115.54	553.66	906.66	172.69
Nilgiris	6.60	83.66	336.26	886.22	1237.67
Karaikal	2.41	9.82	22.70	55.54	69.98

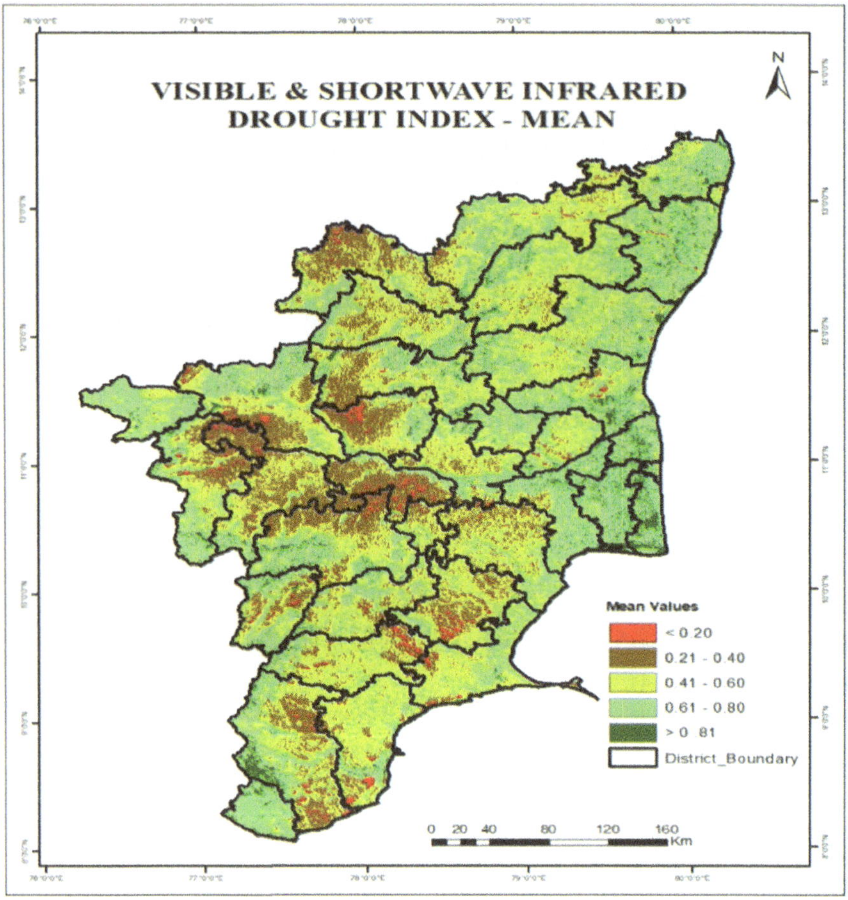

Fig. 18.4 Season's maximum VSDI Mean 2006–2018

results prove that Tirunelveli, Tiruppur, Krishnagiri, Karur and Dindigul are the five districts that are facing acute agricultural drought conditions in the study area.

The recommendations to overcome the agricultural drought in the study area are put forward for consideration.

- Artificial groundwater recharge should be done to eradicate water stress in the study area.
- Proper land improvement and reclamation methods would improve the vegetation cover of the study area.
- Proper soil management is the dire need for increasing the vegetation areas in the study area.
- Soil enrichment can be made by following effective irrigation systems that include drip and sprinkler irrigation, which in turn increase the vegetation cover in the study area.

Table 18.3 Area covered by each vulnerability classes (sq. km) for each district—VSDI mean

District	Very highly vulnerable	Highly vulnerable	Moderately highly vulnerable	Slightly highly vulnerable	Very slightly vulnerable
Karur	1395.82	855.68	288.02	309.00	61.23
Tiruppur	1149.34	2057.93	1069.27	539.22	142.3
Dindigul	1072.06	1415.41	1343.33	1529.55	697.00
Tirunelveli	1028.81	1598.03	1832.33	1550.16	812.11
Krishnagiri	959.85	1793.58	1437.14	890.89	55.54
Erode	772.29	1172.14	1676.22	1811.88	558.54
Namakkal	749.05	974.87	830.14	701.24	169.15
Sivagangai	717.17	1226.13	1741.42	536.16	8.05
Coimbatore	694.74	1038.25	1100.18	1356.16	548.83
Virudhunagar	493.66	967.36	2086.48	601.74	100.89
Salem	461.78	1237.78	1725.37	1379.93	442.57
Pudukottai	428.46	1561.05	1867.64	778.62	20.45
Thoothukudi	411.66	721.90	2437.07	844.84	219.06
Theni	359.61	525.11	676.93	984.75	325.42
Tiruchirappalli	324.40	1319.51	1582.68	1032.67	250.61
Ramanathapuram	284.80	585.32	2042.37	1071.04	134.70
Madurai	235.53	737.35	1532.98	1081.45	127.40
Vellore	195.98	1188.19	2524.39	1858.30	330.90
Dharmapuri	164.48	1356.32	1501.48	1129.10	355.37
Cuddalore	101.05	236.29	913.37	1561.80	799.39
Thanjavur	79.37	228.45	498.44	1807.69	831.10
Ariyalur	73.84	271.49	799.50	670.92	135.13
Villupuram	70.89	753.51	3107.45	3039.67	665.01
Thiruvannamalai	66.97	917.29	2716.08	2002.18	514.27
Thiruvallur	65.63	391.48	1084.56	1406.93	472.95
Kancheepuram	58.39	134.43	866.79	2367.26	1065.30
Perambalur	10.63	102.23	621.44	896.74	130.24
Nagapattinam	7.73	11.59	55.49	894.54	1624.38
Kanniyakumari	5.26	36.33	133.57	992.58	510.51
Nilgiris	3.06	90.91	575.61	1610.58	270.26
Chennai	2.58	34.04	86.78	38.21	8.59
Thiruvarur	0.97	5.10	71.21	943.26	1250.66
Karaikal	0.00	1.72	12.83	73.04	72.88

- Improved farming practices and best-suited technologies like agro-horticulture, silvi-pasture and agro-forestry can be followed to improve the green cover in the study area.
- Dry farming technologies such as soil and water conservation measures are also suggested.
- Water harvesting techniques, irrigation management and moisture retention activities should be followed for the optimum use of the water in the study area.
- Policies should be framed in adopting suitable restriction on the use of vulnerable lands. This can be better achieved through national spirit and co-operation with technical experts, planners, administrators and active participation of farmers.
- Native trees, which need minimum water, may also be planted in fallow and waste lands.
- Increase in the use of organic molasses and adopting dust mulching technology can lead to an efficient water use.
- Jatropha plant is recommended in waste lands of the study area. The oil obtained by crushing its seeds is so ideal for conversion into biodiesel. Apart from this, adding greenery to the study area provides additional income to the farmers.
- Construction of check dams, percolation ponds, infiltration tanks, recharge pits and desilting of tanks is recommended so that the underground water can be regenerated in the study area.
- Ex situ water harvesting practices like farm ponds, community tanks, watersheds and pools should be constructed.
- This study will be helpful to soil scientists and environmentalists in looking for ways to protect soil moisture and soil microorganisms.
- In par with the ecological perspectives, urban designing in the study area is very important. For this, the existing urban land use characteristics should be studied. Under the zoning ordinances, the acts, rules and laws should be framed. The unused open space within the city limit should be mapped using GIS and its suitability for agriculture can be analysed. The degraded open spaces within the city limit should be bought under Urban Agriculture.
- Subsistence-oriented urban agriculture that includes the Urban Agriculture performed as home gardening, community gardening and institutional gardening in school, colleges, offices and hospitals, are recommended, to increase the greening of the study area.
- This study will also be beneficial to the urban planners for future land use planning of the study area.
- To adopt farm practices that include the increased use of organic manures, vermin-culture and agronomic practices like crop rotation measures can increase the efficiency of irrigation water utilization (Bana 2014).
- Indigenous technical knowledge (ITK) on natural resource management can be extrapolated to understand the modern concepts of disaster risk management in terms of early warning, preparedness, mitigation, response and relief (Gupta et al. 2011).
- Adopting integrated watershed approach for maximizing rain water usage would help to overcome the agricultural drought (Bana 2014).

- Crop management practices like sowing of crops at optimum time, securing and maintaining nutrient management, control of weeds, pests and diseases will help in minimizing the agricultural drought (Joshi, 2009).

The Government of Tamil Nadu under "Tamil Nadu Vision 2023" has framed strategic initiatives regarding the eradication of agricultural drought (Vision Tamil Nadu 2023). Those include:

- Promoting market-driven agricultural produce
- Accelerating innovation and extension mechanism
- Functional consolidation of land holdings
- Emphasis on mechanization
- Improving agricultural productivity
- Assurance of timely irrigation
- Creating a robust supply chain
- Skill development in agriculture
- Ensuring timely irrigation by intelligent use of available water by
  - Adopting effective recharging mechanisms,
  - Drip/sprinkler irrigation systems,
  - Timely cleaning and deepening of tanks and canals,
  - Adopting appropriate cropping pattern based on water availability and
- Location-specific programmes include:
  - Integrated Wasteland Development Programme (IWDP),
  - Integrated Watershed Management Programme (IWMP),
  - National Watershed Development Project for Rainfed Areas (NWDPRA) and
  - Drought-Prone Area Programme (DPAP).
- Added to this, the Government of Tamil Nadu's "Environmental Policy 2017" has introduced a programme named, "Integrated environmental well-being into development programme", which has promoted the following techniques (Tamil Nadu State Environmental Policy 2017).
- Preserving Prime Agricultural Land:
  - Government is giving thrust to preserve the prime agricultural lands by recommending suitable and profitable cropping system, reclamation of problem soils, identification and conversion of fallow lands for agriculture.
- Precision Farming and Micro Irrigation Schemes:
  - Government will promote Water Soluble Fertilizers (WSF)/Liquid bio fertilizers for various crops, as this provides optimum quantity of water and nutrients in well-balanced proportion directly to the active root zone.
- Soil Analysis and Database Creation:

- Government is exploring avenues to rejuvenate the soil health and fertility through detailed soil survey, soil sample collection and analysis to identify problematic soils. The recommendations will be recorded in "Farmers Integrated Hand Book" a record for effective planning.

18.8 Conclusion

The assessment of agricultural vulnerability in Tamil Nadu using NDWI, NDVI and VSDI clearly states the existence of agricultural drought between the periods 2006 and 2018. The resultant maps and recommendations would help the decision makers to take steps to overcome the drought, district-wise, in the study area.

References

Akhtar Ali Memon et al (2015), Flood monitoring and damage assessment using water indices: a case study of Pakistan flood. Egyptian J Remote Sens Space Sci 18(1):99–106

Ashraf et al (2015) A comparison of change detection analyses using different band algebras for Baraila Wetland with Nasa's multi-temporal landsat dataset. J Geogr Inf Syst 7:1–19

Bana (2014) Drought in India: Its impact and mitigation strategies—A review. Indian J Agron 59(2):179–190

Chakraborty, Sehgal (2010) Assessment of agricultural drought using modis derived normalized difference water index. J Agric Phys 10:28–36

Chandrasekar et al (2009) Identification of agricultural drought vulnerable areas of Tamil Nadu, India, using GIS-based multi criteria analysis. Asian J Environ Disaster Manag 1:40–61

Chandrasekar et al (2010) Land Surface Water Index (LSWI) response to rainfall and NDVI using the MODIS Vegetation Index product. Int J Remote Sens 31(15):3987–4005

Chen et al (2005) Vegetation water content estimation for corn and soybeans using spectral indices derived from MODIS near-and short-wave infrared bands. Remote Sens Environ 98:225–236

Das Prabal (2017) Analysis of agricultural drought severity using MODIS satellite images. In: Proceedings volume of international conference, ISBN:978-93-5267-355-1, pp 277–282

Delbart et al (2005) Determination of phenological dates in boreal regions using normalized difference water index. Remote Sens Environ 97:26–38

Downing, Bakker (2000) Drought discourse and vulnerability, Chapter 45. In: Wilhite DA (ed) Drought: a global assessment, natural hazards and disasters series. Routledge Publishers, UK

Drought Management Plan (2017) Government of India. Ministry of Agriculture and Farmers Welfare, Department of Agriculture, Cooperation and Farmers Welfare

Fang et al (1997) An EOF analysis on drought effect on agriculture in China during 1978-1994. J Nat Disast 6(1):59–64

Fu, (1991) Analysis of the geographical distribution and disastrous condition of drought in China. J Arid Land Resour Environ 5(4):1–7

Gao (1996) NDWI for remote sensing of vegetation liquid water from space. Remote Sens Environ 58:155–162

Gao, Goetz (1995) Retrieval of equivalent water thickness and information related to biochemical components of vegetation canopies from AVIRIS data. Remote Sens Environ 52:155–162

Government of Tamil Nadu, March (2012) Vision Tamil Nadu 2023-Strategic plan for Infrastructure development in Tamil Nadu

Gu et al (2007) A five-year analysis of MODIS NDVI and NDWI for grassland drought assessment over the Central Great Plains of the United States. Geophys Res Lett 34(6)
Gu Y et al (2008) Evaluation of MODIS NDVI & NDWI for vegetation drought monitoring using Oklahoma Mesonet soil moisture data. Geophys Res Lett 35
Gupta et al (2011) Traditional intellect in disaster risk mitigation: Indian Outlook-Rajasthan and Bundelkhand Icons. Indian J Tradit Knowl 10(1):156–66
Hanqiu X (2006) Modification of normalised difference water index (NDWI) to enhance open water features in remotely sensed imagery. Int J Remote Sens 27(14):3025–3033
He et al (2013) Quantitative assessment and spatial characteristic analysis of agricultural drought risk in China. Nat Hazards 66:155–166
Jackson et al (2004) Vegetation water content mapping using Landsat data derived normalized difference water index for corn and soybeans. Remote Sens Environ 92:475–482
Ji et al (2009) Analysis of dynamic thresholds for the normalized difference water index. Photogramm Eng Remote Sens 7(11):1307–1317
Jiang G et al (2012) An analysis of vulnerability to agricultural drought in china using the expand grey relation analysis method. Proc Eng 28:670–676
Johanna et al (2020) Drought vulnerability in the United States: an integrated assessment. Water 12:2033
Joshi, (2009) Contingency crop planning for dryland areas in relation to climate change. Indian J Agron 54(2):237–243
Kogan (1990) Remote sensing of weather impacts on vegetation on non-homogeneous areas. Int J Remote Sens 11:1405–1419
Krishna Das (2017), NDVI and NDWI based change detection analysis of Bordoibam Beelmukh Wetlandscape, Assam using IRS LISS III data. ADBU J Eng Technol 6:17–21
Legesse, Suryabhagavan (2014) Remote Sensing and GIS based agricultural drought assessment in East Shewa Zone, Ethiopia. Trop Ecol 55(3):349–363
Liu et al (2002) Synthetic analysis and quantitative estimation of agricultural vulnerability to drought disaster in Hunan Province. J Nat Disast 11(4):78–83
Lui et al (2013) Assessing vulnerability to drought based on exposure, sensitivity and adaptive capacity: A case study in middle inner Mongolia of China. Chin Geogr Sci 23:13–25
Maki et al (2004) Estimation of leaf water status to monitor the risk of forest fires by using remotely sensed data. Remote Sens Environ 90:441–450
Matthews et al (1994) Climatic soil moisture deficit- climate and soil data integration in GIS. Clim Change 28:273–287
Mcfeeters (1996) The use of normalized difference water index (NDWI) in the delineation of open water features. Int J Remote Sens 17:1425–1432
MODIS—https://modis.gsfc.nasa.gov/about/
Mohit Singh et al (2016) Normalised difference water index-based water surface area delineation and assessment. Int J Environ Sci 6(5)
Murthy et al (2014) Analysing Agricultural Drought Vulnerability at sub-district level through exposure, sensitivity and adaptive capacity-based Composite Index. In: The International archives of the photogrammetry, remote sensing and spatial information sciences, vol XL-8, pp 65–70
Murthy et al (2015) A study on agricultural drought vulnerability at disaggregated level in a highly irrigated and intensely cropped state of India. Environ Monit Assess 187(3):140
Myneni et al (2009), Potential gross primary productivity of terrestrial vegetation from 1982 to 1990. Geophys Res Lett 22:2617–2620
Nathan (1998) Droughts in Tamil Nadu: a qualitative and quantitative appraisal, drought network news (1994–2001), Digital Commons@University of Nebraska-Lincoln 62 10(3)
Olga et al (2002) Assessing vulnerability to agricultural drought: a nebraska case study. Nat Hazards 25:37–58
Rogers et al (2004) Reducing signature variability in unmixing coastal marsh thematic mapper scenes using spectral indices. Int J Remote Sens 25(12):2317–2335

Rokni et al (2014) Water feature extraction and change detection using multitemporal landsat imagery. Remote Sens 6:4173–4189

Sruthi MA (2015) Agricultural Drought Analysis using the NDVI and land surface temperature data: a case study of Raichur district. In: International conference on water resources, coastal and ocean engineering, Aquatic Procedia, vol 4, pp 1258–1264

Tamil Nadu State Environmental Policy (2017) Department of Environment, Government of Tamil Nadu

Tucker, Seller (1986) Satellite remote sensing of primary production. Int J Remote Sens7:1395–1416

Vaani, Porchelva (2017) GIS based agricultural drought assessment for the state of tamil nadu, india using vegetation condition Index (VCI). Int J Civil Eng Technol 8(5):1185–1194

Vaani, Porchelvan (2017) Assessment of long-term agricultural drought in Tamil Nadu, India using NDVI anomaly. Disaster Adv 10:1–10

Wang et al (2002) Spatio-temporal change of drought disaster in China in recent fifty years. J Nat Disast 11(2):1–6

Xiao et al (2002) Characterization of forest types in north-eastern China, using multi-temporal SPOT- 4 Vegetation sensor data. Remote Sens Environ 82:335–348

Xu (2002) Spatial expansion of urban/town in Fuqing of China and its driving force analysis. Remote Sens Technol Appl 17:86–92

Yu (2001) Study on water bodies extraction and classification from SPOT image. J Remote Sens 5:214–219

Zhang et al (2013) VSDI: evaluation of the visible and shortwave infrared drought index in China. Int J Disaster Risk Sci 4(2):68–76

Zribi Mehrez et al (2016) Analysis of the effects of drought on vegetation cover in a Mediterranean region through the use of SPOT-VGT and TERRA-modis long time series. Remote Sens 8:992

Chapter 19
Impact of Climate Change on Oil and Gas Sector and Mitigation Approach for Sustainable Development

Keshav Singhal, Abhijeet Pawar, and Sandeep Deedwania

Abstract Climate change is one of the most pressing issues that the world is facing. Human influence is believed to be the dominant cause. The impact of climate change is visible in the form of changes in frequency and intensity of extreme climatic events, such as heat waves, cyclones, extreme rainfall, droughts, reduction in water availability, loss of land mass and vector-borne diseases. The adverse effects of climate change impact physical infrastructure and efficiency of processes in manufacturing industry across all sectors including oil and gas. The impacts could be damage to assets, interruption in production, a disruption in the supply chain, availability of water and employees' health etc. All these impacts cause not only physical risks and losses but financial as well. Thus, it is essential for businesses to mitigate risks related to climate change and implement mitigation measures. This paper demonstrates various impacts of climate change on the oil and gas sector and adoption of mitigation measures to be sustainable and build environmental resilience.

Keywords Climate change · Mitigation measures · Oil and gas · Risk mitigation · Risks and losses

19.1 Introduction

Human beings depend on nature for their needs. Our earth has a limited capacity to generate resource and absorb the generated waste (Earth Observation Day 2019). Some of the resources we consume are for generation of energy in the form of heat and electricity. Energy captured gradually by photosynthesis over the course of time

K. Singhal (✉) · A. Pawar · S. Deedwania
Hindustan Petroleum Corporation Limited, Mumbai, India
e-mail: keshavsinghal@hpcl.in

A. Pawar
e-mail: abhijeetpawar@hpcl.in

S. Deedwania
e-mail: sdeeds@hpcl.in

R. K. Mishra et al. (eds.), *Sustainable Climate Action and Water Management*, Advances in Geographical and Environmental Sciences,
https://doi.org/10.1007/978-981-15-8237-0_19

has been stored up as dense reservoirs of energy in the form of coal, crude oil and natural gas. With the ability to generate and control fire and the knowledge on utility of fossil fuels, human beings have been able to release these vast amount of energy in a relatively short time scale. The flip side of this development story is that the usage of energy for human developmental need comes with associated liberation of carbon trapped in biomass, coal, and oil and gas in the form of carbon dioxide, a potent greenhouse gas (Sinha and Nongpluh 2014) (IPCC 2017). The pace of consumption of resources has been ever-increasing; sometimes in mid-1970s, our resource consumption started exceeding the earth's bio-capacity which resulted in ecological overshoot (Earth Observation Day 2019).

With the release of more and more amount of carbon dioxide in the atmosphere, the impacts of climate change have started to become even more evident, with climate change emerging as one of the biggest threats that the mankind is facing (World Economic Forum 2019). The climate change phenomenon is a result of number of factors which can be broadly categorized into natural and man-made. However, since the dawn of the industrial revolution, the impacts of anthropogenic activities on climate are becoming more evident (NOAA ClimateGov 2019).

The oil and gas sector stands at the fulcrum of the debate: development vs climate change. The sector is crucial for India's energy security as well as its growth curve. In 2017, oil and gas constituted over one-third of India's primary energy mix (Dudley 2015). Gas is a fertilizer feedstock for nitrogen fertilizers which lead to increase in agricultural productivity (McArthur et al. 2017). Use of gas leads to environmental improvement through provision of cleaner cooking and transportation fuels. Dependency on oil and gas is the foundation for achieving high economic growth (IBEF 2019).

The energy sector itself contributed to approximately 10% of 2010 global greenhouse gas emissions due to activities such as fuel extraction, refining, processing and transportation (USEPA 2019). The backbone of the National Development Agenda, i.e., the energy infrastructure (Gavin et al. 2018) is vulnerable to the physical impacts of climate change (Cruz and Krausmann 2013). Past experience shows that, of the range of risks posed by climate change, the oil and gas sector is vulnerable primarily to extreme climatic events and slow onset of events. The physical infrastructure and efficiency of processes in various oil and gas sector plants are sensitive to climatic conditions. The impacts can be financial and production losses due to infrastructure damage.

19.1.1 Impact of Climate Change on Oil and Gas Sector

The oil and gas industry may experience the impacts of climate change in the following ways (TERI 2018) (Table 19.1):

With the increased risks due to climate change, the mitigation measures become all the more important. These measures can be categorized into various focus areas,

Table 19.1 Impact due to climate change events

Cause	Example	Impact
Extreme events	Heavy rainfall, Heat waves, cyclones Storm surges Floods	Delay in infrastructure development Damage to assets Shut down of production Disruption of supply chains Health issues for the employees
Slow onset of events	Rise in average temperature, Increased number of dry days Depletion of water resources Salinization of ground water and soil Desertification	Impact on efficiency of units such as cooling towers and compressor stations Risks to pipelines Fresh water scarcity for refineries Water shortage impacting through-put, energy cost, or additional investment in technology

such as energy efficiency, water management and renewable energy; however, there is a need to structure these measures in a framework.

19.1.2 Methodology

This paper proposes an environmental sustainability framework for organisations.

Although comprehensive, the number of topics that can be incorporated into this framework may increase. These topics may vary from organisation to organisation; however, the topic selection in the above framework is made based on sectoral relevance. The weightage and importance of these topics may differ on case-to-case basis. For customizing this framework, materiality analysis could be used by an organisation (Fig. 19.1).

19.2 Interventions

Businesses are increasingly identifying climate change as a key risk. Business strategies are being formulated by giving due consideration to risks and opportunities arising out of climate change. The alignment of business strategies with various national and international frameworks, such as Sustainable Development Goals (17 Goals), United Nations Global Compact (10 Principles) and Nationally Determined Contributions (8 NDCs), is visible in today's context. It is also important to incorporate anticipated changes in climatic events and resulting impacts in regulations and guidelines within the organisation.

There is a clear need of intervention at all three levels (Fig. 19.2).

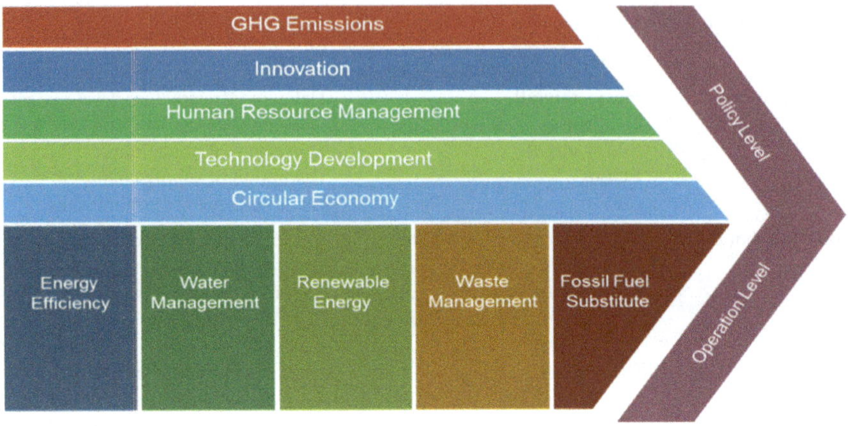

Fig. 19.1 Environmental sustainability framework

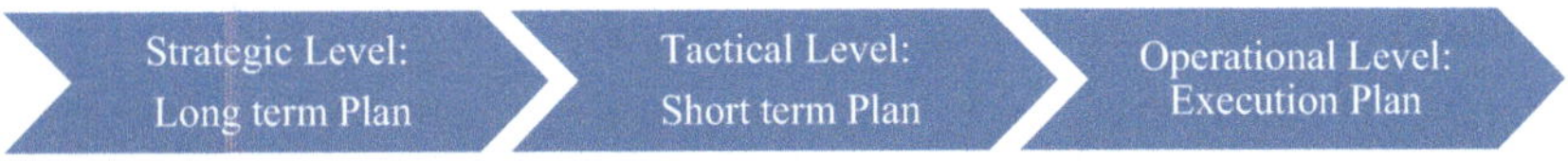

Fig. 19.2 Level of Interventions

At all these three levels, various interventions that may help organisations toward Sustainable Development are as follows:

- Policy intervention: In order to show commitment from senior management and bring out change at mass level, policy-level interventions are required. It makes the efforts coherent and consistent and acts as a guiding principle.
- Focus areas, strategy, targets and action plan: Broad focus areas should be identified as per environmental sustainability framework. A strategy should be developed in line with business strategy and along with targets and action plan across focus areas. The strategy should also incorporate identified climate change-related concern areas.
- Research and development (R&D): A comprehensive R&D strategy is required to constantly innovate ways to combat impacts of climate change.
- Measuring and monitoring system/information systems: There is a strong need of efficient measuring and monitoring systems to monitor environmental parameters of the organisation, particularly in oil and gas sector in Indian context, wherein oil marketing companies have hundreds of operating locations.
- Human resources: The required number of manpower based on organisational context is required to handle suitable portfolios. Constant capability building is also required of the select manpower. The required manpower at field, regional and zonal level will help in driving sustainability agenda at operational level.

- Communication and disclosure: Organisations are required to disclose the information in transparent and effective manner to stakeholders using various available frameworks. The disclosures should be balanced with positive and negative impacts being dealt equally.
- Internal carbon pricing (ICP): ICP is one of the innovative tools to attach monetary value to carbon emissions. This helps organisations to move toward low-carbon projects by attaching a notional value to the projects. Many Indian organisations across sectors have adopted the concept, however, Indian oil and gas sector is yet to make significant strides in this journey.
- Adopt innovative rating frameworks/indices: Organisations are adopting various frameworks such as Green Rating for Integrated Habitat Assessment (GRIHA) by The Energy and Resources Institute (TERI), Green Building framework by India Green Building Council (IGBC) and GreenCo Rating by Confederation of Indian Industry etc. Life cycle assessment (LCA) can also serve as a useful for organisations to measure the total environmental impacts and take steps to reduce it.
- While most of the frameworks are typically focusing more on the environmental pillar (CII 2018) vis-à-vis economic and social pillars of sustainable development, several organisations have developed their own frameworks to measure and benchmark sustainability performance of operating locations. One such sustainability framework is proposed as below (Table 19.2).

The framework is expected to evaluate two years of performance of an operating location. Based on the marks obtained the location is given a level of Leader, Achieve, Challenger and Striver (Fig. 19.3).

This framework will help organisations to benchmark the operating locations and understand the opportunities for improvement. This will also help in creating a sense of competition among rated operating locations to perform better.

19.3 Conclusion

The impact of climate change on the oil and gas sector is pertinent. Considering the capital-intensive nature of the sector and its requirement of vast infrastructure to operate, the measures to mitigate the impacts of climate change need to be imbibed in the operating philosophy of organisations. The paper introduces an environmental sustainability framework for the organisations with probable interventions aimed towards environmental excellence and climate resilience. A new sustainability framework is also introduced which can be used by organisations to measure operating location's performance on three pillars of sustainability.

Table 19.2 Sustainability framework for operating locations

Pillars	Categories	Maximum marks	Parameter	Marks allotment
Profit (100 Marks)	Through-put	20	Increase in through-put	2 Marks per increase % of through-put
	Cost leadership	30	Decrease in cost per KL or MT	3 Marks per % decrease in operating cost per unit
	Innovation	50	Innovation resulting in increase in throughput or cost leadership	10 Marks per innovation
Planet (250 marks)	Energy	30	Reduction in Specific Energy Consumption (SEC)	3 Marks per % reduction in SEC
		20	Energy-saving initiatives	5 Marks per initiative
		10	ISO 50001 certification and energy monitoring system	10
	Water	30	Reduction in specific water consumption (SWC)	3 Marks per % reduction in SWC
		20	Water-saving initiatives	5 Marks per initiative
		10	Rainwater harvesting (RWH)	1 Mark per 5% of RWH potential captured
	Renewable energy	40	% of total energy requirement substituted by renewable energy	0.5 Mark per % substitution
	Green belt	10	% of Green belt area beyond stipulated guidelines or 33% of the total area (whichever is higher)	1 Mark per % of additional green cover

(continued)

Table 19.2 (continued)

Pillars	Categories	Maximum marks	Parameter	Marks allotment
	Waste management	30	Reduction in specific waste generation	2 Marks per % specific waste generation reduction
	Innovation	50	Innovations having significant positive impact on any categories under planet pillar	10 Marks per innovation
People (150 Marks)	Safety	30	Nil Incidents in last two years	30 Marks
	Employee engagement	25	Initiatives taken towards employee engagement	At least one programme on half-yearly basis
	Training and development	25	Average training man-days for employees	3 Marks per one average man-day
	CSR Project	20	Corporate social responsibility (CSR)	10 Marks per CSR Project
	Innovation	50	Innovations having significant positive social impact	10 Marks per Innovation
	Total	**500**		

Fig. 19.3 Levels of indices

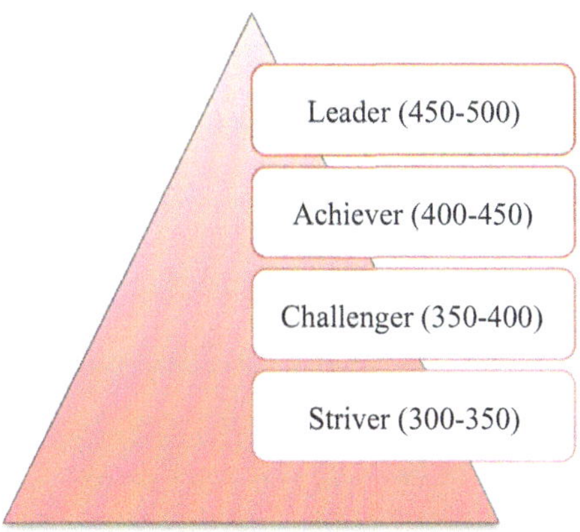

References

Bridge G, Özkaynak B, Turhan E (2018) Energy infrastructure and the fate of the nation: introduction to special issue. Energy Res Soc Sci 41:1–11. https://doi.org/10.1016/j.erss.2018.04.029
CII (2018) Green Company Rating System Reference Guide Version 3
Cruz AM, Krausmann E (2013) Vulnerability of the oil and gas sector to climate change and extreme weather events. Clim Change 121(1):41–53. https://doi.org/10.1007/s10584-013-0891-4
Dudley B (2015) BP statistical review of world energy 2016. London, UK
Earth Observation Day (2019). http://www.overshootday.org. Accessed 20 Aug 2019
India Brand Equity Foundation (IBEF) (2019) https://www.ibef.org/industry/oil-gas-india.aspx. Accessed 20 Aug 2019
IPCC (2007) Climate Change 2007: Impacts, adaptation and vulnerability. In: Parry ML, Canziani OF, Palutikof JP, van der Linden PJ, Hanson CE (eds) Contribution of working Group II to the fourth assessment report of the intergovernmental panel on climate change. Cambridge University Press, Cambridge, UK, 976 pp
McArthur JW, McCord GC (2017) Fertilizing growth: Agricultural inputs and their effects in economic development. J Dev Econ 127:133–152. https://doi.org/10.1016/j.jdeveco.2017.02.007
NOAA Climate.gov (2019) https://www.climate.gov/news-features/understanding-climate/climate-change-atmospheric-carbon-dioxide. Accessed 20 Aug 2019
Sinha KR, Nongpluh CL (2013) Development versus environment. IOSR J Hum Soc Sci 19(8):52–54. https://doi.org/10.9790/0837-19845254
TERI (2018) A study on climate change risks and preparedness for oil & gas sector in India New Delhi. The Energy and Resources Institute, 326 pp
United States Environmental Protection Agency (USEPA) (2019) https://www.epa.gov/ghgemissions/global-greenhouse-gas-emissions-data. Accessed 20 Aug 2019
World Economic Forum (2019) https://www.weforum.org/agenda/2019/01/these-are-the-biggest-risks-facing-our-world-in-2019. Accessed 20 Aug 2019

Chapter 20
Spatio-Temporal Analysis of Climate Change in Belgaum District, Karnataka

Basavaraj R. Bagade

Abstract Belgaum district in Karnataka has been experiencing drought continuously for the past few years. To study the phenomenon of low rainfall in the district, statistical analysis of monthly and annual rainfall data of rain gauge stations located in the district is performed. There was no relation between the rainfall in each of the monsoon months and the monsoon total and among the rainfall of the monsoon months. The serial correlation was poor, indicating the absence of any persistence in the series. Linear regression of rainfall data for Belgaum observatory on split sample data has indicated a falling trend in recent years.

Keywords Rainfall · Climatic fluctuation serial correlation · Linear regression

20.1 Introduction

Belgaum district in Karnataka is one of the districts hard hit by drought in the state. The river flows and reservoir levels have been reported to be diminishing in the last few years. The reservoir at Naviluteerth on Malaprabha has not been able to fill up to the F.R.L. during the last few years.

The Belgaum district has experienced severe drought conditions during 2012–2013 which recurred during 2012–2013, 2014–2015 and is continuing in 2014–2015. Based on rainfall data received up to 15 July, 2015, the Government of Karnataka (2015) has declared eight out of ten talukas in the district as drought affected. Four of these talukas were reported to be under severe drought and two under mild drought. The cropped area in the district was reduced to about 67% of the normal area.

To understand the phenomenon of low rainfall in the district, a climatologically study of the rainfall of Belgaum district and neighbouring area has been undertaken. The study is particularly aimed at identification of the presence of any trend or persistence.

B. R. Bagade (✉)
Department of Studies and Research in Geography, Rani Channamma University, Belgaum, Karnataka, India
e-mail: bagadebr@gmail.com

R. K. Mishra et al. (eds.), *Sustainable Climate Action and Water Management*, Advances in Geographical and Environmental Sciences,
https://doi.org/10.1007/978-981-15-8237-0_20

20.2 Review

Studies for identifying the trend and periodicities in the rainfall series have been carried out by several authors for more than 30 years. Reynolds (1953) and Gregory (1956) had studied the regional variation of rainfall in Britain. Around the same time, Pramanik and Jagannathan (1953) found systematic variations in the annual rainfall over certain parts of India.

Namias (1968) had studied the trends in rainfall of Central Park observatory, New York and related them to general circulation aberrations. Mitchel (1968) had also studied the trends in rainfall of Central Park observatory using power spectrum analysis.

20.3 Problem the Statement

Climatic fluctuation is a complex and vaguely comprehended phenomenon. In general, prediction of climate is possible if:

i. A climatic variable is statistically auto-predictable from knowledge of its own past history,
ii. A climatic variable is statistically correlated with one or more environmental variables, which in turn are statistically and physically auto-predictable.

20.4 Location of the Study Area

Belgaum district is the northern-most district in Karnataka bordering Maharashtra. There are ten talukas in the district, namely Athani, Bailhongal, Belgaum, Chikodi, Gokak, Hukkeri, Khanapur, Raibag, Ramdurg and Saundatti.

The district is bounded on the west by the Western Ghats. The area is more or less undulating one. The elevation of the hills extends to more than 1000 m a.s.l. of the total geographical area of 1,346,348 ha; 192,776 ha is under forest, and 627,328 ha which is nearly 50% of the geographical area is normally under crops.

The Belgaum district is semi-arid in parts of Athani, Raibag, Gokak, Bailhongal taluks and Saundatti taluks. Climate in rest of the district ranges from sub-humid to humid towards west.

Nearly 95% of the annual rainfall is received during the period April to October, because of Southwest Monsoon. Most of the remaining rainfall is received during November and December under the influence of Northeast Monsoon. The normal monthly and annual rainfall (India Meteorological Department, 1962) of rain gauge stations located in the district is given in Table 20.1.

The district has two major river systems Ghataprabha and Malaprabha, which are tributaries of river Krishna. The other rivers Hiranyakesi and Markandeya are

Table 20.1 Normal monthly and annual rainfall 1951–2001 (in MM)

Station	Belgaum	Khanapur	Bailhongal	Saundatti	Ramdurg	Gokak	Athani	Chikodi	Hukkeri	Bijapur	Dharwad
Jan	3.6	1.3	1.3	2.8	2.3	3.6	4.6	2.5	1.8	5.8	2
Feb	1.3	0.8	1.8	3.3	2.5	1.3	1.3	1	2.5	2.3	1.5
Mar	9.9	5.1	8.4	7.4	10.9	5.6	5.6	6.9	8.4	7.1	8.9
Apr	43.2	28.7	36.8	36.1	29	30	26.7	34	32.8	19.6	48.3
May	66.5	53.3	62.7	62.7	51.1	57.1	53.1	51.8	66.5	29.5	74.4
June	190.7	285	100.3	62.5	70.9	68.8	72.9	80.8	82.5	76.2	95.3
July	459	693.9	136.4	86.9	73.7	75.9	75.2	127.3	129.8	58.7	174
Aug	247.1	345.9	87.1	71.4	70.6	63	137.9	86.4	82.8	65.8	121.4
Sep	118.9	119.9	110.7	112	120.4	95.3	93	94.5	103.9	141.5	102.4
Oct	111	102.9	114.3	106.7	83.1	109.5	33.3	98.8	109.2	76.7	125.2
Nov	40.6	37.1	40.6	51.1	38.3	44.2	8.9	38.9	43.7	30.7	48
Dec	11.4	9.7	8.4	7.4	9.7	9.7	9.7	9.9	11.7	6.6	1.9
Annual	1303.2	1683.6	708.8	610.4	572.5	564	587.9	632.8	675.6	520.5	813.3

Normal Monthly and Annual Rainfall and Rainy Days
India Meteorological Department, 2001

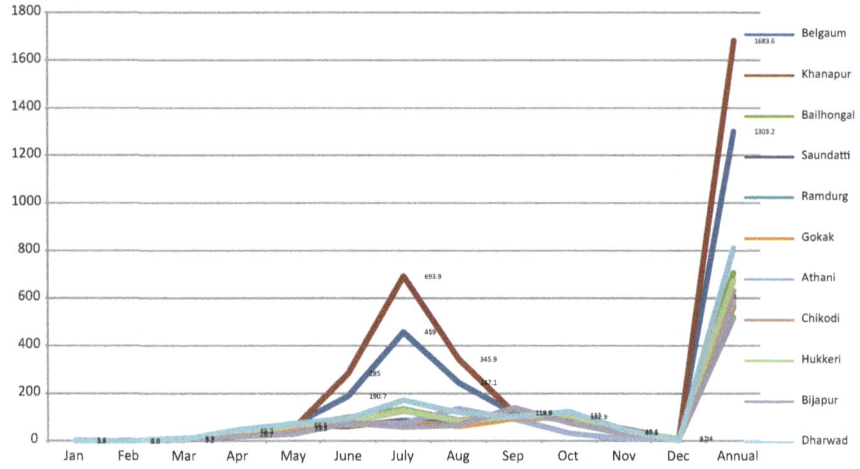

Fig. 20.1 Normal monthly and annual rainfall 1951–2001 (in mm)

tributaries to Ghataprabha. There are two dams on Ghataprabha at Hidkal and on Malaprabha at Naviluteerth (Fig. 20.1).

20.5 Methodology

The average monthly and annual rainfall for the Belgaum district and the catchment of Malaprabha up to Naviluteerth has been computed by Thiessen polygon method. The average for Ghataprabha up to Hidkal could not be computed due to gaps in the data of Chandgad, a rain gauge in the Kolhapur district of Maharashtra which has influence over the catchment.

The statistical parameters mean, standard deviation and coefficient of skewness for rainfall data have been computed for each of the 12 months and also for monsoon (June–Sep), non-monsoon (Oct–May), water year (June–May) and annual (Jan–Dec) periods. Also, the decade means and progressive means for 10, 20, 30,… years have been computed to identify the periods of low rainfall epoch.

20.6 Study of Monsoon Pattern

To study the dependence of the total monsoon rainfall on the rainfall of each monsoon month, and the relationship of the rainfall of each of the monsoon months with one another, cross-correlation using the Eq. (1) has been worked out.

where N is the length of the series and x_i and y_i are two arrays of data. Cross-correlation has also been worked out to see the dependence of non-monsoon rainfall on the previous monsoon rainfall.

20.7 Application

The statistical techniques described earlier have been applied to the monthly, seasonal and annual rainfall data of rain gauge stations in Belgaum, its district average and also catchment average rainfall of Malaprabha river up to Naviluteerth. To visualize graphically, the presence of any trend, plots of annual rainfall of all the rain gauge stations in the district and the average annual rainfall of Belgaum district and of catchment of Malaprabha up to Naviluteerth have been made.

20.8 Trend Analysis

The decade means of each rain gauge station, and district and catchment average have been compared with the mean of the whole period of the respective rainfall series. In case of Dharwad, the mean of the last period was only of eight years, i.e., 1971–1978. The rainfall trend at Khanapur and in the catchment of Malaprabha has been further examined by comparison with the observed discharges at Khanapur and estimated inflows into Malaprabha reservoir at Naviluteerth. Linear regression analysis has been carried out on the monthly and annual rainfall series of the rain gauge stations average of Belgaum district and Malaprabha catchment. Polynomial regression of first and second-order has been fitted to the data series of monthly, monsoon total, non-monsoon total, annual and water year rainfall. The results are presented in Tables 20.2 and 20.3, respectively. The analysis has also been carried out in respect of 3-year and 5-year moving average series of monthly and annual rainfall (Fig. 20.2).

20.9 Results

The geographical analysis of the annual rainfall series and their respective three-year and five-year moving average series did not indicate the presence of any trend for any rain gauge station except Khanapur, Belgaum and Hukkeri. Khanapur and Belgaum show a rising trend until 1971 and 1981, respectively. Since these are located in the catchment of Malaprabha, this trend is also reflected in the Malaprabha catchment rainfall series. While Khanapur does not indicate any falling trend after 1961, Belgaum shows a falling trend after 1953. At Hukkeri also, there is falling

Table 20.2 Decadal Mean & T_k values of annual rainfall series

Station	Khanapur	Belgaum	Bailhongal	Hukkeri	Gokak	Chikodi	Ramdurg	Athani	Belgaum District	Malaprabha	Bijapur	Dharwad
1931–41	1614.1	1218.8	686.9	601	582.7	658.8	552.5	572.6	808.6	1206.3	479.5	766.6
	(−0.51)	(−0.46)	(−0.11)	(−0.66)	−0.15	−0.12	(−0.01)	(−0.006)	(−0.27)	(−0.47)	(−0.50)	(−0.33)
1941–51	1536	1310.9	739.2	716.3	613.7	685.8	637.1	630.7	844.8	1191	529.8	853.3
	(−0.70)	(−0.15)	−0.17	(−0.07)	−0.36	−0.29	−0.56	−0.32	(−0.03)	(−0.52)	(−0.24)	−0.14
1951–61	1695.1	1231.8	686.5	568.3	516.2	596.4	481.5	480.9	773.3	1244.2	496.4	806.5
	(−0.31)	−0.41	(−0.11)	(−0.83)	(−0.29)	(−0.26)	(−0.49)	(−52)	(−0.51)	(−0.33)	(−0.42)	(−0.11)
1961–71	1732.9	1372.4	666.8	748.1	592.5	632.6	596.1	674	857.3	1273.5	559.5	784.7
	(−0.22)	−0.05	(−0.22)	−0.1	−0.22	(−0.04)	−0.28	−0.56	−0.05	(−0.22)	(−0.09)	(−0.23)
1971–81	1841.8	1383	762.2	743.8	515	588.7	595.8	579.1	852.7	1362	536.5	855.9
	−0.04	−0.08	−0.3	−0.07	(−0.30)	(−0.31)	−0.28	−0.03	−0.02	−0.11	(−0.21)	−0.15
1981–91	2084.5	1643.7	756.5	959	592.8	738	538.9	591.1	952.8	1512.6	676.5	970.5
	−0.63	−0.94	−0.27	−1.2	−0.22	−0.61	(−0.06)	−0.1	−0.69	−0.66	−0.52	−0.23
1991–01	2104.9	1469.1	647.2	781.5	563.9	552.7	544.6	522.4	874.4	1471.1	661.1	832.4
	−0.68	−0.37	(−0.33)	−0.27	−0.03	−0.53	(−0.06)	(−0.29)	−0.17	−0.51	−0.44	−0.02
2001–11	1966.6	1336.3	730.3	758.9	533.1	632.5	485.4	565.7	857.9	1414	674.7	730
	−0.34	(−0.07)	−0.13	−0.15	(−0.18)	−0.04	(−0.46)	(−0.04)	−0.06	−0.3	−0.5	(−0.52)

*Value significance at 5% level of significance

Table 20.3 Progressive decade means of annual rainfall series

Station	Khanapur	Belgaum	Bailhongal	Hukkeri	Gokak	Chikodi	Ramdurg	Athani	Belgaum District	Malaprabha	Bijapur	Dharwad
1931–41	1614.1	1218.2	686.9	600.9	582.7	658.8	552.5	572.6	808.6	1206.3	479.5	766.6
1941–51	1575.1	1264.8	713.1	658.6	598.2	672.3	594.8	601.7	826.7	1198.6	504.6	809.9
1951–61	1615.1	1253.8	704.2	628.5	570.9	647	557	561.4	808.9	1213.8	501.9	808.8
1961–71	1644.5	1283.5	694.9	658.4	576.3	643.4	566.8	589.5	821	1228.6	516.3	802.8
1971–81	1683.9	1303.4	708.3	675.5	564	632.5	572.6	587.4	827.3	1255.4	520.3	813.4
1981–91	1750.7	1360.1	716.4	722.7	568.8	650	567	588.1	848.3	1298.3	546.4	839.6
1991–2001	1801.3	1375.6	706.5	731.1	568.1	636.1	563.8	578.7	852	1323	562.8	838.5
2001–2011	1822	1370.7	709.4	734.6	563.7	635.7	554	577.1	852.7	1334.3	576.7	806.7
2011–2012	1824.5	1357.5	707.3	729.4	559.8	638.9	553.9	573.7	849.5	1333.3	–	–

*Mean based on data up to 2011 only

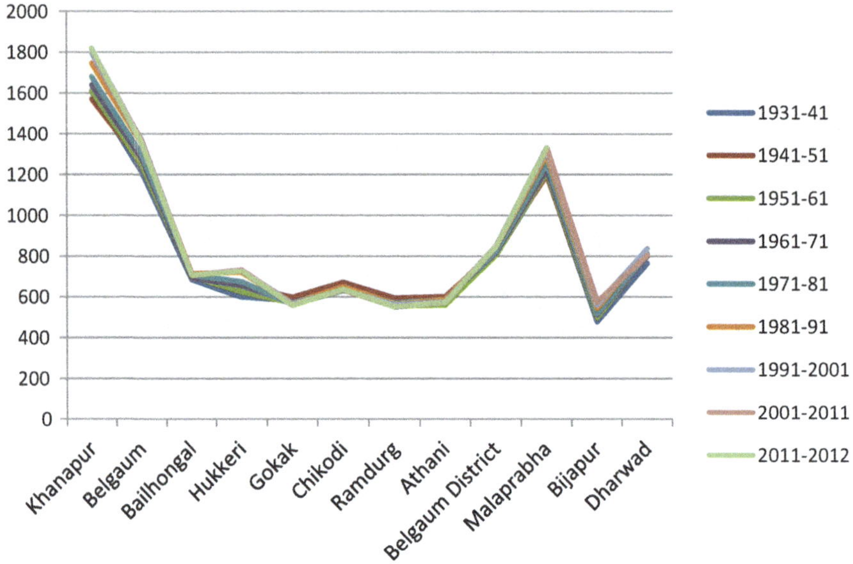

Fig. 20.2 Normal monthly and annual rainfall 1951–2001 (in mm)

trend in the rainfall after 1956. The average rainfall of Belgaum district does not show any trend.

20.9.1 Rainfall Characteristics

The coefficient of variation of rainfall during the non-monsoon months and season is more than that of the monsoon months' and season's rainfall. Rainfall during the monsoon months is less skewed as compared to rainfall during the non-monsoon months. With the exception of Belgaum, Bailhongal and Hukkeri, the seasonal, water year and annual rainfall of all other rain gauge stations, Belgaum district average and Malaprabha catchment average is nearly normally distributed as indicated by the coefficient of skewness during these periods. The coefficient of skewness of seasonal, water year and annual rainfall of Hukkeri is high.

The correlation of the rainfall of monsoon months among themselves and with the monsoon total rainfall is insignificant indicating the general independent nature of rainfall occurrence. The monsoon and non-monsoon rainfall also showed similar behaviour.

20.9.2 Rainfall Trends in Time

From the decadal means of annual rainfall and the progressive means presented in Tables 5 and 6, it may be seen that rainfall during 1901 to 1910 and after 1980 was generally low. The rainfall was high during the decades 1950–1960, 1961–1970 and 1971–1980. It may also be observed that the decadal means are very close to the long period mean, excepting for Belgaum and Hukkeri during the period 1951–1960.

20.9.3 Linear Trend

In general, the monthly rainfall series do not indicate the presence of any linear trend. Linear regression analysis on split samples, however, revealed a significant rising trend in June months' rainfall at Khanapur and falling trend in the July month's rainfall at Belgaum in the last 25–30 years. The effect of the rising trend in June at Khanapur is also noticed in the catchment rainfall series (1962–1984) of Malaprabha. However, the decreasing trend in July noticed for Belgaum is not significant in the Malaprabha series.

20.9.4 Trend in Annual Series

Linear regression analysis of the annual rainfall series has revealed significant rising or falling trend at Khanapur, Belgaum, Hukkeri and Bijapur. While a rising trend has been noticed in the annual rainfall series at Khanapur and Bijapur, a falling trend is noticed in the series of Belgaum (1953–1984) and Hukkeri (2012–2013).

20.9.5 Trends in the District and Catchment Average Series

The presence of rising trend noticed in the monthly or annual rainfall series of Khanapur and the falling trend noticed in the monthly and annual rainfall series of Belgaum and Hukkeri is not reflected in the monthly or annual rainfall series of the district averages of Belgaum because of the relatively small influence that each of these gauges have in the determination of the district averages and the nullifying effect introduced by the rising and falling trends. As such, the Belgaum district averages do not show any linear trend either in the monthly or the annual rainfall series.

20.9.6 Nonlinear Trend

The continuous monthly rainfall series of Bailhongal, Bijapur and Hukkeri reveal some trend as indicated by the polynomial regression coefficients in Table 8. While the trend at Hukkeri and Bijapur is nonlinear (second order), the trend at Bailhongal is linear (first order). The nonlinear trend at Bijapur is significant at the 5% level of significance while the trend indicated at Bailhongal and Hukkeri is not significant.

20.9.7 Nonlinear Trend of Seasonal and Annual Series

In addition to Bailhongal, Hukkeri and Bijapur, the monsoon and water year rainfall, the polynomial regression analysis also indicate the presence of nonlinear trend at Khanapur. While the trend at Bailhongal is linear (first order) and significant at the 5% level of significance, the trend at the other three stations is nonlinear and is significant at the 1% level of significance.

The non-monsoon rainfall series do not show any trend and the polynomial regression of the annual series show that the trend is present at only Khanapur, Hukkeri and Bijapur and this is significant at the 1% level of significance.

20.9.8 Nonlinear Trend of District and Catchment Average Rainfall Series

The polynomial regression of the district averages of Belgaum do not indicate the presence of any form of trend in either monthly, seasonal or annual rainfall series.

Monsoon, annual and water year series of Malaprabha catchment, however, reveal some trend of which only the trend in the annual rainfall series is significant at 1% level of significance.

20.9.9 Regression of Moving Average Series

The linear and polynomial regression of the 3-year and 5-year moving average rainfall series of the rain gauge stations, district and catchment average rainfall has indicated the same trends as was obtained from similar analysis with the actual rainfall series. Significant trends are noticed in the 3-year and 5-year moving average series of Khanapur, Hukkeri, Bijapur and Malaprabha catchment.

20.10 Conclusions

From the results of the analysis, it could be broadly concluded that:

i. There is significant rising trend in the rainfall at Khanapur during the period 1901 to 1961 and at Bijapur from 1901 to 1980.
ii. There is a falling trend in the rainfall at Belgaum and Hukkeri especially for the period 1971–2001.
iii. Because of averaging of point values, the trends noticed in the series of Khanapur, Belgaum and Hukkeri are evened out in the average rainfall series of Belgaum district which did not indicate the presence of any trend.
iv. The flow data at Khanapur and the inflow into the reservoir on Malaprabha are in good agreement with general pattern of rainfall. The trend noticed at Khanapur is also reflected in the rainfall series of Malaprabha catchment.

The trends noticed in this study in the rainfall series for Khanapur, Belgaum, Hukkeri and Bijapur rain gauges may be part of a long-term cycle. This, however, needs further examination by studying the behaviour of rainfall series of the stations in the contiguous region. The effect of changes in land use over the years on rainfall patterns in the region would also have to be considered in this connection.

References

Bhargava BN, Bansal RK (1969) A quasi-biannial oscillation in precipitation at some Indian stations. Indian J Meteorol Geophys 20(2):127–128

Government of Karnataka (1985) Drought Prone Areas Programme, Belgaum District, Annual Plan 1985-86 and 1986-87

India Meteorological Department (1970) Monthly and Annual Rainfall and Number of Rainy days 1901-50, vol IV-A

Mitchell JM (1968) Some remarks on the statistical predictability of drought and related climatic anomalies. In: Proceedings of the conference on the Drought in Northeastern United States (15–17 May 1967), pp 129–159

Winstanley D (1973) Recent rainfall trends in Africa and the Middle East and India. Nature 243(5407):464–465

World Meteorological Organizations (1966) "Climatic Changes", Technical Note 79

Part III
Sustainable Development: A Way Forward

Chapter 21
Status of Drinking Water Sustainability Using GIS in Rangareddy: A Case Study

B. Srinagesh and K. Sudarshan

Abstract A persistent problem in India is the availability of rural drinking water. Access to clean drinking water is largely seen as a responsibility of the government. Water is the most precious and valuable resource and vital for the socio-economic growth and sustainability of the environment. Public supply of drinking water is mainly carried out by tap, while private supply is well through tube and later in most states has now become the predominant source. Groundwater in the region has been utilized through wells tapping various water-bearing formations to provide more fresh water to supply the increased demand for water supply and irrigation. This practice negatively affected the whole of the hydro-geological setting of the basins. There are two major water bodies in Ranga Reddy District: Osman Sagar and Himayat Sagar, which cater to the needs of Hyderabad, the capital city. The district of Ranga Reddy was in second place, with 72 per cent access to drinking water; however there was a sharp disparity between urban and rural districts of Ranga Reddy, with appalling figures for the latter. The study seeks to highlight the potential water resources available and the sustainable management of the water resources.

Keywords Ground water potential · Water grid · Drinking water and Water sustainability

21.1 Introduction

Water is a primary source for development, health, and human survival. Water is essential for human development, and without water there is no forward movement of human and human development. Sustainable development of water and utilization of water sources are very much important in the present-day conditions. Water is

B. Srinagesh (✉)
Department of Geography, UGC-HRDC, Osmania University, Hyderabad, India
e-mail: geosri74@gmail.com

K. Sudarshan
Department of Geography, Osmania University, Hyderabad, India

R. K. Mishra et al. (eds.), *Sustainable Climate Action and Water Management*, Advances in Geographical and Environmental Sciences,
https://doi.org/10.1007/978-981-15-8237-0_21

the most precious and valuable resource and vital for the socio-economic growth and sustainability of the environment. In some arid countries, water resources are limited, scarce, and mainly sourced from groundwater. In some parts, surface water is limited to a few river systems and intermittent streams that are associated with rain. Precipitation is vital and the primary source of recharge for various groundwater aquifers in these regions. Groundwater in the region has been utilized through wells tapping various water-bearing formations to provide more fresh water to supply the increased demand for water supply and irrigation. This practice negatively affected the whole of the hydro-geological setting of the basins.

Management of water resources has become a major effort for the Telangana government in the region. Various water institutions and private companies are using GIS as a tool to manage water resources. The GIS can be used to capture data and develop hydrologic dataset for all components of water resources; understand the region's hydrology; map sources of contaminations; prepare water quality and water–rock interactions maps; delineate the watershed areas; and much more.

In developing places like India, it is very important to improve drinking water facilities to populated areas, because of huge population development in urban as well as in rural India. It is necessary to improve drinking water facilities to overcome all kinds of health issues related to drinking water.

21.1.1 Study Area

The study area falls near and around HMDA area and it lies between 77°21′ and 78°51′ longitude and 16°54′ and 17°48′ N latitude. The Rangareddy district falls on the western and south-western part of the Telangana state, with the area covering 7500 km^2, and population 5,296,741, which is 15.05% of total population. The Hyderabad urban city is surrounded by Rangareddy district. Because of developmental activities, it is spreading in all directions in Rangareddy. There is a huge amount of expansion in and around Hyderabad city. Most of the expanding urban centres are in Rangareddy district, because of expanding of all kinds of industries like pharma city, IT sector, recreational centres like Film city, Adibatla water park, Aviation Park, other health care centres like NIMS, and it is due to this Rangareddy district has gained prominence. Rangareddy district is the fastest developing district of Telangana with a decadal urban growth of 48.16 and 70.21%. Anathagiri hills is the main water source for Osmansagar and Himayathsagar, and is the major water source in Rangareddy district. Rangareddy district has 37 mandals and 870 revenue villages. The district receives 835 mm rainfall and also has a high frequency of droughts owing to poor distribution of rainfall (Fig. 21.1).

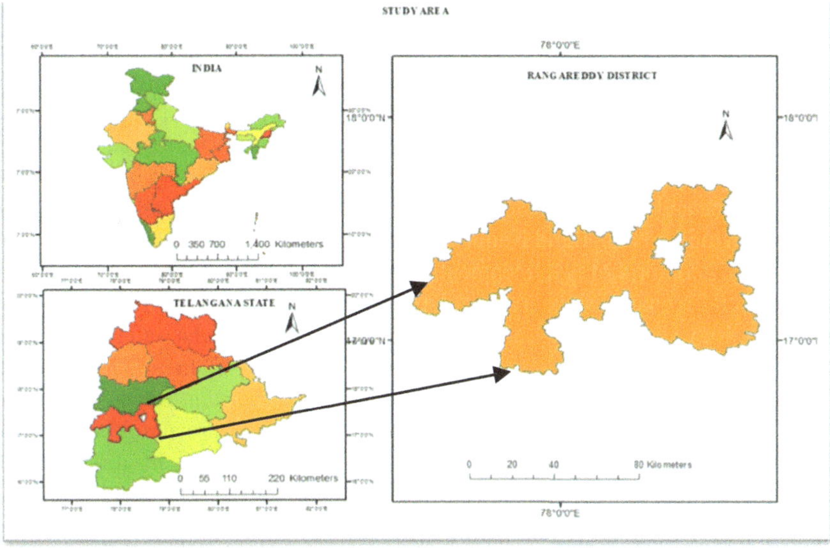

Fig. 21.1 Location map of study area. *Source* TRAC

21.1.2 Aims and Objectives

(1) Preparing potential surface water resources and groundwater resources area maps.
(2) Preparing drinking water grid in the study area.
(3) Sustainable management and development of water resources in Rangareddy district.

21.1.3 Review of Literature

There are large number of journals, articles, and related books referred to in this field. Review of the literature is a vast and most important element to explore and make an awareness to find the output of the research given by so many authors on this sustainable development. Nowadays, satellite remote sensing techniques are used for groundwater exploration, especially for delineating hydro-geomorphological units. Recently, remote sensing with multi-temporal high-resolution satellite data has become a strong tool for monitoring aspects such as vegetation cover, soil degradation, urban expansion, and more generally for most types of land-cover/land-use (LC/LU) changes. GIS has been found to be one of the most powerful techniques in assessing the suitability of the land. The interest of hydrogeologists in remote sensing is shifting towards the dynamic aspects of water balances and contamination, for which an appropriate instrument is provided by combining image processing, GIS procedures, and numerical modelling (Meijerink 2009). The integrated groundwater investigations and management using satellite remote sensing and GIS application in parts of Musi river basin, Nalgonda, and Rangareddy districts, Telangana state have been studied. Sreedevi et al. (2005) have delineated groundwater potential zones based on the study of morphometric studies in Pageru river basin, Cuddapah district of Andhra Pradesh. Water is a scarce resource in the world, and it is classified as a groundwater and surface water. For better understanding it has taken Sivaganga district as a study area in Tamil Nadu. The main objective is to create thematic maps like geology, geomorphology, lineaments, drainage density, and land-use/land-cover data. Artificial recharge sites are interdependent on various layers like soil, lineaments, lithology slope geology, and geomorphology (Balachandar and Eaton 2010). Machender et al. (2014) have studied the assessment of trace element contamination in soils around Chinnaeru River Basin, Nalgonda District, India. Development and growth of population on the earth create imbalance in the nature and need to create infrastructure to developing population. There are no proper plans in storing of rain water and sustainable development (Jeevananda Reddy 2018). The water level data of 810 groundwater level in March 2018 is 11.90 m bgl, whereas it was 10.96 m bgl in 2017. The rise in water levels is 0.15 m in Bhadradri-kothagudem district and 2.28 m in Rangareddy district. The 161 mandals showing deep water levels have to improve. It must sustain these resources by adopting proper steps and giving higher priority on water conservation and management of groundwater (Government of Telangana Ground Water Department 2018).

21.2 Methodology

For studies like this, identification of review of literature is very crucial. Multi-thematic mapping is done in the study area using primary and secondary data, which included maps related to geology, soil, climate, demographics, and groundwater-related data. Database and thematic maps are generated using GIS and RS. Overlapping of all thematic map layers and analysis of data make final the output.

21.2.1 Geology and Lineaments

The area predominantly exposed rocks of peninsular gneissic complex (PGC) along with endues of schists (Older metamorphics), basic dykes (proterozoic), Bhima group comprising upper proterozoic sedimentary formations, and a thin cover of Deccan traps (cretaceous Paleogene) and laterite (Pleistocene). The PGC comprises two variety of older granite and younger granite. The older metamorphics are trending in NNW-SSE to NW-SE direction. Granite is distributed all around the area; it is shading grey to pink, medium to coarse-grained. It is forming higher topographic levels and observing denudational hills, dome-shaped mounds.

Table 21.1 Depth range and yield ranges in basalts and granitic terrain

Depth range (m)	No. of EW		Yield range (lps)	Basalts	Granites
	Basalts	Granites			
20–30	0	1			
30–40	1	20	1–3	8	18
40–50	2	15	3–5	1	2
50–60	4	4	>5	4	2
60–70	6	2			
70–100	14	2			
100–150	5	1			
150–200	2	0			

Source CGWB

The basaltic lavas of the Deccan traps cover either the Bhima sediments or the granitoids, around Vikarabad, Tandur, and Pargi. These include different flows of basalt and intertrappean beds. The thickness of each flow varies from 15 to 20 m. The lineaments are trending in NW-SE and NNE-SSW direction and faults are common in this area. A series of WNW-ESE trending faults are seen in the southeastern part of the area.

21.2.2 Geomorphology and Geohydrology

In the western part, around Pargi and Vikarabad, a plateau of Deccan basalts is formed. The groundwater in granitic terrain is fracture-controlled and the water table varies between 35 and 70 ft. The fractures are trending NW-SE and NS and it is controlling groundwater (Table 21.1).

21.2.3 Land Use and Soil

Northwest and southwest areas are the grass and scrublands with small agricultural patches observed. The southeast is covered by forest land, and in the east part Rangareddy covers Hyderabad which is predominantly urban centre and extends radial pattern from centre to all parts of the city. The Rangareddy district comprises red sandy soil with northwest part occupying laterite soil and medium black soil. A part of mixed red and black can be seen in the southeast.

21.3 Drainage

The river Musi is draining from west to east in this region, and Osmansagar and Himayatsagar are the two reservoirs constructed on Musi river which are supplying drinking water facilities to Hyderabad, Secunderabad, and surrounding Rangareddy. The Musi river comprises Trellis type of drainage pattern where granitic rocks are observed. The chief source of irrigation is tanks, wells, and canals in this area.

21.4 Results and Analysis

21.4.1 Geology

Rangareddy district has 37 mandals. The geology patterns in these mandals are almost similar but in some areas like in the western part, it showed little bit variation in terms of geology. There are mandals like Tandur and Basheerbhag covered with sandstone, purple shale, and limestone. Almost 70% of the area in the Rangareddy district covers granite and gneisses type of rock formation, but some areas in western part cover basalt and intertrapped areas like Vikarabad, Chevella, Pudur, Pargi, Peddemul, Bantwaram, Marpalle, Moninpet, Nawabpet, Shankarpalle. The laterite is like a belt from north to south covering areas like Mominpet, Vikarabad, Dharur, Pudur, Pargi and parts of Marpalle, Peddemul.

21.4.2 Soils

It is a typical combination of soils observed in Rangareddy. In the western part, it is shallow gravelly red soils and also shallow loamy to gravelly. In the eastern part, it is covered by deep wet dark brown soils in areas like Keesara, Shairpet, Ghatkesar, Hayathnagar, Manchal Yacharam, parts of Kandukur, Maheshwaram, Uppal, Qutbullapur, Kurtanchal, Gandeed, Yelal, Dharur, Nawabpet. The areas like Shamshabad, Maheshwaram, Saroornagar, Balanagar, Qutbullapur, Shabad are expanding of settlements because of urbanization. Yelal, Dharur, Vikarabad, Nawabpet, Kurtanchal, Gandeed areas are covered by deep wet soils and dark brown soils. The large waterbodies like Osmansagar and Himayathsagar are located in Rajendranagar and Moinabad mandal only.

21.4.3 Lineaments

Lineaments are the features or faults in geological structures. According to the Rangareddy district, these lineaments are distributed in southeast and south direction of the district. The mandals such as Manchal, Ibrahimpatnam, Kulkacharla, Ganded, Yacharam, Shankarpalle, South of Chevella, Rajendranagar, Keesara, Shameerpet are the areas or spot heights that are having lineaments. These areas are favourable for surface water infiltration into the ground.

21.4.4 Streams

The spot heights like Uppal, Ghatkesar, Qutbullapur, Pargi, Pudur, Chevella, Mainabad, Maheswaram, Manchal are the mandals which are having high stream density. These areas are considered to be a high surface water flow and lead to a rise in groundwater levels.

21.4.5 Water Grid

Water grid is one of the drinking water programmes to serve drinking water to all the households in the Telangana state. To execute the project and for effective service to the people, the government has set up Telangana Drinking Water Supply Corporation Limited. It has taken nearly one lakh acres of forest land and provided alternate land. The objective of this water grid is to provide 100 l per person of drinking water to rural and 150 l per person in urban households. Godavari and Krishna are the two perennial rivers and these are the resources supporting the Telangana water grid. The state has 26 internal grids that are supplying water through pipelines (Table 21.2).

Water management and water development are the main questions in society. Everybody uses water for household, drinking, industrial, and agricultural purposes. The source of water is from rains through surface waterbodies and river bodies and groundwaters. It is a process where no one is involving to increase the water table or conservation of waterbodies in and around the city to have a natural storage to get infiltration into the groundwater table. For developing the city, it is our duty to use water in a proper way and we should not waste it. Day-to-day the population is increasing and the requirement of water is also increasing in double (Tables 21.3 and 21.4; Figs. 21.2, 21.3, 21.4, 21.5, 21.6 and 21.7).

Table 21.2 Water grid of Rangareddy district

Dist	Fully covered				No safe source				Partially covered				Total habitats	Total population
	Habitats	%	Population	%	Habitats	%	Population	%	Habitats	%	Population	%		
RangaReddy	384	23	2,44,297	11	176	11	2,47,406	12	1082	66	16,34,180	77	1,642	21,25,883

Source TRAC

Table 21.3 Year-wise population and water per capita

Year	Population (lakhs)	Per capita (cubic meter/year)	%
1951	3610	5177	31.53
1955	3950	4732	28.82
1991	8460	2209	13.45
2001	10270	1820	11.08
2025	13940	1341	8.17
2050	16400	1140	6.94
Total		16,419	100.00

Source Government of India 2009

Table 21.4 Waterbodies in Rangareddy district mandal-wise

S. no.	Name of the mandal	No. of waterbodies
1	Balanagar	17
2	Ghatkesar	37
3	Hayathnagar	84
4	Ibrahimpatnam	56
5	Kandukur	118
6	Keesara	47
7	Maheshwaram	84
8	Malkajgiri	16
9	Menchal	4
10	Medchal	124
11	Moinabad	20
12	Qutbullapur	77
13	Rajendranagar	46
14	Saroornagar	60
15	Serilingampally	60
16	Shahabad	63
17	Shamirpet	138
18	Shamshabad	71
19	Shankarpally	10
20	Uppal	16
21	Yacharam	13

Source Based on HMDA

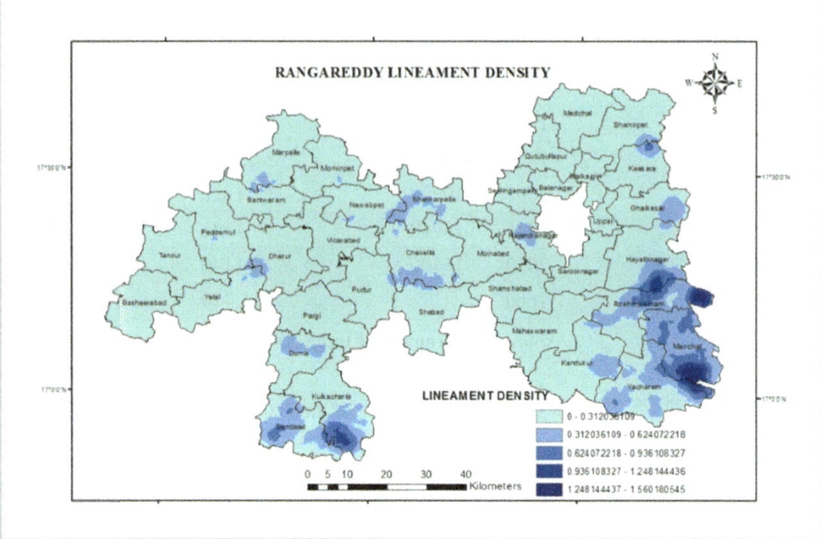

Fig. 21.2 Rangareddy mandal-wise density. *Source* TRAC

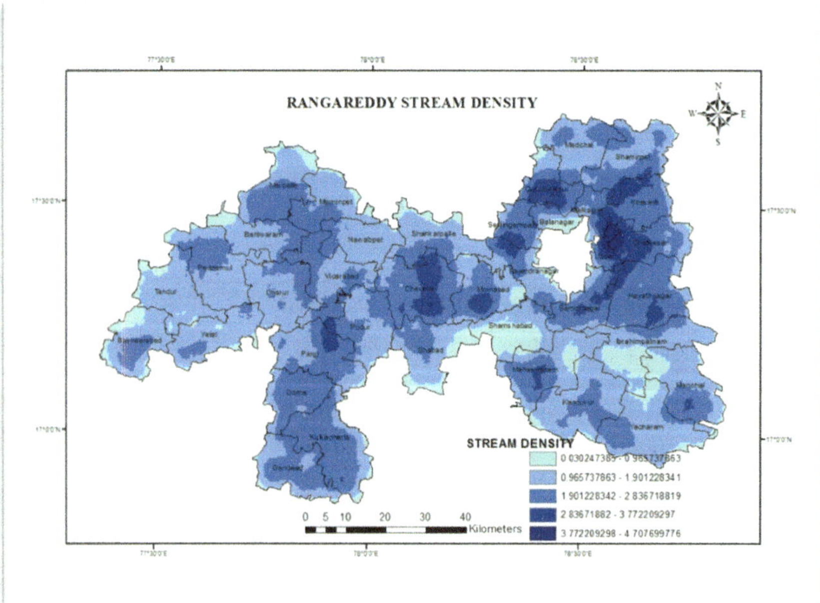

Fig. 21.3 Rangareddy streams density. *Source* TRAC

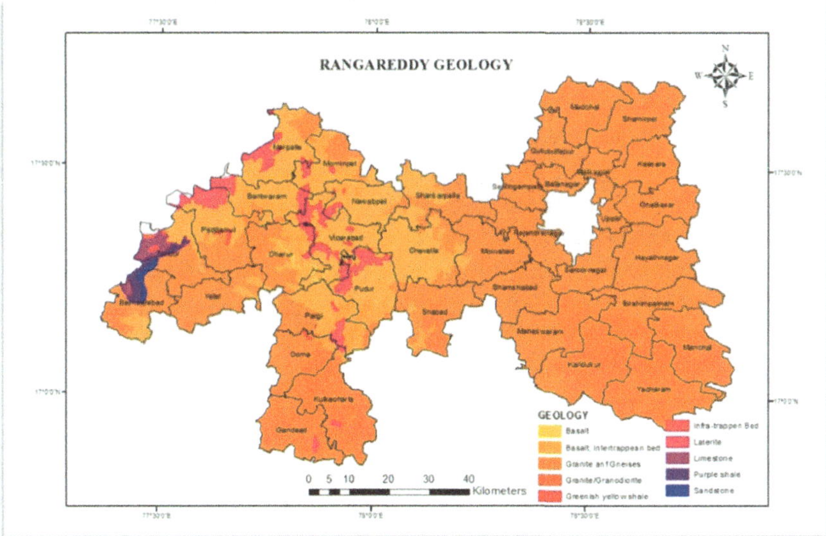

Fig. 21.4 Rangareddy geology. *Source* TRAC

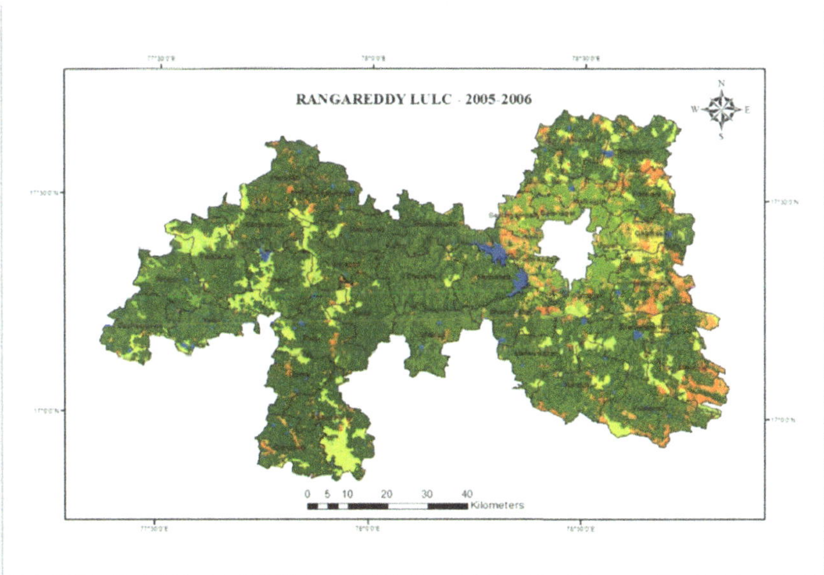

Fig. 21.5 Rangareddy LULC 2005–2006. *Source* TRAC

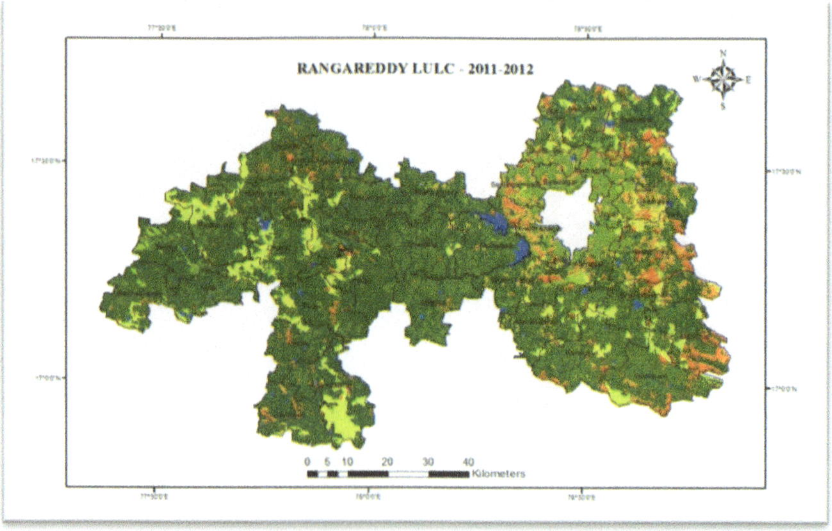

Fig. 21.6 Rangareddy LULC 2011–2012. *Source* TRAC

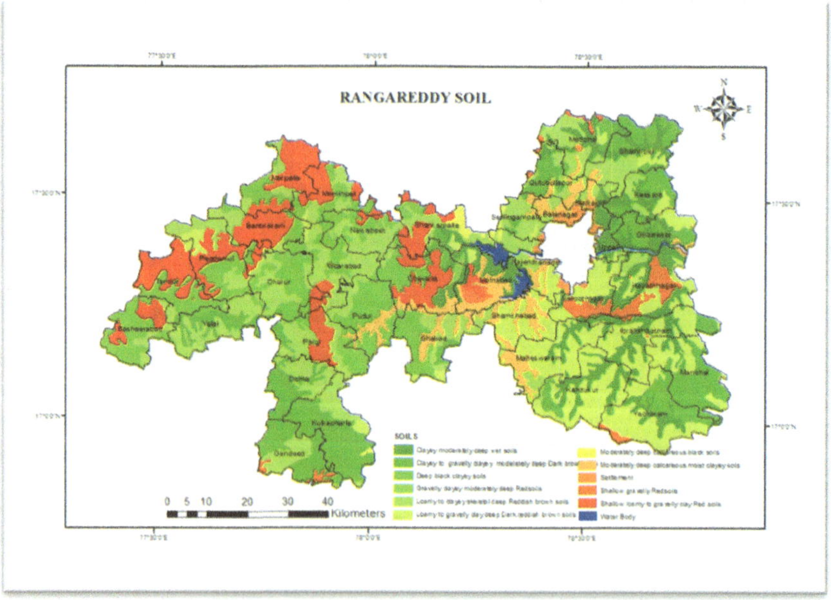

Fig. 21.7 Rangareddy soil. *Source* TRAC

21.5 Conclusion

Sustainable development of water resources refers to reducing the usage of water and recycling of waste water for different purposes, such as cleaning, manufacturing, and agricultural irrigation in such a way that the water demands of future generations are not hampered. Today, the concept of sustainability and the sustainable development has become one of the main issues in the development of contemporary society, economic, and technology. In this regard, it is possible to refer to the case of water use in the contemporary world because water is the major source of life on the earth and the quality of life of people and survival of many species depends on the water. The sustainable development being applied to the use of water can make the use of water more efficient and environment-friendly. In reality, water is used inefficiently. At any rate, the water pollution and the exhaustion of water resources of the earth become the major challenges for the future of the mankind. In this context, the sustainable development is impossible without the development of efficient policies concerning the water use. Today, agriculture is one of the major consumers of water resources that has a negative impact on water resources and supply of potable water.

Acknowledgements The authors are extremely thankful to ICSSR for assigning project and providing financial assistance to carry out this work.

References

Balachandar S, Eaton JK (2010) Turbulent dispersed multiphase flow. Ann Rev Fluid Mech 42(1):111–133

Central Ground Water Board (CGWB) http://cgwb.gov.in/

Government of Telangana Ground Water Department (2018) http://gwd.telangana.gov.in/

Jeevananda Reddy S (2018) For a workable green revolution for India. Curr Trends Biomed Eng & Biosci 13(2):555857. https://doi.org/10.19080/CTBEB.2018.12.555858

Machende G, Dhakate R, Prasanna L, Govil PK (2014) Assessment of heavy metal contamination in soils around Balanagar industrial area, Hyderabad, India

Meijerink AMJ (2009) Remote sensing applications to hydrology: groundwater. Hydrol Sci J 41(4):549–561

Sreedevi PD, Owais S, Khan HH, Ahmed S (2005) Morphometric analysis of a watershed of South India using SRTM data and GIS. J Geol Soc India 73:543–552 (2009). Published 20 May 2009

Chapter 22
Cultural Landscapes of Kakatiyas and *Mission Kakatiya* Scheme Under Government of Telangana

G. S. V. Suryanarayana Murthy, Vaibhavi Dhote, and Ashita Agarwal

Abstract Since the known history of India, numerous cultural landscapes have flourished throughout this civilization, but the awareness to identify these settlement geographies as cultural landscapes has been missing for decades leading to binary of operational confusion during management, conservation and administration of these geographies. Natural settings have always been woven into the cultural fabric of this nation with such an ease that nature and humans coexist. This paper examines one such cultural landscape in India. Kakatiya dynasty that hailed from Warangal and ruled the Telugu speaking region in early medieval eastern Deccan happened to raise rulers, whose course of action flowed together with the existing topography of the region. Kakatiyas are remembered for their monumental contribution to the Telugu culture as well as agriculture, irrigation and building technology. A study of Telugu past under Kakatiyas reveals the agriculture, irrigation, settlement patterns and art bearing strong roots from cultural knowledge for adjusting, understanding and camouflaging with the topography. The adamantine relation between existing topography, the flow of water, erection of massive temple complexes, foundations of villages and adjacent farmlands is what makes this assembly a cultural landscape unique to Kakatiyas. Numerous cultural landscapes throughout the globe have been inscribed under UNESCO as World Heritage Sites, one of them being from India—the *Bhimbetka* Rock Shelters, and there are many like the Kakatiyan built environments, which fall into different categories meeting the tangible with intangible heritage. For the sustainable land use of these landscapes, the Government of Telangana has launched *Mission Kakatiya* to protect and manage the irrigation and groundwater in these areas. This approach ensures better management of these landscapes.

G. S. V. Suryanarayana Murthy · V. Dhote (✉) · A. Agarwal
IHSRDM, Hyderabad, India
e-mail: dh.vaibhavi@gmail.com

G. S. V. Suryanarayana Murthy
e-mail: ihsrdm@gmail.com

A. Agarwal
e-mail: arashitagarwal@gmail.com

R. K. Mishra et al. (eds.), *Sustainable Climate Action and Water Management*,
Advances in Geographical and Environmental Sciences,
https://doi.org/10.1007/978-981-15-8237-0_22

Keywords Cultural landscapes · Kakatiyan irrigation systems · Settlement patterns · Sustainable land use · Harmonious coexistence of culture and nature

22.1 Cultural Landscapes as Defined by UNESCO

In 1992, the World Heritage Convention became the first international legal instrument to recognize and protect cultural landscapes. According to guidelines and format issued by UNESCO, cultural landscapes are ***combined works of nature and humankind***; they ***express a long and intimate relationship between peoples and their natural environment***. In cultural landscapes there are three major categories, namely ***clearly defined landscape designed and created intentionally by man***, ***organically evolved landscapes*** and ***associative landscapes.***

22.2 The Kakatiyan Interventions Over the Natural Profile of Earth

This peninsular region of Deccan plateau (Fig. 22.1) has monotonous peneplain embellished with stony ridges over granite mounds and black stone dykes occurring regularly. The mounds being not higher than half a kilometre and mostly enveloping a piece of land provide picturesque landscapes. Kakatiyan engineers used these mounds and the sloping terrain to build thousands of *Cheruvus* (lakes) which were connected to each other by means of manually dug canals, throughout their dominion. The agricultural land, nearby which these canals rode, became fertile enough to yield thrice a year. Building embankment wall and joining two adjacent ridges allowed rainwater to cease flowing, thus, creating a lake. Sluice gates were built and distribution of water was regulated.

An elaborate irrigation system was developed, taking care that the natural setting is not hurt; rather it is benefited by such lakes and brooks. Today after seven centuries the relation of people with these waterworks is as veritable as it was when the systems were being laid. A close interaction between nature and culture is visible. The lakes and canals are still in use and irrigate the paddy fields (Fig. 22.2).

Heritage, be it built or un-built, has coexisted harmoniously with nature and this is the key to sustainability. Culturally significant spaces need not be always built in locations of great aesthetic or scientific importance, but its merging with whichever context available/chosen is necessary. This is exactly what Kakatiyas did. The vibrancy of Kakatiyan cultural landscapes is an outcome of cohesion between the built and un-built, tangible and intangible (Fig. 22.3).

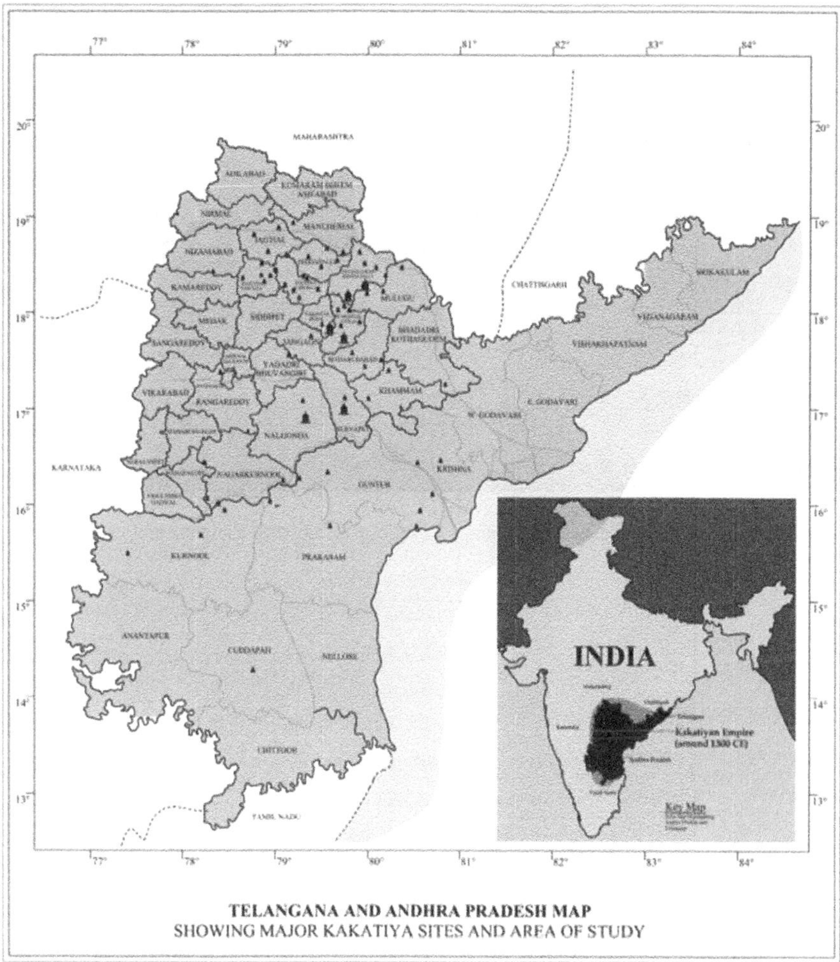

Fig. 22.1 Map—Major Kakatiya sites in Telugu speaking Region *(Present day Andhra Pradesh and Telangana)*

22.3 Sustainable Land Use by Kakatiyas

Building of irrigation system included construction of dams/lake bunds. These man-made structures were planned in a way that it does not hamper the existing ecological balance of the region. There are thousands of dams built by Kakatiyas, not a single one is on a river. Building dams on rivers is disastrous from a sustainable point of view as it obstructs the flow of the river causing silting and devastating the marine life there. Kakatiyan irrigation engineers were sensitive towards nature, which is evident from their interventions.

Fig. 22.2 Arial view: Rudreshwara Ramappa Temple at Palampet, Telangana and its environs

Fig. 22.3 Landscapes of Kakatiyas comprising majorly of Paddy Fields and Granitic Mounds. *Image Source* Author

22.3.1 Silting of the Lakes

1. The water tanks must have got silted and needed regular de-silting. In that case, the sludge removed from the lakebed would have been deposited in the fields nearby, enriching fertility of the soil, *or*

Fig. 22.4 Ramappa Cheruvu (Lake) and Lake Bund

2. As evident from first-hand observations of local people, most of these tanks have never been de-silted in the last seven centuries. Still, the need to de-silt any of these tanks was never felt as their efficiency never decreased (Fig. 22.4).

22.4 Spiritual Affinity Towards Nature

> The gods always play where groves are near, rivers, mountains and springs, and town with pleasure gardens.
>
> ...*Brihat Samhita,* LV. 8
>
> ...*Bhavisya Purana*, I. CXXX. 15

Religious scriptures and treatises on Indian Temple Architecture specify suitable locations for erecting consecrated spaces. To satisfy these scriptural norms, the rulers in the past built reservoirs near hills and farmlands. Later villages with mighty religious edifices adorned the landscapes and registered human presence.

In case of *Palampet* village in *Telangana* which houses the great *Rudreshwara Ramappa* Temple, there are *Shivalayams* built atop Lake bund and the Lingams in these temples face water. The temple is built over the dam to purify water which runs down into the lake from forests and wild-lands (Fig. 22.5). It is the natural entity-water, which is not physically but psychologically treated with a sacred presence near the waterbody. Such a setting where waterbodies are edged with consecrated places is common in Indian cultural landscapes (Fig. 22.6).

Like water and *cheruvus*, hills and stones also have a prominent role to play in cultural landscapes of Kakatiyas.

Many of the temples are built in the low-lying area at the foothills of the granitic mounds and thus have farmlands in the foreground and mountains in the background. The cultural wealth of Kakatiyas is reflected in these temples which are built either

Fig. 22.5 Shivalayam atop tank bund facing Ramappa Lake, Palampet, Telangana. *Image Source* Author

Fig. 22.6 Shivalayam atop Lake Bund, Palampet

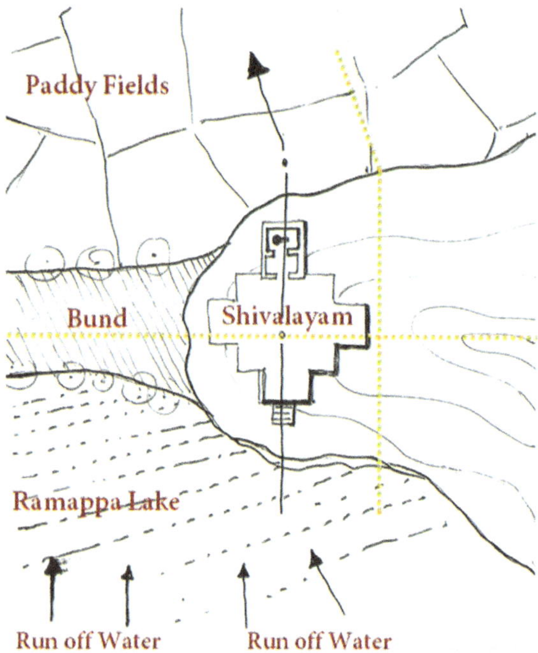

of granite or sandstone with sacred spaces built in black dolerite. The change in colour and texture (properties) of stone at specific locations in temples is symbolic. The combined effort of Kakatiyan *Sthapatis* (architects) and *Shilpis* (sculptors) in framing the culturally and spiritually pious atmosphere in a natural geographical setting is evident from the existing heritage sites of Kakatiyan era.

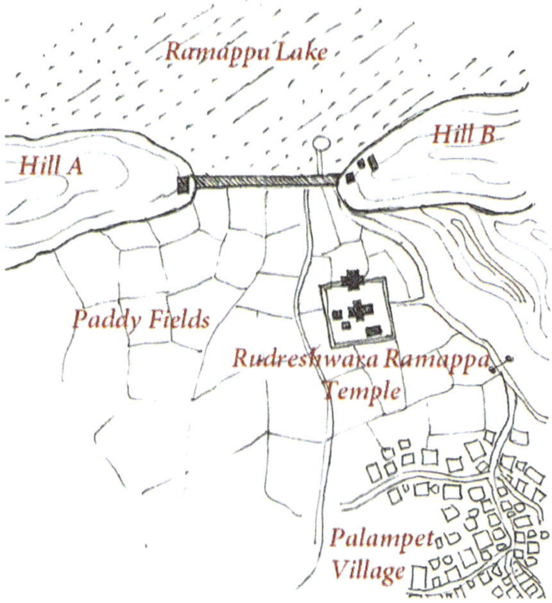

Fig. 22.7 Base Map, Ramappa Temple and its environs

22.5 Settlement Pattern of Kakatiyan Cultural Landscapes

The inter-relation between elements of naturally occurring geography and human contribution to this landscape is adamantine. It is a two-way relation. Even though settlers had a specific way to lay their footprint on the existing topography, the settlement pattern has to be understood and studied as a whole. The location of village and irrigated cultivation lands is actually a response to the context. The context which has monotonous peneplains and slopes with undulations is pierced with water channels that connect other Kakatiyan lakes creating a series of chain link tanks (Fig. 22.7).

22.6 The Skyline of Kakatiyan Cultural Landscape

A striking feature of Kakatiyan cultural landscapes is the common skyline which is visible in almost all the Kakatiyan sites. The skyline comprises culturally vibrant centre (temple complex) with paddy fields and palm trees, village huts, and a hillock in the background. Farmers in fields, men climbing the straight wine pine trees, grazing cattle herds, and water from the nearby lake being distributed in the fields via channels is what is seen in the living cultural landscapes of Kakatiyas. This setting is not temple-centric and the capacity of the villagers to self-organize and enact to the existing settlement is what fosters the link between culture and nature (Fig. 22.8).

Fig. 22.8 Skyline of a typical Kakatiyan cultural landscape

22.7 Protection of Degrading Cultural Landscapes

The closeness and dependence of human civilization on geographical assemblage and natural elements have always been established by the man around the globe and it is this relationship that has always ended up drastically, once the civilized man starts over-exploiting the existing resources which he had yet relied on. The sustainability quotient keeps falling and cultural landscapes transform into deserted ruins. To avoid such disasters, awareness regarding recognition and protection of cultural landscapes is necessary. Concepts like '*Eco-Cultural Landscape*' are needed to be emphasized on. Institutes and NGOs throughout the world have stepped forward in taking up this task and Kakatiya Heritage Trust, Warangal is one such body. The existing Telangana state government has also launched a public welfare movement called **Mission Kakatiya,** which aims primarily at rejuvenating water tanks and reservoirs throughout the state.

22.8 Mission Kakatiya

Mission Kakatiya is a programme for restoration of all the minor irrigation tanks and lakes in Indian state of Telangana. The programme helps in rejuvenating 46,531 tanks and lakes, storing 264 TMC water across the state in five years.

22.8.1 *Objective*

To enhance the development of agriculture-based income for small and marginal farmers by accelerating the development of minor irrigation infrastructure, strengthening community-based irrigation management and adopting a comprehensive programme for restoration of Kakatiyan Tanks.

Under Mission Kakatiyas, irrigation projects like lift irrigation, canals, aqueducts, barrages and flood banks are undertaken throughout the Indian state of Telangana.

Acknowledgements The authors acknowledge the work of Prof. M. Pandu Ranga Rao, coordinator of INTACH- Warangal chapter for last three decades and founder of Kakatiya Heritage Trust (a non-profit trust that endeavours to preserve, promote, protect and propagate the contribution made by the Kakatiyas to Telugu culture) along with retired IAS officer Mr. B. Papa Rao in 2009; and acknowledge the assistance of Mr. Manthan Joshi, Architectural intern, M/s Kshetra, Hyderabad in preparing the illustrations.

References

Balla SR. Chain Link Tanks- Kakatiya Dynasty
Chiba M, Jain S, Ghosh S, Mathur VB (2018) Cultural landscapes of Asia
Eck DL (2012) India: a sacred geography
Grover N, Singh KN (2004) Cultural geography—From and process
Pletcher K (2011) The geography of India—Sacred and historic places
Rao M, Pandu Ranga. Geo-technical splendour of the Kakatiyas
Reddy E. Sivanagi. Irrigation Technology of Kakatiyas
Silva KD, Sinha A (2017) Cultural landscapes of South Asia—Studies in heritage conservation and management

Chapter 23
Collection and Quantification of Leaf Litter from Urban Street Plantation: A Case Study on South-West Delhi, India

Deepanshi Gandherva and Prodyut Bhattacharya

Abstract Plantation of urban street trees is being actively promoted as a forecast to facilitate urban areas to adapt and mitigate against climate change, augment urban sustainability, reduce air and noise pollution and avenue beautification. Urban biomass from green areas is a potential resource for bioenergy recovery, which is widely unused in India. Different types of organic material (e.g., grass, leaf litter) produced play an important role in ecosystem services after decomposition in soil, which usually occur in mixtures due to common collecting practice. Leaf litter produced and shed from roadside trees are an increasing concern and burden to the city administration, which has the responsibility to dispose the autumn foliage. Despite being banned (Municipal Solid Waste (Handling and Management rules) 2000), this leaf litter undergoes illicit burning in the city which causes pollution. Delhi itself is experiencing a very bad phase of air pollution, hence, such activities pronounce the effect. Hence, this study focused on methods of collection and quantification of the leaf litter produced from selected road transect in South-West Delhi. The methodology utilized here in the collection of the leaf litter was the sweeping method accompanied by weighing at the collection centre. The collection was conducted in collaboration with South Delhi Municipal Corporation in the post leaf fall period (March to mid-May). The total length of the transect covered was 11 km and the average leaf litter collected per day was 68.7 kg during the peak time with maxima and minima being 105.5 kg and 25 kg, respectively. Since leaf litter produced by roadside trees is an increasing burden to the city administration, the problem could be tackled effectively if the collected leaf litter can be converted to useful manure and solid fuel.

Keywords Urban leaf litter · Collection and quantification methods · Solid waste management · Bioenergy · Urban street plantations

---

D. Gandherva (✉) · P. Bhattacharya
University School of Environment Management, Guru Gobind Singh Indraprastha University, Sec 16 C Dwarka, Delhi, India
e-mail: deepanshi.ganderva93@gmail.com

P. Bhattacharya
e-mail: prodyutbhattacharya@yahoo.com

R. K. Mishra et al. (eds.), *Sustainable Climate Action and Water Management*, Advances in Geographical and Environmental Sciences,
https://doi.org/10.1007/978-981-15-8237-0_23

23.1 Introduction

Leaf litter characteristics and its decomposition is well-studied in natural ecosystems because of its role in ecosystem carbon budgets, Soil Organic Matter (SOM) formation and nutrient cycling (Hobbie et al. 2013). But, leaf litter decomposition has been studied less in urban environments, and past terrestrial urban decomposition studies mostly emphasized on comparisons along urban to rural gradients (McDonnell et al. 1997) and on the effects of urban pollution on decomposability and litter chemistry (Fritz 1988; Cotrufo et al. 1995; Carreiro et al. 1999). Lesser well known is how decomposition profits in the environments shaped by urbanization. Urban areas are characterized by high impermeable surface (Nowak and Greenfield 2012), such that substantial leaf litter in the urban areas falls on streets/parking lots/blown away. Solid waste management is a major problem in most of the developing countries. The organic portion of the solid waste can be decomposed or recycled into beneficial products. Dumping off the huge quantity of leaf litter accumulation in the forests, along the streets, rail lines, etc. has always been a major problem. Leaf litter is often piled up and set on fire in India and several other countries in the southern hemisphere (Singh et al. 2017) and it adds on to air pollution (Abbasi 1999).

Urban biomass from green areas is a potential resource for bioenergy recovery, which is widely unused. Different types of organic material (e.g., grass, leaf litter) usually occur in mixtures due to common collecting practice. Leaf litter produced and shed from roadside trees (an integral part of the urban ecosystems) are an increasing concern and burden to the city administration, which has to dispose the autumn foliage (Nurmatov et al. 2016). The disposal is not only cost-intensive but also an organizational challenge. Once urban leaf litter is collected, it is considered as a waste. Leaf litter is usually collected by municipalities for composting, burial or direct incineration (Springer 2012). Despite being banned (Municipal Solid Waste (Handling and Management rules) 2000), this leaf litter undergoes illicit burning which causes pollution. Delhi (study area) itself is experiencing a very bad phase of air pollution, hence, such activities pronounce the effect. Some alternatives must be adopted to reuse the leaf litter beneficially and hence reducing the pollution. There are various studies conducted in Delhi in reference to urban biodiversity, green space, in relation to pollution abatement, Policy, law and legislation, societal contribution and cultural aspect (Bhattacharya and Nigam 2010). Preferred species of trees are planted on the roadside by Delhi Government which are large canopied, tall and doesn't require significant pruning (Nagendra and Gopal 2010; Bhalla and Bhattacharya 2015).

Conversion of the accumulated leaf litter into high-quality solid fuel would be an appealing option as this would offer two main advantages to municipalities: leaf litter would be converted from waste into a product, which could be sold and thereby reduce the costs for leaf litter removal and also energy is provided for local demand (Hensgen et al. 2011). Evidence of leaf litter use for energy production is scarce as yields are low with conventional anaerobic digestion techniques (Liew et al. 2012). But efficiency can be increased by using some other methods as discussed in this proposal. The utilization of urban leaf litter from roadside plantation has many other advantages, like it reduces the volume of waste to be incinerated, reduces adverse environmental impacts, decreases disposal costs, etc. (Department of Environmental Protection, Boston 2016).

The forests in Delhi comprise of Delhi Ridge, mini forests maintained by government agencies and a larger fragmented green area that is non-forested but tree dominated. Such green area occurs in patches which include trees in parks, gardens, avenue trees, trees along streets, trees within the premises of historical monuments, University/Institute/Educational campuses, landscapes, posh office areas, places of beautification, etc. (Vegetation of Delhi 2017).

Considering the harmful effects of increased carbon emissions in Delhi, particularly, vehicular pollution, it became imperative to evaluate the role played by such fragmented green areas. It has been discussed that leaf litter produced by roadside trees is an increasing burden to the city administration bodies. The problem could be tackled effectively if the collected leaf litter can be converted to useful manure and solid fuel.

23.2 Methodology

23.2.1 Study Area

The geographical area of Delhi is 1483 sq. km which is 0.05% of the total geographical area of the country. The river Yamuna, the Aravalli range and the plains in between both of these form alluvial deposits of recent formation and dominate the physiography of Delhi (State of Forest ReportS 2017). The rainfall of Delhi ranges between 400 mm to 600 mm. During the summer months of April, May and June, the temperature can rise up to 40–45° C while winters are typically cold with temperatures during December and January falling to 4–5 °C. It has 9 districts and none are hill or tribal. Figure 23.1 gives the study area map. In this study, the Dwarka zone of South-West Delhi was covered and three sectors, Sectors 13, 14 and 16, were undertaken.

23.3 Collection of Leaf Litterer

The method used for the collection of leaf litter from the selected sectors of Dwarka was the sweeping method. The sweeping technique is usually applied on compacted surfaces. The collection was conducted in collaboration with the concerned civic authority of the area, South Delhi Municipal Corporation (SDMC). The leaf litter was collected from the ground of the selected roads in three sectors of Dwarka (Sector

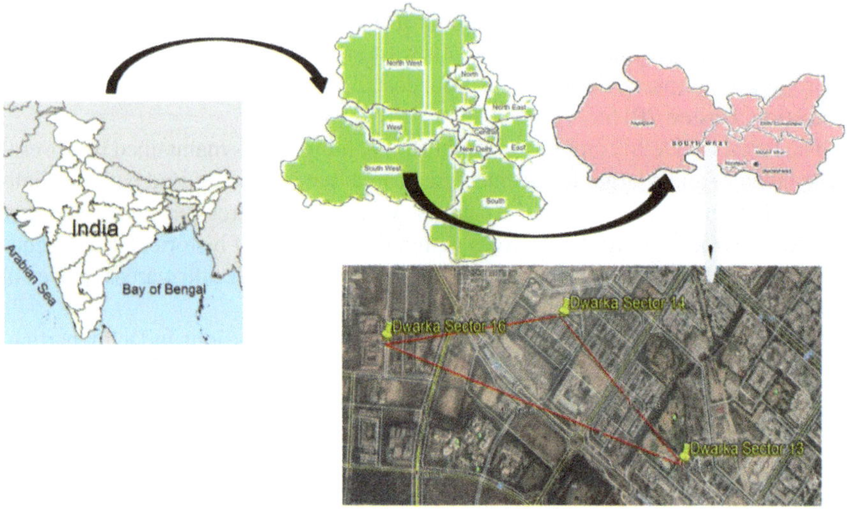

Fig. 23.1 Study

13, 14 and 16). The collection took place every Friday at the designated collection centre of the SDMC. The collection period was March 1 to May 17, 2019.

23.4 Result and Discussion

Collection rounds involved multiple road types, major roads, pavements, residential streets and few minor roads. Table 23.1 shows the road types in the sampled area and the percentage sample (leaf litter) collected from these roads.

Collections most commonly swept from the road surface, around 70%, followed by residential streets. Samples of street leaf sweepings were submitted for analysis of physical contaminants. The collected leaf litter consisted of contaminants like glass, plastic, metal, stones, etc. These contaminants were removed manually from the sample. The major contaminant in the collected sample was plastic. The samples were measured at the collection centre using a weighing machine. For the conservation of the sample, they were filled into airtight polyethylene barrels for processing at a later stage (Fig. 23.2).

Table 23.1 Road type and percentage leaf litter collected from these roads

Road type	Percentage leaf litter collected from these roads
Major roads	48%
Residential street	31%
Minor Road	21%

Fig. 23.2 Exemplary display of leaf litter on the streets and its collection

The total length of the road trail covered was 11 km which included major-minor roads and residential streets. The leaf litter was collected for eleven consecutive weeks covering from March 1 to May 17, 2019 depicts the amount of leaf litter collected during this period. This clearly shows how the quantity of collected leaf litter declines from March to May. The maximum leaf litter was sampled during the month of March with total 371.6 kg of leaf litter. The week which showed maximum leaf litter collection was the second week of March with 105.5 kg of collection, and the least being collected in the last week, i.e. eleventh week, from May 11 to May 17 with 25 kg of total leaf litter collection. The total leaf litter collected during the sampling period was 668.4 kg which is quite significant. A longer winter has delayed the shedding of leaves in Delhi this year (2019). The process of shedding of leaves should have been over by March end but this year the roads were littered with leaves for a longer period. Temperature and humidity play a major role in the shedding process of leaves.

Graphs 23.1 and 23.2 shows the leaf litter collected during the months of March, April and May. Clearly, March month shows the maximum amount of leaf litter collected, followed by April (194.9 kg) and May (101.9 kg).

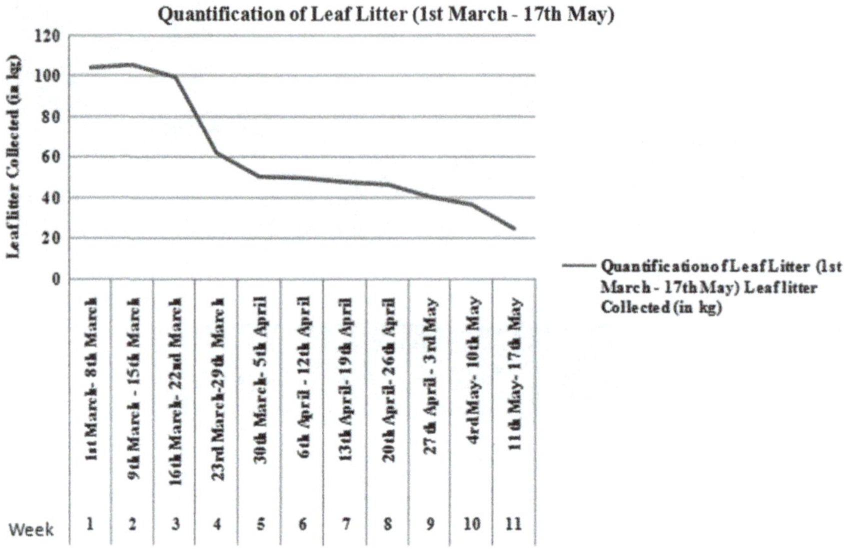

Graph 23.1 Quantification of leaf litter from March 1 to May 17

Graph 23.2 Total leaf litter collected during the months of March, April and May

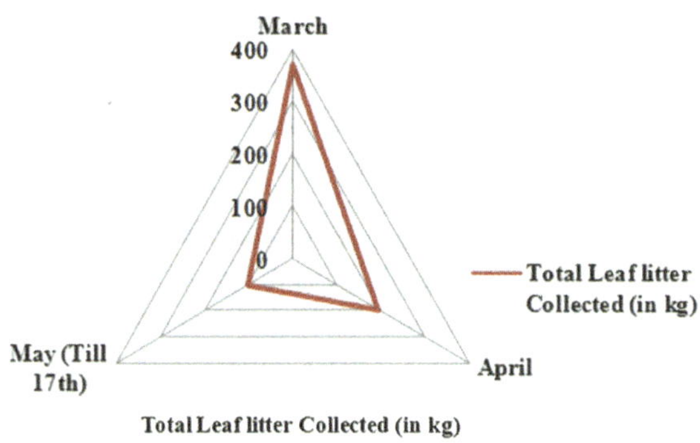

23.5 Conclusion and Recommendation

Urban biomass from green areas is a potential source for bioenergy recovery which is widely unused. By investigating leaf litter from a full season of leaf fall, it is found that urban environments produce a significant amount of leaf litter which can be utilized for various useful purposes. This study only conducted collection followed by quantification of the leaf litter from the roadside plantations and residential streets. The further analysis (both physical and chemical) of this collected leaf litter must be carried out. This collected biowaste can be converted into compost which can further be utilized by city dwellers for organic farming. Managing leaf litter by composting is an alternative approach to waste management since the leaf litter is neither burnt nor landfilled. This would also prevent from illicit burning of leaf litter, and hence reduce the air pollution. Cities lack such quantification studies which should be carried out. In general, leaf litter is clubbed up with other types of biodegradable wastes which leads to improper quantification of this important unused biomass. The economic, social and ecological implications of such implementations should be taken into account sincerely.

Acknowledgements The authors thank the institutional support provided by the Dean, University School of Environment Management, Guru Gobind Singh Indraprastha University, Delhi. Thanks are due to the DST, New Delhi, for financial support under INSPIRE Junior Research Fellowship Scheme to Deepanshi Gandherva. The authors are also thankful to SDMC authorities for giving the required permissions to carry out the work. The contents and views expressed in this research paper are the views of the authors and do not necessarily reflect the views of the organizations they belong to. The author declares no conflict of interest.

References

Abbasi SA (1999) Environmental pollution and Its Control. Cogent Int, Philadelphia and Pondicherry

Bhalla P, Bhattacharya P (2015) Urban biodiversity and Green Spaces in Delhi: a case study of new settlement and Lutyens's Delhi. J Hum Ecol 51:83–96

Bhattacharya P, Nigam V (2010) Status Paper on Urban Forest and Biodiversity. Paper presented in the International Conference on Urban Forestry and Biodiversity organized by AFE, New Delhi, India, February 25 to 26, 2010

Carreiro MM, Howe K, Parkhurst DF, Pouyat RV (1999) Variation in quality and decomposability of red oakleaf litter along an urban–rural gradient. Biol Fertil Soils 30:258–268

Cotrufo MF, De Santo AV, Alfani A, Bartoli G, De Cristofaro A (1995) Effects of urban heavy metal pollutionon organic matter decompositio in Quercus ilex L. woods. Environ Pollut 89:81–87

Department of Environmental Protection (2016) Leaf and yard waste Composting guidance document. Bureau of Waste Prevention, Boston

Fritze H (1988) Influence of urban air pollution on needle litter decomposition and nutrient release. Scand J For Res 3:291–297. State of Forest Report 2017. Executive summary. http://fsi.nic.in/cover_2017/executive_summary.pdf. Accessed 29 Aug 2019

Hensgen F, Richter F, Wachendorf M (2011) Integrated generation of solid fuel and biogas from green cut material from landscape conservation and private households. Bioresour 102:10441–10450

Hobbie SE, Baker LA, Buyarski C, Nidzgorski D, Finlay JC (2013) Decomposition of tree leaf litter on pavement: Implications for urban water quality. Urban Ecosyst 10(9):1–17

Larondelle N, Haase D (2013) Urban ecosystem services assessment along a rural–urban gradient: A cross-analysis of European cities. Ecol Indicat 29:179–190

Liew LN, Shi J, Li Y (2012) Methane production from solid-state anaerobic digestion of lignocellulosic biomass. Biomass Bioenergy 46:125–132

McDonnell MJ, Pickett STA, Groffman P, Bohlen P, Pouyat RV, Zipperer WC, Parmelee RW, Carreiro MM, Medley K (1997) Ecosystem processes along an urban-to-rural gradient. Urban Ecosyst 1:21–36

Municipal Solid Waste (Handling and Management rules) 2000, Ministry of Environment and Forest Notification. http://www.moef.nic.in. Accessed 12 April 2019

Nagendra N, Gopal D (2010) Street trees in Bangalore: density, diversity, composition and distribution. Urban For Urban Green 9:129–137

Nowak DJ, Greenfield EJ (2012) Tree and impervious cover change in U.S. cities. Urban For Urban Green 11:21–30

Nurmatov N, Gomez DAL, Hensgen F, Buhle L, Wachendorf M (2016) High quality solid fuel production from leaf litter of urban street trees. Sustainability 8:1–13

Singh S, Yadav I, Juneja SK (2017) Leaf litter utilization through composting: a review. J Agrofor Nat Resour Manag 4(5):398–402

Springer TL (2012) Biomass yield from an urban landscape. Biomass Bioenergy 37:82–87

Vegetation of Delhi, Forests of Delhi. https://delhi.gov.in. Accessed 19 May 2019

Chapter 24
Enablers of Sustainable Development of Randullabad Watershed in Satara District of Maharashtra State, India

Bharat Kakade

Abstract The Watershed development approach has proved to be successful in conserving the natural resources for better livelihood of people in rainfed areas of India. The impact of different programmes funded through Government, International agencies or corporates is good; however, the sustainability issues have remained unaddressed. There have been some successful examples of sustainable development of watersheds, and the Randullabad watershed happens to be one such successful case. The Randullabad watershed project in Satara district of Maharashtra state, implemented under Indo-German Watershed Development Programme (IGWDP) Phase-III, was assessed for sustainability using new methodology suggested by Kakade (2017). The new methodology based on the principles of 'strong sustainability' of FAO is advocated for sustainability assessment of 'new generation holistic watershed development projects' in India. Under the social domain, sustainability scores were 0.45 at baseline (2008), 0.66 at project completion (2013) and 0.79 after five years post-project (2018); under economic domain, these scores were 0.50, 0.87 and 1.27 while under ecological domain, scores were 0.46, 0.86 and 1.07 respectively. These scores are against the threshold level of 1.00 for each domain. Watershed Sustainability Index (WSI) improved from 0.47 in 2008 to 0.79 in 2013 and to the level of 1.02 in the year 2018 against the threshold of 1.00. Since the WSI in 2018 was very high, it was thought necessary to ascertain the rationale of high level of sustainability of watershed. This paper throws light on the factors contributing to the sustainability of the Randullabad watershed. This analysis of the parameters contributing to the sustainability of watershed revealed the dominance of dynamic and holistic project design, continued investment, local leadership and institutions, role of external facilitating organizations, creation of remunerative value chains, hybrid model of politicization of watershed governance and shift towards the circular economy.

B. Kakade (✉)
BAIF, Development Research Foundation, Warje, Pune, Maharashtra, India
e-mail: bkkakade@baif.org.in

R. K. Mishra et al. (eds.), *Sustainable Climate Action and Water Management*, Advances in Geographical and Environmental Sciences,
https://doi.org/10.1007/978-981-15-8237-0_24

Keywords Sustainable watershed · Social · Economic and ecological indicators of sustainability · Participation of people · Politicization of watershed governance · Circular economy

24.1 Introduction

Sustainable Watershed Development is the "Development and management of watershed ecosystem and the use of watershed services for socio-economic benefits to community without depleting the resources through socio-cultural and institutional mechanism."

Many frameworks have been developed and used to assess the watershed sustainability across the globe. The approaches have been focussing on a certain sector, say for example, some focussed on the livelihood or economic growth of communities, some on ecological or environmental conservation and some on hydrological aspects. Some frameworks have adopted Human Development Index (HDI) for assessment. Water Poverty Index (WPI), Environmental Sustainability Index (ESI) and UNESCO's international Hydrology, Environment, Life and Policy (HELP) to define watershed sustainability index (WSI) (Chaves and Alipaz 2007) are the few other frameworks. Some researchers used some indicators based on the project objectives. The design and composition of conservation measures, the mode or approach of implementation, resources and its user interactions, governance mechanism and longer term continuation need to be taken into account while understanding the sustainability. The facets of sustainability should reflect in the assessment framework.

Therefore, a necessity was revealed for a methodology encompassing different facets of ecosystem, socio-economic aspects and externalities to have sustainable watershed design, execution and sustainability assessment approaches of watershed development projects. With this background, Kakade (2017) proposed a new methodology for sustainability assessment of recent holistic watershed development projects in India. This methodology has been adopted in assessing the sustainability of the Randullabad watershed project.

24.2 Profile of Randullabad Watershed

Randullabad village is located between Latitudes and 17°50′ E to 18°05′ E and Longitudes of 74°10′ N to 74°20′ N in Satara district of Maharashtra state in India. Figure 24.1 gives the location of the Randullabad watershed.

Randullabad is a drought-prone village and the rainfall pattern is irregular. The annual average rainfall of Koregaon station, which is 38 km away, for the period 1999 to 2017 was 668.49 mm. The rain gauge station was established in Randullabad village in the year 2010. The records of the Randullabad station from 2010 until 2017

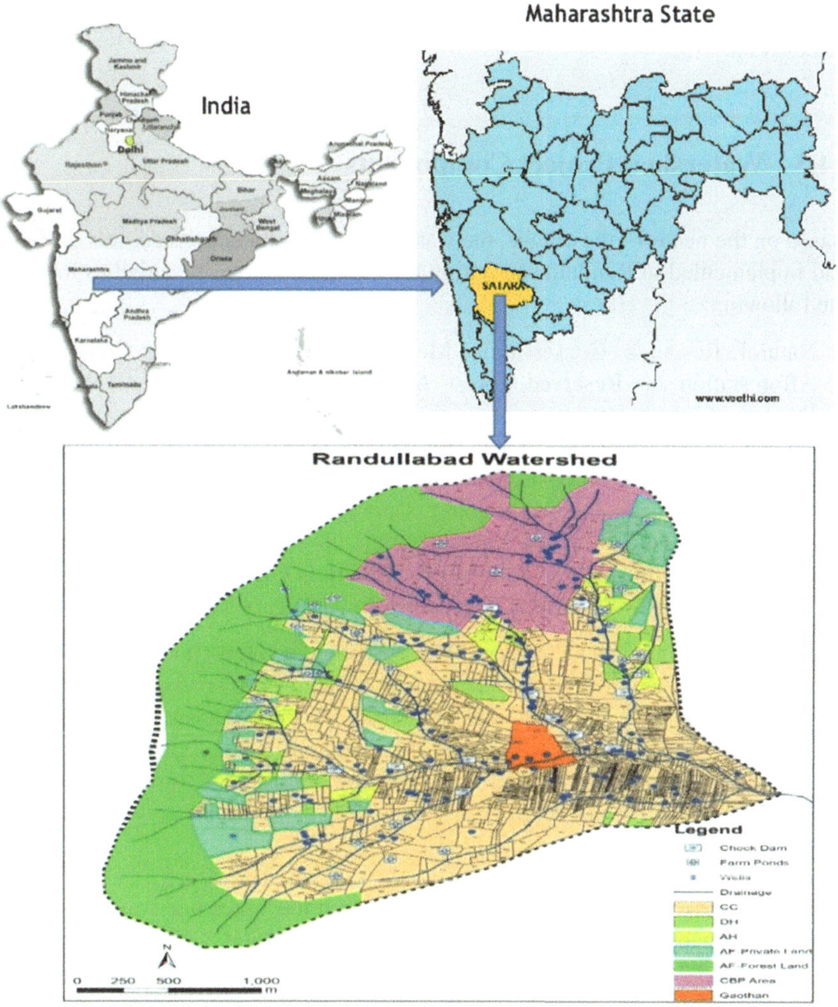

Fig. 24.1 Location Map of Randullabad Watershed

show the average annual rainfall of 648.40 mm, slightly less than that of Koregaon station.

Randullabad is a medium size village with an area of 836 ha and a total of 395 families. It has a population of 1857 of which 914 are males, while 943 are females as per Population Census 2011. Landless (3%) and small and marginal landholders form 85% of the population. The literacy rate was 86.94%, which was slightly better than the Maharashtra state average of 82.34%. The male literacy was 92.71% while the female literacy rate was 81.29% (Census 2011). There were just two farmers holding above 8 ha land. In the year 2018, the Schedule Castes (SC) constituted 5.3% while Schedule Tribes (ST) were 0.76% of the total population. Other Backward Caste communities were about 12.18%. There was also a single household of Nomadic Tribe (NT).

24.3 Watershed Project Components

Based on the needs of the village, the watershed development project was designed and implemented in Randullabad during the year 2008–13. The activities included the following:

- Natural Resource Conservation Measures: Soil and Water Conservation, Afforestation on Reserved Forest Area, Crop Cultivation (CC) Measures, Drainage Line Treatment
- Livelihood Measures: Dry Land Horticulture (DH), Agro- Horticulture (AH), Dairy Husbandry, Home-based income-generation activities
- Social (Awareness and institutions): Community awareness and stakeholder engagement, Formation and Strengthening of Village Watershed Committee (VWC), Formation and Strengthening of Self-Help Groups (SHGs) and *Sanyukta Mahila Samiti* (SMS)—Women's Federation, Drudgery Reduction Activities.

Project was implemented during 2008–2013 by the Gram Panchayat of Randullabad village through Village Watershed Committee (VWC) named '*Jai Hanuman Panlot Vikas Samiti*' and with the facilitation of BAIF Development Research Foundation, Pune as Resource Support Organization (RSO). Financial support to the project was through National Bank for Agriculture and Rural Development (NABARD) under the Indo-German Watershed Development Programme (IGWDP) Phase-III.

As already mentioned, a sustainability assessment of the project was conducted during the year 2018 using a new framework, and the watershed was found to be highly sustainable for five years after the project completion.

24.4 Objective of Study

The objective of this paper is to ascertain the enabling factors of sustainability of the Randullabad watershed.

24.5 Methodology of Study

The research design and indicators used for the study were adopted from the sustainability methodology suggested by Kakade (2017). The field study was conducted during November 2017 to June 2018. Based on the data requirement and expected sources of information, a combination of different tools was used, which include the following:

i. Secondary data collection through *Grampanchayat*, Agriculture department, Census 2011.
ii. Household socio-economic survey based on the pre-decided questionnaire. Out of 394 households, 125 households were randomly selected from the list of households available with the Village Watershed Committee (VWC). This is about 30% of the sample size. The data analysis has been done for these 125 households; however, the income data could be analyzed only of 110 households due to inaccuracy and gaps in the data of 15 households.
iii. The baseline socio-economic data of the year 2008 of the households selected for the survey in 2017–18 was available with BAIF. This was used for comparison of the status of different indicators pre-project and present.
iv. Site visits were undertaken to capture and verify the status of water harvesting structures, farm ponds, check dams, vegetation and other soil and water conservation measures.
v. Focus Group Discussions (FGD) with members of Village Watershed Committee (VWC), dairy cooperative, *Grampanchayat*, Sanyukt Mahila Samiti (SMS) and SHGs, villagers and students.
vi. The records of VWC, including the proceedings in a minute book maintained by VWC, transactions of maintenance fund, the records of baseline surveys, village profile and watershed treatment measures were studied and obtained for onward analysis.
vii. BAIF and VWC records on the project activities, convergence activities, meteorological data, groundwater data and other study reports available with *Grampanchayat* and BAIF.

The detailed socio-economic survey was conducted during the period from November 2017 to June 2018 capturing the data of all three agriculture seasons of the year 2017–18. Several rounds of group discussion with community members were conducted during this period in addition to household surveys. The group discussions were organized with the members of *Grampanchayat*, 'Milk Cooperative', Village Watershed Committee, 'Samyukt Mahila Samiti' and other key local leaders from 31 August to 1 September, 2018 in the village. After the entire data was compiled and analyzed, the interaction with key persons of Randullabad was arranged on 10 December, 2018 at the BAIF office in Pune to validate the information, to fill up the missing information and also to inform the villagers about the primary outcomes of the study.

The sustainability indicators have been classified into three key domains viz. social, economic and ecological. The methodology of data analysis followed is as under:

- The sustainability indicators have different units. For independent analysis and comparison across the years, these units could be used, but for comparison of indicators on a common platform, say for web diagram within the groups as well across the year, all the indicators had to be converted at a normalized scale using distance to target method of normalisation. The indicator targets of this study are based on the targets of Sustainable Development Goals (SDG)

- Weights of the indicators as derived through the participatory weighting method has been adopted to ensure strong sustainability.

Finally, for each of the domain, the indicator-wise comparative sustainability analysis for the years 2008, 2013 and 2018 is presented using a web diagram.

24.6 Sustainablity of Randullabad Watershed

Since the paper aims at identifying and critically commenting on the factors, which contributed to higher level of sustainability, it is necessary to first present the base on which the discussion has to happen. Therefore, the overall sustainability analysis considering the measurable indicators, which was done by Kakade (2019) under the three main domains of social, economic and ecological, in the form of a web diagram is given below. Parameters used for the development of web diagram are as follows:

i. **Social**—Community awareness, Participatory approach, Access to resources, Equity (resource class, social and gender), Education, Conflict management and Institutions (capacities of institutions, water use efficiency, Sustainability funds)
ii. **Economic**—Family income, Distribution of benefits, Migration, Human-induced capital or assets, Farm production and diversification, Access to credit
iii. **Ecological**—Soil conservation (wasteland management), Siltation in water reservoirs (condition of reservoirs), Water (storage capacity, area under irrigation in Rabi), Water use efficiency, Groundwater status, Forestry (vegetation cover). Based on the field study observations, the factors contributing to the higher level of sustainability scores in the year 2018 have been identified and critically commented in the succeeding section.

The weighted values of indicators are aggregated at the level of Social, Economic and Ecological domains. The sustainability scores of each domain are derived through weighted arithmetic mean. The web diagram constructed using the aggregated sustainability scores of social, economic and ecological domains is as given in Fig. 24.2.

The web diagram in Fig. 24.2 shows a higher level of sustainability in all three domains, social, economic and ecological, in the year 2018. Watershed has weak sustainability scores in 2008 when the project was initiated. The values in the year 2013 appear to have achieved a medium level of sustainability. The overall sustainability index (the geometric mean of social, economic and ecological domains) of the year 2008 is 0.47, for 2013 it is 0.79 and for the year 2018, the sustainability index is 1.02 against the threshold sustainability index of 1.00.

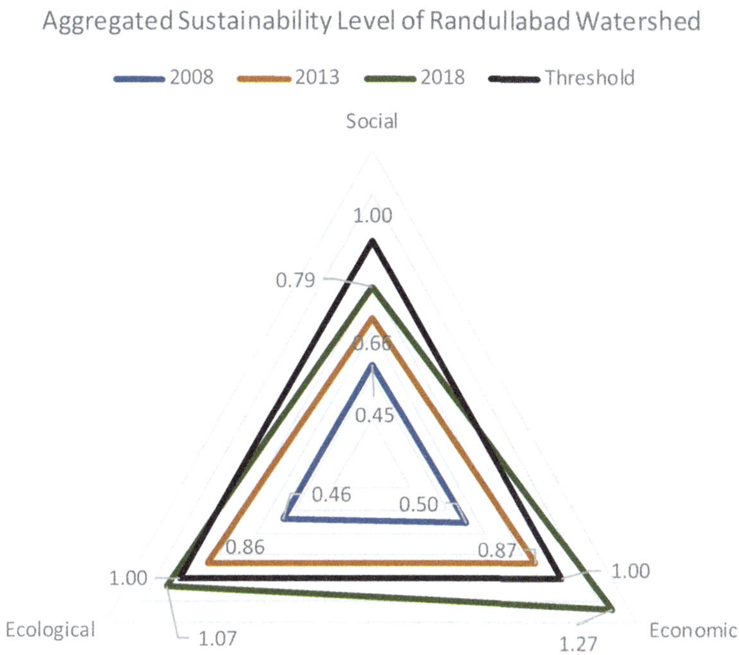

Fig. 24.2 Web diagram—Randullabad Watershed Sustainability in different years

It is evident that the economic growth is non-exploitative, as the economic growth has been achieved along with the improved performance of ecological indicators. The improvement in the natural resource base, judicious use of the resources and good governance has ensured non-exploitative economic growth.

Sustainable development of Randullabad continued as the resource base created continued to provide additional productivity without depletion, rather the natural resource base appears to be enriched post-project. The Watershed community is able to have enough livelihood generation through the watershed. The migration is restricted only for skilled employment outside, while there has been in-migration of labour as well as creation of wages for surrounding villagers.

It is necessary to know the factors contributing to the synergy among sustainability domains and continued improvement in sustainability levels after the project. The following section critically looks at the enabling parameters of sustainability of the Randullabad watershed.

24.7 Enablers of Sustainability

The enabling parameters have been identified based on the field studies conducted. The parameters are classified into the broad categories of Social (including institutions), circular economy, politicization of watershed development, role of external organizations and project design. Although these are interrelated, but for the sake of organizing the discussion, they are presented under each of the category.

24.7.1 Social Parameters

The aggregated sustainability score of the social domain as seen in Fig. 24.2 is very strong after the project. Participatory processes in project implementation and post-project management, good governance systems through capable institutions and local leadership may have contributed to enhanced social capital in the village.

24.7.2 Participation of Community and Equity

Initially, it took a lot of time and effort for BAIF (as RSO) and Randullabad *Gram Panchayat* (as PIA) to mobilize the community and Village Watershed Committee (VWC), but it gained momentum over a period and continued even after the project completion. The question is what led to the generation of impetus and such continued participation? As seen from the project reports and minutes of VWC meeting, one of the factors appears to be initial awareness generation measures and exposure visits of farmers by GP. During the project implementation, since most of the land-owning community had direct benefits of soil and water conservation and some livelihood activities and skill training to landless, both came forward and the project could be implemented well. However, a major portion of the development of the village took place after the completion of IGWDP. The interests of all sections of the community were addressed through various measures during this phase and these measures were not only related to natural resource conservation and management but also to other developmental measures for improving the overall well-being of people and ecosystem. There were tangible benefits to all sections of the community in terms of access to drinking water, sanitation, water for agriculture, housing scheme, local employment in farms and sources for higher income.

The overall annual average income was increased from ₹97,431.82 in 2008 to ₹175,174 in 2017–18. The important fact to be noted is that the benefits were for both resource-less and resource-rich households. The primary source of income of the landless household was shifted from wages and service to business, while that of small and marginal farmers shifted from service to agriculture. Overall, there was a shift from service as a major income source to agriculture. This could be one of the major reasons for people supporting the programme and also helping to maintain the resources.

Dr. Mahesh and Dr. Prajakta Jagtap, being the health practitioners, were able to provide health services in the village and also extend the services through medical camps. This helped Prajakta Jagtap to influence the village women to undertake drives such as weekly village cleanliness camps and make women participate in all mainstream community initiatives. The women were actively involved in providing the toilets to all the families to achieve ODF (Open Defecation Free) village status.

The representation of women is noticeable in the key village institutions. All major institutions including GP, VWC, SMS and Cooperatives had women members. It was, however, strange to note that landless and Scheduled Caste people were not part of the VWC and SMS. During the field study by the author, it was brought to the notice of the leadership in the village and tried to motivate for inclusiveness in membership. The result was that the landless were included in the committees, but the representation of SCs could not be done both in VWC and SMS. Thus, the issue of true democratic representation in the village institution could not be addressed. The deep rooted caste system and discrimination in India were also seen in Randullabad and the complexity in slackening the same was experienced in Randullabad.

Except for this caste-based complexity reflecting in decision making through village institutions, a substantial degree of equity in accessing the resources seems to have been achieved in Randullabad. Thus, the equity in sharing the benefits seems to be more through add-on development interventions or 'watershed-plus' interventions. Sangameswaran (2006) also drew a similar conclusion regarding equity in the village Hiver Bazar in the state of Maharashtra (India). As per Sangameswaran's observation, the equity outcome in Hiver Bazar was better than many other watershed programmes, mainly due to the use of watershed-plus and other attenuating measures.

24.7.3 Local Leadership

The developmental initiative by GP and VWC and the consultative approach of the members in planning and executing the development activities were quite evident in the village. In the process of implementation of interventions, the interactions of people taking lead were increased with other farmers as well as with people settled in nearby cities. The visible impact of the project could be one of the contributing factors towards the inspiration of lead members as well as the community. These members in managing project were also inspired due to encouragement from the people in general. Unlike in past, when there was only one local institution, i.e. *Gram Panchayat*, during and post-project, there were several local institutions created within the village. VWC, SMS, development society, credit cooperative and milk cooperative were there for enthusiastic members to join and take an active part in the development process. In addition, SHGs were there for women to join. All this helped flourish the leadership among many men and women in the village.

Over a period of ten years (from 2008 to 2018), about 25 to 30 young villagers had become active in the development process. It included both men and women. The education in the village was another factor in favour of nurturing such leadership. The GP *Sarpanch* (*Head of Gram Panchayat*) in 2018 was an advocate, while SMS chairperson (Dr. Prajakta Jagtap), a doctor by profession, was later elected as *Sarpanch*. Both had already a good exposure to social service. The local teams helped a lot in mobilizing the resources from outside for the village. The leadership skills and the number of skilled village leaders were evident from the way the conflicts were handled throughout the project period and post-project and the conflict management systems laid down.

The latent potential among the local progressive community members got an opportunity to excel during the project phase and they could also deliver the results after the project.

24.7.4 Local Institutions

VWC implementing the watershed development activities is a routine and normal practice. However, the implementation by VWC under the leadership of GP was exceptional and successful too. VWC could do a remarkable work in the implementation of watershed development activities with good quality. They together did an exceptional contribution in mobilizing the funds for complimenting the conservation measures under watershed development project, enhancing the watershed productivity and water use efficiency. Moreover, several development initiatives were taken up during a short span of four years after the completion of watershed project. The funds mobilized by VWC and *Gram Panchayat* during the project period 2008–13 itself was ₹8,192,478/- (Indian Rupee = US $ 69 in July 2019) and post-project period until 2018 was ₹10,252,878/-. Overall, the resources worth ₹18,445,356/- were mobilized. Both VWC and SHGs had acquired the capacity to mobilize resources for development in the village. The maintenance of the resources developed and its judicious use was the most important aspect for ensuring the longer term benefits, which were also well attended by the GP and VWC.

Since the upper mountainous catchment belongs to the forest department, Joint Forest Management Committee (JFMC) was formed in the village. JFMC's role was to plan and execute the soil and water conservation and plantation in the forest area. The efforts of JFMC in conservation and adoption of LPG by most of the village families together had impacted positively in improving the vegetation. As it was observed from the satellite images, vegetation cover was increased from 1.03 sqkm in 2008, 1.55 sqkm in 2013 to 2.62 sqkm in 2018. The major increase was found to be in the forest area, while there was some change in riparian vegetation in the downstream area.

In addition to the able statutory institution—GP, JFMC— and watershed project related institutions—VWC, SMS— the other institutions such as dairy cooperatives and credit cooperatives had also complimented the developmental initiatives as well as economic growth in the village. As mentioned under the section of 'local leadership', these institutions also had provided the platform for the local leadership to grow.

The role played by facilitating organization—BAIF, was also crucial in resource mobilization. It is also important to note that the BAIF continued its association with Randullabad even after the project period to facilitate the overall development. This is elaborated further in the section under role of external organizations.

24.7.5 Regulatory Measures

In addition to the implementation of project, the village already had a culture of tapping and using only the groundwater from a shallow aquifer through open wells. The control on over-exploitation of groundwater was followed strictly, as no single borewell was allowed for extracting the water for agriculture. Curtailment of water-intensive crops, managing the drinking water supply throughout the year and introduction of water-saving devises—drip and sprinkler— were undertaken after project completion. The control on over-extraction, improved water use efficiency and control on the cultivation of water-intensive crops were the key contributors in ensuring the access of water for all the farmers in the watershed.

Any developed watershed can face water shortage in summer if there is no control on the exploitation of aquifers. Although it is the most difficult, but very important measure of water regulation in the village, which actually helped prohibit tanker-supply for drinking even in drought years. Training by the organization ACWADAM on groundwater budgeting and its controlled use had helped villagers acquire the skill of managing the available water resources.

24.8 Towards Circular Economy

Circular economy aims at providing the societal benefits in a restorative or regenerative way by design. First, the biological ingredients or nutrients of consumables can be safely returned to the biosphere, either directly or in a cascade of consecutive uses. Second, the durable assets are designed from the start for reuse or upgrade, and third, the energy required to fuel this cycle should be renewable to increase the resilience (WEF 2019). With the shifting of sources of income from secondary to primary source (agriculture) and increased use of local resources, recycling and renewable energy along with improved ecosystem, the Randullabad watershed appears to have moved towards circular economy.

A major shift was observed in the income source of small and marginal farmers in Randullabad from 'service' based to 'agriculture' based post watershed project. While income source of landless families was shifted from 'service' and 'other sources' to small businesses. The major source of income in village was found to be agriculture in 2018. In alignment with the principles of circular economy, the villagers had used the local resources (land, water and livestock) for economic growth, but at the same time, the resource base had been also strengthened or enriched. Due to more biological inputs in soil (using biogas slurry), increased number of dairy animals leading to higher amount of manure availability and balancing the groundwater recharge and use of LPG in place of firewood for cooking together improved the resource regeneration. People may have been attracted to manage their resources better due to an increase in income through agriculture. Another reason for farmers agreeing to groundwater regulation appears to be the fact that all the farmers have better access to water for irrigation, which enabled them to grow more crops. An increase in groundwater levels and vegetation cover has been reported five years after project completion. So the economic growth has taken place along with ecological well-being.

Dairy, in addition to crop, had also emerged as another major source of income after project. The productive livestock holding had been sharply increased. The asset holding of both agriculture and transport had been also enhanced substantially. Credit uptake was increased from ₹16,227 per household in 2008 to ₹56232 in 2018 due to the higher level of economic activities. Banking, transport, dairy value chain, potato and other vegetable value chains together led to the rapid emergence of the service sector. The services required for accelerated production, including agriculture operations and transportation, were provided locally. Thus, the support enterprises were another income-generating sector that emerged after project. Although, these support services, especially the transport, consume the external energy contributing to the linear economy and going against the circular economy; however, the services were provided by the locals. Dairy enterprise has contributed in terms of recycling waste back to agriculture. There was also a shift from grid-supply of power to solar power for drinking water in village again increasing the renewable energy consumption.

Due to a major rise in employment from agriculture in the village, over 100 labourers had returned to village from cities. While there was a reduction by 60% in labour going out for the work, and out of those who had continued to go out, most of them had gone for skilled jobs. The rise of about ₹150 per day for women and ₹200 per day for men was also seen in the village due to more demand for labour. This shows the achievement of self-sufficiency in terms of employment generation.

The increased income, increased productive assets and services have accelerated the economic growth. At the same time, the local resources—land and water—had been judiciously used, rather enriched. The local prosperity, through local resources without its over-exploitation, has shown the signs of circular nature of economy.

24.8.1 Politicization of Watershed Governance

Randullabad watershed project had a unique arrangement for implementation. Project Implementing Agency (PIAs) of all the ‘Indo-German Watershed Development

Projects' (IGWDP) supported through NABARD were NGOs. PIA for Randullabad project was '*Gram Panchayat* (GP)' of Randullabad, which implemented the project through 'Village Watershed Committee' (VWC). BAIF Development Research Foundation, Pune was RSO (Resource Support Organization). The role of RSO was to train the PIA and VWC and facilitate the project implementation.

The Watershed development programme funded through IGWDP had depoliticized the watershed governance. The funding had been directly routed through NABARD to the Non-Government Organizations (NGOs) and VWCs for implementation. Being a political body, and the area of GP being in non-alignment with the boundaries of watershed, *Gram Panchayat* was usually kept away from the implementation. This has proved to be a successful approach in implementing the watershed projects in a transparent and democratic manner, as VWC is the representative body of *Gram Sabha*. However, the watershed development projects were announced by the PIA (NGO) in the village as a technical and apolitical activity, hence actively pursued a dissociation from village politics or any political ideology. "In case of '*Hiwre Bazar*' and '*Ralegan Sidhi*', the project was undertaken by the village panchayat in a politicized manner. The development here could be inferred as more sustainable than in many IGWDP villages because of the constitutional recognition of institutions. The approach followed by IGWDP also has its successful examples of *Darewadi, Mendhwan* and many others but the sustainability of the institutions and maintenance of self-help groups has been a constant problem in most of the projects" (Mishra 2010) visited by Neeraj Mishra in Ahmednagar district. Nevertheless, overall sustainability was much better in IGWDP, Maharashtra compared to many other projects.

On the other hand, under *Hariyali* guidelines of 2003 by Govt of India, there was complete politicization of the watershed programme (GOI 2003). The *Gram Panchayat* was the executing body (somewhat similar to the VWC) and the PIA was *Panchayat Samiti* (Block level Panchayat). *Zilla Parishad* (District Panchayat) through District Rural Development Agency (DRDA) was Nodal Agency for implementation. Although these are constitutionally recognized bodies, due to the issues faced in the implementation, Govt. of India had to change these guidelines in 2008 and again had to go back to the implementation through VWC under either GO or NGO as PIA (GOI 2008). This also means both over politicization and over de-politicization may have been the issues in implementing sustainable watershed development projects.

Randullabad appears to have achieved a balance between two extremes by adopting a hybrid model of governance. In Randullabad, the key actor in implementation was the *Gram Panchayat*. Randullabad GP played a role of Project Implementing Agency (PIA) in place of an NGO (or taluka panchayat in case of '*Hariyali* programme'). So at the level of village institution, a successful attempt of politicization of watershed governance was made by NABARD and BAIF. Moreover, the VWC had representation from both *Gram Panchayat* and others suggested by *Gram-sabha*. (*Gram-sabha* includes all the adults in the village). The chair-person of VWC was the *Sarpanch* (Head of *Gram Panchayat*). This kind of model integrated very well the two bodies and the coordination was very well achieved. The project was made as a platform and was mainstreamed into the overall development of the village. This also helped *Gram Panchayat* to leverage the additional resources to implement other developmental interventions.

Due to political influence and social pressure, *Gram Panchayat* and VWC could approach *Zilla Parishad* (District level political body under *Panchayati Raj* system of local self-governance in India), Different Govt. Departments viz. Revenue, Forest, Agriculture and Irrigation to tap various development schemes of these departments. They could also mobilize funds from the Member of State Legislative Assembly.

Post-project completion, *Gram Panchayat* continued the rigour of development process initiated under the watershed project. The 'Pani Foundation' an organization, which has created a movement of water conservation in the state of Maharashtra through an annual competition of villages, was also approached by the villagers. Some conservation works were identified and implemented through 'Pani Foundation'. There was still some gap in saturating the soil and water conservation in the village, which was then filled up through a funding support from UPS Foundation, Pune.

All was not well in this arrangement of project implementation. Being a political and elected body, Randullabad *Gram Panchayat* was not an exception to the party-politics. Before elections, the party-politics used to gain momentum leading to strained relations among different groups aligned to political parties. It used to slow down or even stop the development projects in the village for over a period of a couple of months pre-election of *Gram Panchayat*. However, some of the VWC members, who were not very strongly aligned to the political parties, used to neutralize the extremities in the relations among the leading team of village. It is also to be noted that some five to six households had kept their opposition to GP alive even after ten years down the line. Due to this opposition, GP and VWC were not able to control 100% the cultivation of sugarcane—the water-intensive crop. Of these, two to three families continued sugarcane cultivation even after the strict ban in the entire village. This means, even in this model of governance, 100% participation and cooperation could not be achieved.

There appeared to be very less impact on villagers during the electoral process of state (provincial) and national assemblies.

24.8.2 Role of External Organizations

In addition to the financial support, National Bank for Agriculture and Rural Development (NABARD) was responsible for overall project monitoring in addition to RSO. Any unaddressed intervention of project used to be pointed out by NABARD. The whole approach of IGWDP in implementation of three steps is also one of the important contributors to the success of project– viz. 'Qualifying *shramdan*' (voluntary labour)—after qualification of the village for the project, the 'Capacity Building Phase' enables the PIA and VWC to be ready for the final and main 'Full Project Implementation Phase' (NABARD 2014). This ensures first the willingness of villagers to participate proactively in the implementation and then the capacity of implanting agency at various stages of project.

The role of BAIF as Resource Support Organization (RSO) was crucial, as it was a risk of selecting this village despite knowing the fact that an attempt by other organization in selecting the village in past was failed due to the non-cooperation of villagers. After the selection of village, BAIF took another 'risk' of choosing Randullabad *Gram Panchayat* as PIA. There was a special effort by BAIF deputing a strong facilitator right in the village, backed up by the senior RSO team in organization's Headquarters in Pune. The awareness generation, training of GP, VWC and motivation of progressive farmers in joining hands for bringing change in the village helped develop a good team of villagers, who later took lead in implementation.

BAIF had previous experience of 'Landscape Approach' in holistic development of a village or cluster of villages. In this approach, organization enables the community to come together for a week to first assess their resource and then develop a 'Participatory Village Development Plan' (PVDP). In this process, a group of multidisciplinary team of BAIF, *Gram Panchayat*, other local institutions, forest department as well as district-level representatives contribute to the planning process. Then the local institutions are empowered to mobilize resources through Govt. and non-Govt sources for implementation of the plan. This experience of overall development planning and empowering people to mobilize resources may also have played a role in Randullabad. This has directly resulted in improved performance of institutions and enhanced natural capital.

In addition to core watershed development activities, BAIF facilitated the community to initiate add-on livelihood measures such as dairy development and skill-based employability. The ideation for the holistic development of village was initiated by BAIF, which was seriously pursued and realized by the GP and community. Unlike other projects, the association of Randullabad GP and BAIF continued after project closure, mainly due to the enthusiasm of villagers and also the vicinity of the area to BAIF Headquarters in Pune. BAIF also facilitated introducing 'Integrated Renewable Energy and Sustainable Agriculture' (IRESA) model in the village. This model has an integration of pre-fabricated biogas unit, slurry filtration, use of dry slurry for vermicomposting, reuse of filtered water in digester or use for vegetable cultivation. In the year 2015, UPS Foundation with the facilitation by BAIF supported the village to implement soil and water conservation works worth ₹2,800,000/-.

During the fieldwork of this research, some weak areas were observed mainly in terms of equity in the representation of all sections of the community in the village institution. This was brought to the notice of VWC and GP, which subsequently was addressed by VWC by expanding the membership to uncovered sections. However, it is necessary to note that the inclusion of the scheduled caste (SC) community did not happen. That indicates a sign of barrier between SCs and other sections of the community. Nevertheless, there had been serious efforts by GP increasing the access to water by most of the households, which helped bring most of the people into the mainstream development process.

BAIF had also invited ACWADAM, an organization having expertise in groundwater management for incorporating the groundwater management strategy in the watershed development. ACWADAM established the weather monitoring system in the watershed. The interventions included geological and hydrological investigations, aquifer delineation, formation of water user groups, support in developing water sharing mechanism for these groups, working towards drinking water security and awareness and sensitization of the community (Dhavan et al. 2013).

The villagers were educated on recording the weather data and its application for agriculture as well as crop-disease prediction. ACWADAM also studied the fluctuations of groundwater and its effect on crops. As per Dhavan et al (2013), 'the village proves that communities, if provided with the necessary information, are able to and willing to manage their water resources in a participatory and negotiated manner.' This has increased the water availability for a longer period and access to all.

In the year 2011, with the persuasion from *Grampanchayat* and VWC, *Zilla Panchayat* (District Council), Satara came forward and supported for construction of drinking water well, solar pump and other development activities in the village. The major support was also received to Randullabad between 2013 to 2016 from Dept. of Agriculture, Govt. of Maharashtra. Post-project completion, *Grampanchayat* continued the rigour of development process initiated under the watershed project. The 'Pani Foundation', an organization which has created a movement of water conservation in the state of Maharashtra through an annual competition of villages, was approached by the villagers. Additional water conservation works were identified and implemented through 'Pani Foundation' in the year 2018. Figure 24.3 shows the timeline of ten years and the engagement of external orga-

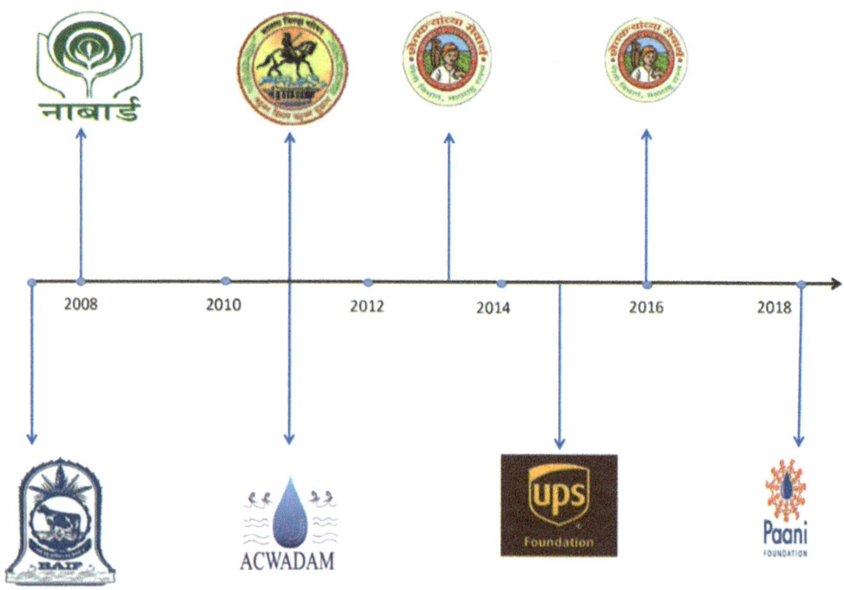

Fig. 24.3 Randullabad watershed project timeline with starting point of engagement of external organizations

nizations. NABARD and BAIF continued till 2013, while others were engaged for shorter durations.

Thus, the concerted efforts by these external organizations and cooperation of local community, GP and VWC worked quite well in creating a strong foundation of the sustainable development process. The facilitation by these organizations has helped improve the performance of many indicators, such as community awareness, institutions, participatory processes, access to water, water use efficiency and forestry and biodiversity.

24.8.3 Project Design, Plan and Finance

Initially, composition of project activities in the Randullabad watershed project was the same as of any other IGWDP supported projects. It had a major component of soil and water conservation and small support for livelihood and women empowerment measures. After gaining mutual trust and rapport, the RSO, i.e. BAIF and Randullabad *Gram Panchayat* together broadened the development agenda for the holistic development of the village.

24.8.4 Demand Driven Planning

In addition to the core soil and water conservation measures, several other developmental activities were initiated towards the end as well as after the closure of IGWDP project. These include drinking water supply and sanitation, community health awareness, biogas, village cleanliness with the initiative of women members, lined farm ponds, water regulation and efficient use through micro-irrigation, desiltation of old water reservoirs, credit linkage, village roads and saturation of soil and water conservation measures in the village. The project activities were thus need-based or demand-driven and not restricted only to soil and water conservation. Most of the above-mentioned activities were taken up through convergence of various schemes and programmes. Market-oriented production system was a special feature of Randullabad agriculture.

The performance of all three sustainability domains, viz. social, economic and ecological, which was average at the end of the watershed project, was enhanced to its maximum level after the project. This result could be due to the combined effect of multi-faceted and dynamic project design and its serious implementation as well as other social, equity, political, organizational, institutional and economic factors. It is evident from the web diagram that the higher level of sustainability was in 2018, which is associated with the add-on activities post-project—both core watershed activities, agriculture activities and other quality of life initiatives. A study of 405 sample watersheds in all five regions of Maharashtra by SOPPECOM (2009) covering eight different modes of implementation revealed that the distribution and likely continuance of benefits vary according to the design (mode) of programme and its actual implementation. Modes like IGWDP and AGY (*Adarsh Gao Yojna*—An Ideal Village Scheme) had dedicated institutional or organizational mechanism for management and support from below which were found to be performing better than the modes of implementation through departments (Samuel et al. 2009). The results in Randullabad appear to be quite similar to the observations of this study.

This also means that implementation of only the conservation measures may not assure sustainability of project. The empowerment of people, nurturing the local leadership and local institutions, active engagement of *Gram Panchayat*, ensuring benefits to most of the families in village and convergence of production, production efficiency, water efficiency regulations for resource management and control on over-exploitation can lead to sustainable development of the watershed area.

24.9 Project Investment

The basic project investment through IGPWDP was **₹8,763,103/-** over a geographical area of 836 ha. It comes to ₹10,482/- per ha, which was below the norm of ₹12000/- per ha as per national guidelines. The additional investment mobilized by *Gram Panchayat* and BAIF together in water conservation, agriculture (₹15,797,530/-) and other development activities in village (₹2,647,826) was **₹18,445,356/-.** Of this ₹8,192,478/- was mobilized during the project period and ₹10,252,878/- was mobilized after the project period till 2018. The additional funds mobilized was higher than the project funds. So the project was just a stimulant in the development process. Thus, the overall investment including project investment and investments through convergence during the ten-year span was **₹27,208,59/-.** Of the overall investment in village, ₹24,560,633/- was on Natural Resource Management (NRM), drinking water supply along with solar energy and agriculture and allied interventions. With project investment and the additional investment on NRM and agriculture, per hectare investment comes to ₹29,378/-, which is much higher compared to the investments made in the country on watershed development projects.

The investment flow was observed to be consistent across this ten-year span of time, starting from the year 2008. The sources of funds were diverse including Govt. mainstream programmes, District level schemes and CSR (Corporate Social Responsibility) funding. Both economic benefits and ecological base kept on enhancing due to continued investments. The economic gains were also fuelled with the credit availed by individual families through formal financing sources. The average annual credit uptake per household had gone up from ₹16226.80 in the year 2008 to ₹56232/- in 2018. It would be however debatable whether this level of investment is possible and appropriate, but it also indicates both the ability of locals to mobilize this level of funding as wells as the availability of funding sources if tapped for the development of a village.

Thus, the amount of investment and continued flow of funds during and post-project period was another important parameter of sustainability of watershed.

24.10 Conclusion

Although the key factors as classified and explained above show their influence on the sustainability independently, there appears to be a great level of interplay among these factors. The basic project design and implementation methodology of IGWDP itself was very robust. There was an in-built process of graduation of community and implementing agency for successful implementation of the project through a 'qualifying *shramdan*' (voluntary labour) and 'Capacity Building Phase'. On top of it, BAIF facilitated the local institutions to enlarge the project design towards holistic development of watershed. The role of external organizations played a key role in constantly facilitating the local institutions initially in implementation and quality assurance and later in accessing the funds, ideation of new interventions and upskilling the institutions. The empowerment of both *Gram Panchayat* and VWC has also played a key role in broadening project design, convergence of many developmental initiatives in the village mobilizing a substantial amount of funding.

Watershed and watershed-plus activities were coupled with the equity aspect in its distribution, resulted in a wider level of cooperation of the community. The sense of circular economy could also be seen in the whole development process. The organic linkage of community institutions, leadership and economic activities was observed in the village. The value chains around dairy, potato, pomegranate and credit flows were clearly seen growing, fuelling the growth of service sector and overall economic growth. People were sensitized and empowered enough to ensure that sustained gains could be harvested only through enriched resources and well-governed watershed.

A hybrid model of politicization of watershed governance coupled with a robust and holistic project design and resource regulation has helped to enhance the sustainability of the Randullabad watershed.

Acknowledgements This paper is an outcome of the PhD research of the author at 'Centre for Technology Alternatives for Rural Areas' (CTARA), IIT Bombay, with the guidance of Prof. N.C. Narayanan and co-guide Mr. G.G. Sohani. The project was implemented by *Gram Panchayat* of Randullabad village through 'Jai Hanuman Panlot Vikas Samiti'—Village Watershed Committee. BAIF Development Research Foundation, Pune facilitated the *Gram Panchayat* and VWC in project implementation and the project was financed by 'National Bank for Agriculture and Rural Development' (NABARD)—Regional Office, Maharashtra, India, under Indo-German Watershed Development Project. Along with the *Gram Panchayat* of Randullabad, officers of BAIF, Mr. Balwant Thorat and Ms. Sneha Shinde, played a critical role in facilitating the community in the development process.

References

Census (2011) District census handbook Satara. Village and Town wise Primary Census Abstract (PCA), Directorate of Census Operations, Maharashtra, India

Chaves HML, Alipaz S (2007) An integrated indicator for basin hydrology, environment, life and policy: the watershed sustainability index. Water Resour Manag 2(5):883–895

Dhavan H, Jagtap P, Lokgariwar (2013) Why participatory groundwater management is no longer an oxymoron: The story of Randullabad. India Water Portal. Accessed 5 April 2013

FAO, ND. Sustainability, accounting for natural resources and environmental degradation. http://www.fao.org/3/x5318e/x5318e04.htm. Accessed 15 Sept 2019

Govt. of India (2003) Guidelines for Hariyali. Department of Land Resources, Ministry of Rural Development, Government of India

Govt. of India (2008) Common guidelines for watershed development projects. National Rainfed Area Authority Planning Commission, Government of India. NASC Complex, D.P. Shastri Marg New Delhi, 110012. www.nraa.gov.in

Kakade BK (2017, 2019). Sustainability assessment of watershed development programme in India. Thesis (draft) submitted in partial fulfilment for the Degree of PhD. Center for Technology Alternatives for Rural Areas (CTARA) Indian Institute of Technology, Bombay

Luukkanen J, Vehmas J, Panula-Ontto J, Allievi F, Kaivo-oja J, Pasanen T, Auffermann B (2012) Synergies or Trade-offs? A new method to quantify synergy between different dimensions of sustainability. Environ Policy Gov 22:337–349

Mishra Neeraj (2010) A Watershed in Watershed Governance' Democracy and (De) Politicization of Development Projects in India. Inaugural-Dissertation, Rheinischen Friedrich-Wilhelms-Universität, Bonn

NABARD (2014) Main strategies adopted under Indo-German Watershed Development Programme (IGWDP).National Bank for Agriculture and Rural Development (NABARD). https://www.nabard.org/english/Indo_German_WDP2.aspx. Accessed 27 Oct 2014

Samuel A, Joy KJ, Suhas P, Ishwar K, Raju A, Ravi R (2009) Watershed development in Maharashtra: a large scale rapid assessment. Society for Promoting Participatory Ecosystem Management (SOPPECOM), Forum for Watershed Research and Policy Dialogue, Pune. India

Sangameswaran P (2006) Equity in watershed development. Economic Times 41(21)

WE Forum (2019) From linear to circular—Accelerating a proven concept. Ellen MacArthur Foundation, and World Economic Forum. http://reports.weforum.org/toward-the-circular-economy-accelerating-the-scale-up-across-global-supply-chains/foreword-ellen-macarthur-foundation/. Accessed 8 July 2019

Chapter 25
Futuristic Urbanism of Eco-Sensitive Zone of India: A Case Study of Kolkata Region

Tazyeen Alam

Abstract The Bengal Delta is one of the ecologically sensitive zones of the Indian sub-continent. This sensitive zone is facilitated by water, which has an entry in all parts of people's lives through seasonal monsoon flooding, and cyclones under extreme weather conditions. The topography determines the flow of drainage, which slopes to the East. The characteristics of this water network as well as a number of other aspects of water resources can be linked to the physiography of the region. One of the most critical elements of this water system is the low-lying areas of the region, which creates a number of economic opportunities at the local level as well as regional level. The paper will understand the significance of such low-lying areas. The intent of the paper is to provide an approach for understanding the hydrological process and its relation with the existing morphological elements. It also aims to understand the spatio-temporal impact of anthropogenic activities on the low-lying areas. The author identifies the different elements of the water system to delineate the study towards creating a balance with these settlements. This will depend on the social and economic dependencies of the communities and the spatial configuration of the area which will be analyzed through primary and secondary surveys. The research will focus on the improvement of the environmental and social conditions through regeneration strategies to integrate the economic and physical resourcefulness of these communities. Out of all the different components of water systems, the intent of the research will be limited to the case of low-lying areas only.

Keywords Bherries · Economic dependency · Encroachment · Low-Lying areas · Urbanization

25.1 Aim and Objectives

This paper aims to understand the spatio-temporal impact of human activities on the low-lying areas considering the following objectives

T. Alam (✉)
Indian Institute of Technology, Kharagpur, Kharagpur, India
e-mail: tazyeen.alam@gmail.com

R. K. Mishra et al. (eds.), *Sustainable Climate Action and Water Management*, Advances in Geographical and Environmental Sciences,
https://doi.org/10.1007/978-981-15-8237-0_25

- To understand the different types of anthropogenic activities that disturb the water ecology in the region
- To study the different types of spatial organization in the region
- To analyze the need for urbanization in the current scenario
- To identify the major component responsible for the disappearing low-lying areas of the region
- To identify the futuristic approach towards the growing population.

25.2 Methodology Framework

The conceptual framework of the research is divided into four broad categories

1. Building an ecological research platform for the study
2. Identification and study of the most critical zone
3. Role of Regional Landscape of Infrastructure
4. Identification of Urban Riparian Zone (URZ)
5. Study of Inadequate Land Cover Theories
6. Understanding the Relationship between the vegetation structure and social characters across the region
7. Understanding the role of Complex Household Structures
8. The Function of the Watershed
9. Research-Decision Making
10. Management and Policy Framework.

25.3 Scope and Limitations

The study will explore the loss of low-lying areas in the region with respect to the trend of urbanization. It will identify the strategies for a resilient and sustainable growth of the region with respect to the population projection of the intervention area. This will include the moderately critical zone along the Bagjola Khal of Kolkata that has the potential to convert into a populous section in the coming years.

The study will be limited to the activities of settlements along the low-lying areas only. This will be covering a pilot project of Jatragachhi to demonstrate the need of the people and the need of the climatic condition of the area with respect to the topography of the region. The paper will not consider the socio-political conditions within the region.

25.4 Introduction

The geography of human settlements has always been shaped strongly tied to the distribution of water and therefore, to a certain sense, distribution of water and urbanism are completely inseparable. Water is a life-giving element, as we see, it is also more life-threatening (Pierce 2009). Many of the rise and fall of great civilizations has a lot to do with their relationship with great water systems. The concept of Ghats in the region plays a vital role in creating a connection with water. It is a kind of mediator between land and water that often become dynamic public spaces. Amidst this relation of water with man and these water-based societies and hydraulic civilizations, things are changing rapidly. There are new ecologies emerging in the region.

The Kolkata Metropolitan Development Area (KMDA) is the most urbanized area within the region. It is a junction of major fluvial and consequential changes over the past 500 years (Pai 2016). Built on a comparatively higher ground of the river banks, it is surrounded by an immense area of low-paddy fields. The topography has also played a vital role in the expansion of the region from its initial point of settlement to the present. The region has witnessed a smooth expansion from the Eastern bank of the Hooghly River to the Eastern part of the region. The local people have activated a number of lakes, ponds and East Kolkata Wetlands (EKW) which provide a number of benefits including biologically purified water, agriculture, aquaculture, employment generation, food generation, trade and commerce and monsoon flood mitigation(Annabel Cremer 2008). The water logistics of Kolkata support the exponential growth of the population in the region (Fig. 25.1).

25.5 Major Findings

The three watersheds within the delineated boundary show the growth of human activities and settlements in the present scenario. Almost 80, 60 and 40% of the first, second and third watersheds region is occupied by settlements which have covered the low-lying areas of the region. The map below clearly depicts the amount of low-lying areas left in each of the greater watersheds of the region. This can be clearly determined that the growing population and the authorities managing the growth of the city are paying no heed to the significance of these low-lying areas. Thus, they seem to be blindly encroaching on them in the greed of land requirement for more buildings and services for the people (Fig. 25.2).

The management authorities and the decision-makers somehow fail to realize the dependency of the local communities on these low-lying areas for their daily survival and the needs of the other urban settlements. This can be understood as we understand the declining agricultural, floriculture and pisciculture activities in the entire region. According to the visual analysis of the region, it can be clearly estimated that the majority of the population dependent on these areas includes the rural sectors of the

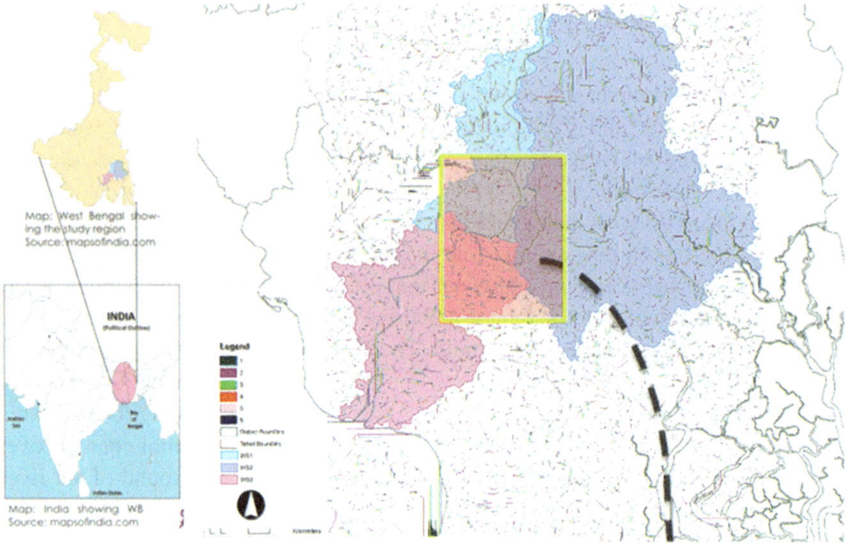

Fig. 25.1 Location Map of the Region (left), Watershed of the Region (right) (Maps of India, GIS, Author 2019)

region which are often left unattended while planning for the futuristic urban growth of a region despite their high impact on the urban settlements (Fig. 25.3).

25.5.1 Ecological Dependency on Low-Lying Areas

The two main economic interdependencies identified in the region include the pisciculture and floriculture which play a major role in shaping the different anthropogenic networks within the region. The study is limited to these two major dependency linkages within the region.

25.5.2 Ecological Dependency on Low-Lying Areas

The economic dependency activities mentioned above are directly associated with the existing ecology in the region. Therefore, there is a direct link between the flora, fauna and the anthropogenic activities that benefit the people economically. Hence, there is a requirement to preserve and maintain the ecological essence of the region that sustains the other important activities within the region.

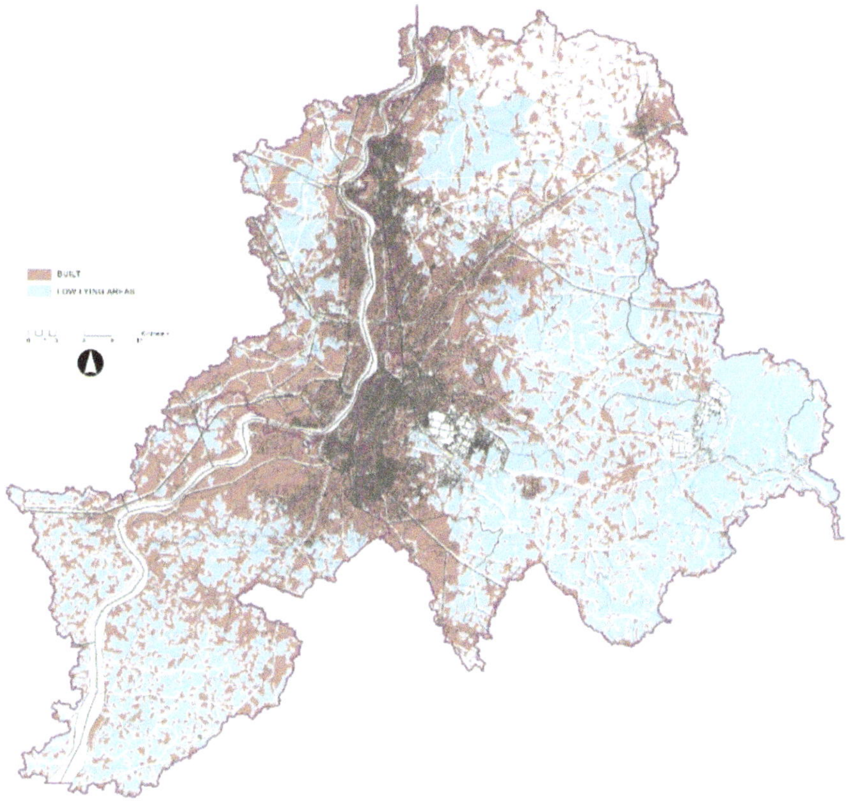

Fig. 25.2 Map showing built and unbuilt (low-lying areas) within the major watersheds

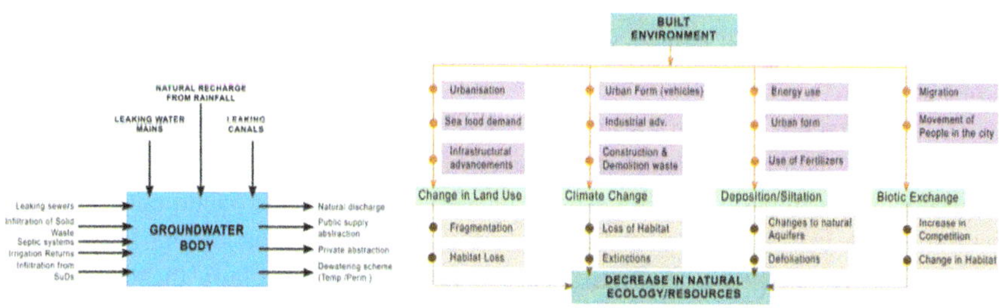

Fig. 25.3 Impact on groundwater (left) and Impact of Built Environment (right) (Author 2019)

25.6 Recommendations

1. Proposal for making the settlements along water bodies interact with the natural elements by facing them as their front yard.
2. Identify and demarcate the area for future urbanization and growth of the city.

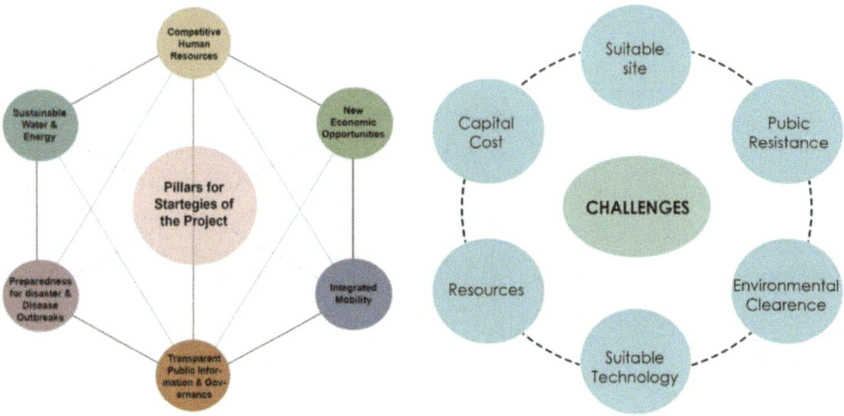

Fig. 25.4 Pillars for strategies (left) and challenges in implementation (right) (Author 2019)

3. Proposal to define the active and passive circulation zone for the city's movement.
4. To propose roads along those edges that have the potential for urbanization.
5. Reclamation of low-lying areas for future development and preservation of bherries culture.
6. Removal of illegal encroachment.
7. Development policies for dense housing that would restrict the F.A.R. Policies to restrict encroachment of low-lying areas and provision of housing for all the communities in its precinct.
8. Propose a policy for protection and conservation of natural and manmade water systems.
9. To propose Discharge norms for the water system (Fig. 25.4).

Project: Reclamation of Low-Lying Area in Jatragachhi and Restoring the Natural Land Drainage.

The Jatragachhi area in New Town was entirely a low-lying plain that was completely in the past 20 years. It was an agricultural patch of land which included seasonal and permanent bherries that were crippled by the encroachments of urbanization. Thus, under the current trend of urbanization, the area suffers from waterlogging and minor flooding. Therefore, the proposal will reclaim these low-lying areas and create a resilient environment for the drainage pattern and the settlements simultaneously. This will define the integrated existence of the present condition along with directing the growth for future expansion (Fig. 25.5).

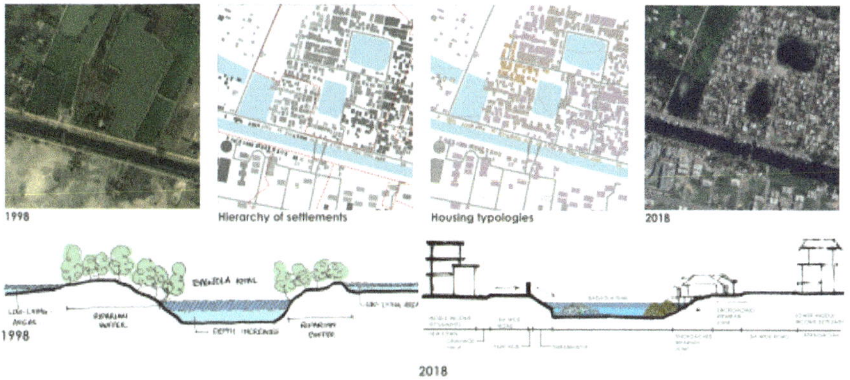

Fig. 25.5 Critical zone A for analysis (Author 2019)

References

Baghchi K (1944) The Ganges Delta, pp 39–49

Bandyopadhyay S et al (2014) River systems and water resources of West Bengal: a review. Special Publication, Geological Society of India

Bhattacharyya A (1975) History of the Bengali Mangalkāvyas (in Bengali), 6th edn. A Mukherjee & Co. Pvt. Ltd., pp 11–14, p 20, pp 61–64, pp 127–130, p 467

Biswajit Nath SA (2013) Urban municipal growth and landuse change monitoring using high-resolution satellite imageries and secondary data. Stud Surv Mapp (SSMS):13

CCWRD (2011) India Vulnerability of Kolkata Metropolitan area to increased precipitation in changing climate. CCWRD, South Asia Region, p 101

Doheny EJ, RJS, Striz EA, Mayer PM (2006) Watershed characteristics and pre-restoration surface-water hydrology of mine bank Run, Baltimore County, Maryland, Water Years. Virginia, Scientific Investigations Report

Eva Hasíková MR, Hume Shane (2012) Hydro-morphological alterations & pressures in European Rivers, Lakes. Transitional & Coastal Waters, European Topic Centre on Inland, Prague

Ganga Action Parivar (2015) Annual Report, Parmarth Niketan Ashram

Ghosh AK (1988) Urban Ecology—A case study of Calcutta. Kolkata, Institute of Local Government and Urban Studies

Sundararaman B, KGV, Lalwin M (2012) Urban Hydrology—a case study on water supply and sewage network For Madurai Region, using remote sensing & GIS techniques. Int J Eng Sci:7–12

Chapter 26
Water Governance for Sustainable Development: Role of Community Participation, Opportunities and Challenges

M. Venu Madhav

Abstract Water is one of the indicators for sustainable development, it plays a vital role in human health and wellbeing. Water governance is an effective tool between communities and institutions for the effective implementation of water services to the community. Sustainable Development Goals (SDGs) targeted that, "By 2030, achieve access to adequate and equitable water, sanitation and hygiene for all and end open defecation, paying special attention to the needs of women and girls and those in vulnerable situations", to achieve the universal access of water services, involvement and participation of community play an important role. Community participation means the participation of people in the administrative process, policy formulation, implementation of the programmes, the evolution of the programmes and sharing the benefits. It implies citizen's control over administration or public influence on the administration. It secures public support to the government policies and makes them a success and constitutes an important means of enforcing administrative accountability. This paper provides a detailed review on various governance issues involved in the drinking water sector and the role of community participation to ensure better service delivery; to substantiate this review, the Gollapally and Kodimial mandals in Telangana State has been taken as a case study.

Keywords Water governance · Sustainable development · Community participation

26.1 Introduction

Water and sanitation play a vital role in human health and wellbeing. Water is a crucial element for the development across the board, in terms of food production, energy,

---

This article is the part of author's Ph.D. work.
M. Venu Madhav—ICSSR Fellow at CESS-Hyderabad.

M. Venu Madhav (✉)
Centre for Economic and Social Studies (CESS), Hyderabad, India
e-mail: venumadhavsharma@hotmail.com

R. K. Mishra et al. (eds.), *Sustainable Climate Action and Water Management*,
Advances in Geographical and Environmental Sciences,
https://doi.org/10.1007/978-981-15-8237-0_26

education and health. Yet billions are still faced with daily challenges accessing even the most basic services. Sustainable Development Goals (SDGs) (2015) targeted that "By 2030, achieve access to adequate and equitable water, sanitation and hygiene for all and end open defecation, paying special attention to the needs of women and girls and those in vulnerable situations", to achieve the universal access of water and sanitation services, involvement and participation of community play an important role.

Community participation means the participation of people in the administrative process, policy formulation, implementation and evolution of the programmes and sharing the benefits. It implies citizen's control over administration or public influence on administration which is essential for the smooth and effective performance of the administrative machinery. Making the administration responsive to the needs of the people, it secures public support to the government policies and makes them a success continuing an important means of enforcing administrative accountability.

26.2 Literature on Water Governance

Water governance is "The range of political, social, economic and administrative systems in place to develop and manage water resources, and the delivery of water services, at different levels of society" (Rogers and Hall 2002). Reasons for water governance are many; water-related problems can be attributed to governance failure rather than to the resource base itself. Increasing complexity in better service delivery in the water sector of real-world processes needs water governance.

A number of authors discussed and detailed the need for ensuring a broad representation of interests in Community participation efforts (Zeithaml and Berry et al. 1993; Korfmacher 2001). The primary advantage of ensuring representative participation is for decision-makers to get an accurate picture of the range of knowledge, thoughts about a particular issue (McComas 2001). Ahmed (2008) identified while community participation has been an integral component of decentralized water management, efforts are still required to address accountability, transparency, capacity building and facilitate participation to the people.

26.2.1 Water Governance in Indian Scenario

Verma et al. (2015) observed institutions as a whole, that the national level framework gives guidance on the sector institutions and their role. According to the author, there are gaps in the bottom level (1) community management (2) community participation with private sector (3) officials participation in O&M process (4) Public-Private Partnership (PPP). Drinking water is the most decentralized service, with panchayats in charge of investment and maintenance. Inadequate access to services can arise because of problems with either the delivery side or the implementation side and the limitations discussed here may not always be unique to rural areas.

Nayak and Saxena (2002) has explained that poor governance practices lead to inefficiency of service delivery. There are too many players involved with an unclear demarcation of job responsibilities. Local authorities in all parts of the world play an increasingly important role in the delivery of fundamental basic public services but authorities also face great challenges, adequacy of own revenues is the key to an improved ability to deliver needed goods and services and to better accountability of local officials to their constituents.

Trivedi et al. (2013) mention that governance is the major key component for the development of society. Without a systematic governance system, goals cannot be achieved. In the case of the water sector, the appropriate governance system the scientific and technological aspects of water-related work are unlikely to be effectively implemented at the policy and management levels. Tropp (2007) observed that the water sector has undergone similar changes as other development sectors regarding for example decentralization and privatization yet uniquely the water sector has lagged behind in addressing explicitly governance challenges.

The above literature reveals that various governance issues are involved in the water sector in rural areas; major problems are availability, accessibility, quality and equity of water services to the Community. This is most important in the context of New Public Management (NPM), which emphasizes four 'Es' (economy, efficiency, effectiveness and equity) and it also emphasizes accountability, transparency and people participation. Hence, the present study is a modest attempt in that direction.

26.3 Trends and Challenges in the Water Sector

The growth of three dimensions of sustainable development social, economic and environmental is depended on the availability, accessibility of water resources and the way these resources are managed to provide services and benefits. According to the Ministry of water resources, India and Central Water Commission (2013) projected estimations on the water sector, there is an enormous decline of per capita water availability in India. It is estimated that inclining of water stress and scarcity from the availability of 1545 per capita in the year 2011 to 1174 per capita in the year of 2051, it was 5178 per capita water availability in the year of 1951 (Fig. 26.1).

Reduction in water per capita is observed over a period of time. This provides valuable information on the behaviour of the water availability regime, which is constantly subjected to changes due to recharge and discharge phenomena. The decline in water level may be due to an increase in the draft (for different purposes) or a decrease in precipitation (less recharge to groundwater) (Central Ground Water Board 2017). According to the Central Ground Water Board report (2017), groundwater availability in Telangana State is reducing over the period.

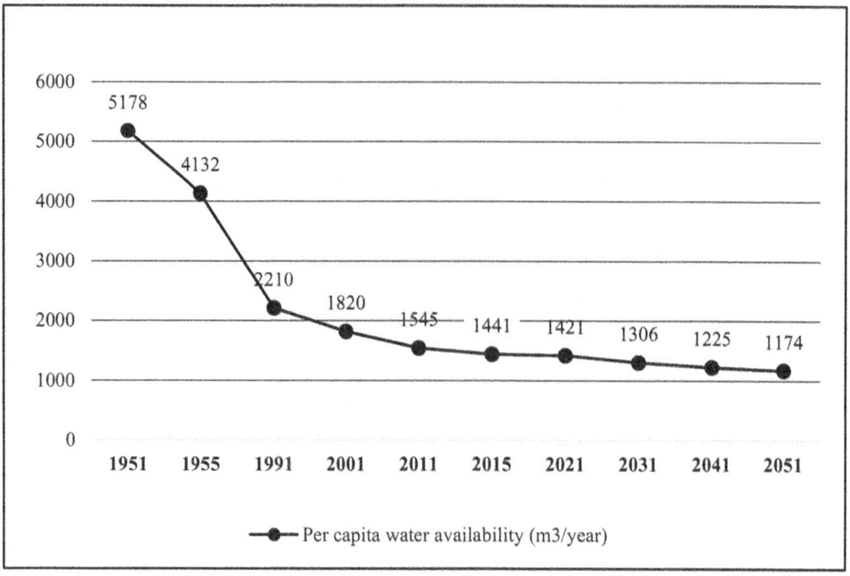

Fig. 26.1 Reducing per capita water availability in India. *Source* Ministry of Water Resources, RD & GR. *Note* 1. The per capita availability from 2015 onwards has been calculated from the 2017 WRA estimate. 2. m3/year = cubic meters its equals 1000 L[1]

26.4 Empirical Strategy

26.4.1 Description of the Study Area

The study examines the availability and accessibility of water for domestic use and the involvement of the community in the governance process to avail better service delivery at four villages of Gollapally and Kodimial mandals in Jagtial district, Telangana State (see Map 26.1). This study adopts the community participation perspective. In its analysis look into the reasons for non-participation of the community and how community involvement, participation leading to successful participation for water governance.

26.4.2 Methods and Materials

For the purpose of the study, two mandals from Jagtial district in Telangana State have been selected. While selecting the study areas, due care has been taken regarding

[1]Cubic meter (m3). BusinessDictionary.com. Retrieved September 11, 2019, from BusinessDictionary.com. website: http://www.businessdictionary.com/definition/cubic-meter-m3.html.

Map 26.1 Map of the Gollaplly and Kodimial mandals in Jagtial district of Telangana State, Sample district and mandals highlighted in the *darker colour*

the variations in terms of geographical area, population, the socio-economic status of communities, size of the mandals and villages, devolution of powers, functions, functionaries and funds by the state government. The study has employed a multi-stage purposive sampling method in selecting mandals, villages and respondents. Two mandals (i.e. Gollapally, Kodimial) have been selected based on the level of development (developed and underdeveloped in terms of water services delivery). There are 15 mandals in the Jagtial district. Rational behind selecting these mandals, in domestic water service, is that Gollapally has 29.7% fully covered habitations and Kodimial mandal has 100% fully covered habitations. In Gollapally mandal, people are mainly depending on alternative sources for domestic water. It may lead to the privatization of water services.

Similarly, from each selected mandals, two villages were selected by giving weightage to the geographical location that is one village from proximate to the mandal headquarter and another village from a distant place to the mandal headquarter. Respondents are selected from the sample villages on the basis of 5% of the total families of all social groups, thus, in Jagtial district, two mandals, four villages, 5% of the total households of the village have been taken as a sample in selected villages. This 5% of households are divided among all social groups of the village such as SC, ST, BC, Minority and Others. 224 households were selected for the study. The data was collected through a structured schedule to assess the household's participation in Gram sabha meetings, awareness of VWSCs and Government-funded programmes (i.e. Swachh Bharat Mission).

26.5 Results

26.5.1 Drinking Water Situation in the Sample Villages

Everyone has the right to water services that are physically accessible within to ensure safe and sustainable piped drinking water supply from surface water sources. Government of Telangana started the Mission Bhagirtha project, and it aims to provide safe drinking water for every village and city household in Telangana State. In sample villages, the Mission Bhagirtha project was partially covered. Water is supplied by Gram panchayat on alternative days. Among four villages, Gollapally and Nachupally are not fully covered with tap connections (covered with 62% and 64%, respectively). It is reported that there is no tap water connection coverage in SC colony of Nachupally village where one of the respondents reported that "they have approached Gram Panchayat for water connections but, they were not able to provide water connection because of the high altitude location of the colony. In all four sample villages, respondents are depending on purchased water supplied by private Reverse Osmosis (RO) water plant for drinking purpose keeping in view the poor quality of public water supply and 'demonstration effect'. Respondents in all villages are depending on alternative water sources for domestic uses (Fig. 26.2). Whereas in the case of Gollapally village 27% of respondents are depending on purchased drinking water and are using open wells for other purposes.

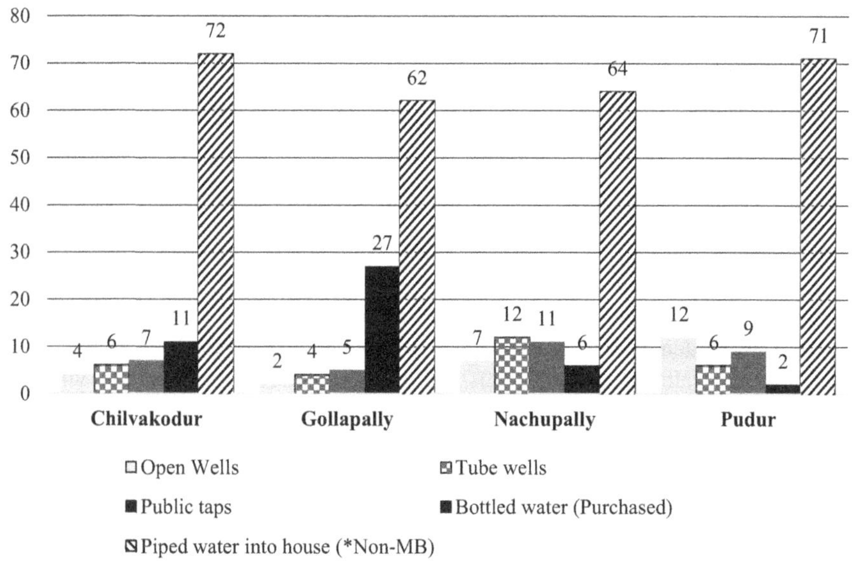

Fig. 26.2 Availability of water according to sources (%). *Source* Based on the primary survey. *Note* Non-MB = Non-Mission Bhagiratha Water

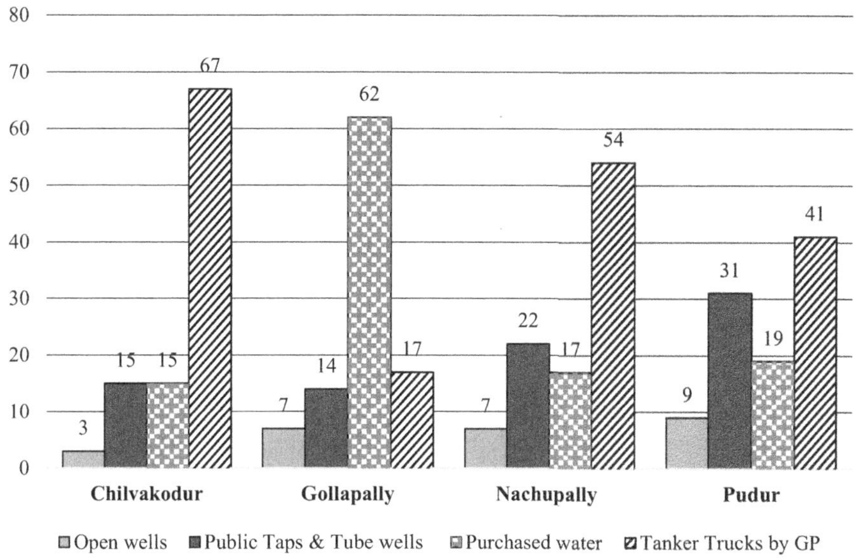

Fig. 26.3 Main source of drinking water during water scarcity (%). *Source* Based on the primary survey. *Note* GP = Gram Panchayat

26.5.2 Sources of Water During Scarcity

There is water scarcity in the summer season due to low water availability and supply from main sources insisting the households to depend on alternative sources. In all the sample villages, households are mainly depending on water tankers provided by the Gram panchayat for a nominal price, which is becoming a financial burden on the respondents, particularly in Nachupally village (Fig. 26.3). Multiple sources of water (i.e. open wells, tube wells, also purchased water, etc.) are being used by respondents.

26.5.3 Community Participation

In India '*Panchayat Raj*' institutions have greater importance to discharge devolution of powers, funds and functionaries which expected to reduce transaction cost, lead to more efficiency in the demand for and delivery of public services and generate greater efficiency in the process of governance at the grass-roots level. For the implementation and success of policies, schemes and programmes, there are two different routes: 1) Local Governments and 2) Community involvement. The community can involve either directly through local governments (by attending Gram Sabha, Ward Sabha meetings) or through the pressure groups like community-based organizations (CBOs), non-governmental organizations (NGO's), etc.

26.5.4 Awareness and Participation of Gram Sabha

During the survey, respondents were found aware of Gram Sabha but are ignored of their role in it. Most of the respondents are in the opinion that Gram Sabha belongs only to political activities where political leaders have more voice. Due to lack of policy awareness, some of the households missed the opportunity of government-funded schemes like toilet construction and household's tap water connections, etc., and few members of the community are managing and diverting funds according to their wish.

Members who attend Gram Sabha meetings have not raised water-related issues even when they suffer from the low-pressure water supply, breakdowns of water pipe connections, less frequency of water supply, etc., because they did not get an opportunity to talk/raise the issues. Other issues like the "Mahatma Gandhi National Rural Employment Guarantee Act" (MGNREGA) works, house constructions, etc., were given priority. A few respondents expressed that they had been involved in Gram Sabha meetings and had participated in water-related issues. In Nachupally village, 12% of the respondents said that they helped in the Operation and Maintenance (O&M) issues in water supply in favour of SC colony demanding for better service delivery which was partially solved being the best example of the result of community participation for better service delivery. 51% of households in Pudur and 62% in Gollapally village regularly participate in Gram Sabha meetings but no positive impact was reported (Fig. 26.4).

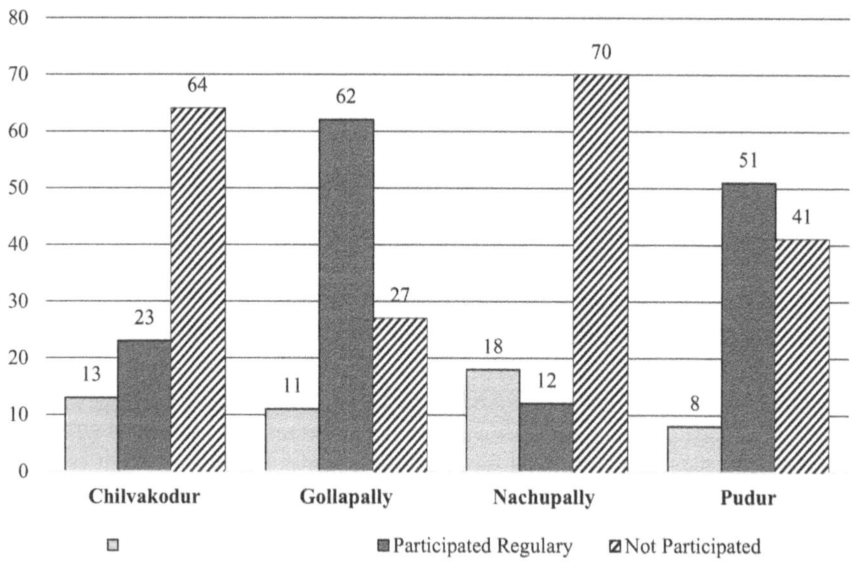

Fig. 26.4 Participation of households in Gram Sabha meetings (%). *Source* Based on the primary survey

26.5.5 *Role of Village Water and Sanitation Committees (VWSCs)*

Village Water and Sanitation Committees (VWSCs) is a subcommittee of panchayats that should actively be involved in the operation and maintenance of drinking water supply schemes. Aims and functions of VWSCs are planning and implementation of water supply schemes of the village for achieving health-based targets that focus on water quality. In the sample villages, VWSCs including women and other members are constituted. Due to ignorance of their responsibility due to political differences, regular meetings are not held for the last year which deprived them of their influence in resolving water-related issues and government machinery.

26.5.6 *Challenges of Community Participation*

Lack of continuity of community involvement was found to be the major challenge in sample villages for better service delivery. In sample villages, there are issues related to the availability of water, access to information about Gram Sabha, VWSCs meetings and its minutes coupled with lack of interest of members to attend the meetings, lack of awareness of the households on government-provided water services, poor maintenance of O&M in water sector, social and cultural practices, privatization of drinking water facilities, lack of political awareness, groups and cast politics, often visits of Gram panchayat members to their houses to collect the required quorum signature of the members for the record sake, etc., are identified as reasons for low community participation. All these issues led to the de-motivation of the community in general particularly women, SC and ST members.

26.6 Conclusion

Sustaining available resources in the water sector is a critical component to ensure impact. This becomes even more significant in rural settings as getting quality and sufficient water supply is a major challenge for them. In the sample villages as a result of irregular water supply, people are depending on the Reverse Osmosis (RO) water plant that provides paid service which a normal man cannot afford out of his low income. Respondents have low awareness of Gram Sabha meetings and their benefits, due to ignorance and unwillingness of the political system, they are partially participating in the governance process; for better service delivery, stakeholders have to involve in the service delivery process by attending Gram Sabha meetings and VWSC meetings or create pressure through participating or involving in community-based organization (CBOs) added with the insights and involvement of academicians,

NGOs, Youth which will make the difference to give sustained water resources to generations ahead.

References

Ahmed S (2008) Sociable happiness. Emotion Space Soc 1(1):10–13
Central Ground Water Board (2017) Ground Water Year Book 2016-17 Telangana State. Government Press
Korfmacher KS (2001) The politics of participation in watershed modelling. Environ Manag 27(2):161–176
McComas KA (2001) Theory and practice of public meetings. Commun Theory 11(1):36–55
Ministry of Water Resources, RD & GR http://mowr.gov.in/, Accessed 20 Sept 2019
Nayak R, Saxena NC, Farrington J (2002) Reaching the poor: The influence of policy and administrative processes on the implementation of government poverty schemes in India. Overseas Development Institute, London
Rogers P, Hall AW (2003) Effective water governance, vol 7. Global Water Partnership, Stockholm
Sustainable Development Solutions Network. (2015) Getting started with the sustainable development goals: a guide for stakeholders. Sustainable Development Solutions Network
Trivedi N, Mau TA, Vajpeyi D, Dwivedi SP (2013) Perspectives on governance and society: essays in honour of professor o.p.dwivedi. Rawat Publications
Tropp H (2007) Water governance: trends and needs for new capacity development. Water Policy 9(S2):19–30
Verma K, Bisht BS, Cronin A (2015) Decentralised governance in water and sanitation in rural India. Academic Foundation
Zeithaml VA, Berry LL, Parasuraman A (1993) The nature and determinants of customer expectations of service. J Acad Mark Sci 21(1):1–12

Chapter 27
Settlement Development of Indigenous People in the Context of Coastal Area Development in Kozhikode

K. K. Nishan Nazer and P. J. Nidhish

Abstract The project aims to prepare a settlement development plan for fishermen community living in the coastal stretch of Kozhikode city of Kerala in the context of the urban development that will be triggered due to the amendments in the CRZ rules in 2019. Several restrictions in built use, land-use, FSI etc that regulated urban growth in CRZ-2 is either withdrawn or relaxed in the 2019 amendments which create challenges of displacement of native population and removal of their traditional land rights. The project involves primary and secondary study about the amendments in CRZ 2019, coastal area and the native fishermen community. Case studies are done to derive the techniques and approaches that can be used in the sustainable settlement development of the native communities living in urban areas. The technique of land pooling and reconstitution is used to unlock the land value which can finance the settlement development process. Land value estimation before and after land pooling is assessed by using weighted overlay analysis in GIS to understand the economic feasibility of the project. Land is pooled and redeveloped under the Fishermen Cooperative Society and the surplus value generated in the process is proposed to be used for settlement development through improving housing and infrastructure. FCS can invest in fishing related activities like production, processing and marketing which can generate both revenue and employment. Unlocking the value of land in a fast urbanizing city like Kozhikode through land pooling can help native fishermen communities to generate finance for mechanizing fishing, diversifying fisheries to secondary and tertiary sectors, recapture profit from middlemen and corporates, reestablish sustainable fishing practices and create an adaptive and resilient fishermen settlement.

Keywords CRZ amendments 2019 · Fishing community · Sustainable settlement development · Land value · Land pooling

---

K. K. Nishan Nazer (✉)
Department of Architecture and Planning, NIT Calicut, Kozhikode, India
e-mail: nishannazer@gmail.com

P. J. Nidhish
Department of Architecture, College of Engineering Trivandrum, Thiruvananthapuram, India
e-mail: ar.nidhishjoseph@cet.ac.in

R. K. Mishra et al. (eds.), *Sustainable Climate Action and Water Management*,
Advances in Geographical and Environmental Sciences,
https://doi.org/10.1007/978-981-15-8237-0_27

27.1 Introduction

The CRZ amendment of 2019 is perceived widely as a pro development initiative to favour tourism and other development activities in the coastal belt. Relaxations made on the earlier restriction will favour development in CRZ areas unseen like before. Such development will trigger an increase in the land value and will intensify the economic activities along the coastal belts. This poses a huge danger to the indigenous communities living on the coastal line with livelihood highly dependent on fishing-related activities. An increase in tourism along with capital-intensive fishing industries by private players will result in the displacement of the native population. It may also result in the loss of traditional livelihood of the native population. In such a scenario, the native population can make advantage of the land they have as a resource to adapt themselves to the new growth trends so that the native communities also benefit from the development activities.

27.2 Methodology

The study involves the primary and secondary study of the amendments in CRZ 2019. Three fishing villages identified by the Kerala Fisheries Welfare Board—Vellayil, Puthiyakadavu and Thoppayil of Kozhikode district—are identified for the study purpose. This coastal stretch and its native population are studied in detail through random assessment techniques and literature studies. Literature case studies are done to derive the techniques and approaches that can be used in the settlement development of the native communities. Based on the case studies, the technique of land pooling and reconstitution (LPR) was identified as the tool to unlock the land value which can finance the settlement development through better housing and infrastructure. Land value estimation before and after land pooling is studied using weighted overlay analysis in GIS to understand the economic feasibility of the project. A master plan is prepared to guide the settlement development of the fishermen communities in the study area. The land is pooled and redeveloped under the fishermen cooperative society which works as an independent entity by handling the surplus value generated in the process.

27.3 Concerns of CRZ Amendments 2019

CRZ amendments of 2019 for CRZ 2 has taken away the concept of hazard line from the HTL as mentioned in the 2011 regulation. It also relaxed the FSI limit to the limit that is prescribed in the municipal laws as on the date of publication (in the case of Kozhikode, FSI is 4). Construction of buildings towards the landward side of the existing roads and existing authorized structures in the seaward of the existing road give immense investment potential in the earlier restricted zones. Construction of pathways and temporary tourism facilities likes shacks, toilets, etc. are permitted in the area up to 10 m from the HTL which belonged to NDZ as per 2011 notification. All these relaxations will exert huge development pressure on the coastal land. This may result in the displacement of the native population and the deterioration of the coastal ecology.

27.4 Development Pressure

The development pressure in the identified fishing villages—Vellayil, Thoppayil and Puthiyakadavu—is analysed using image processing in ArcGIS. A comparative image analysis of the site in 2002 and that of 2018 is carried out to understand the pattern of development in the area. For that purpose, google earth images for the site for respective years were analysed. The analysis was done for three factors—land cover, built use and beach (Fig. 27.1).

In the analysis, it was found that beach has eroded over time and beaches have been converted to built uses. A new harbour has been constructed at Vellayil and it is presumed that the erosion is triggered due to the construction of harbour. It is also evident that built use has increased immensely on the non-CRZ zones of the region which indicates that CRZ regulations have been beneficial in restricting development in the coasts to a great extent. Change in the land cover analysis indicates that any relaxation in the CRZ would induce development in coastal areas. Currently, change in land cover in the coastal area is present only in fishermen community. The development pressure is high within the settlements due to the unaffordability of land in the beach area. Growing fishermen have to subdivide their existing land to accommodate more houses.

27.5 Land Valuation

A primary study was conducted to understand the factors that determine the land value of the study area. The land value of the same area was generated using weighted sum analysis using the parameters that determine the value of land as per the study done by Yomralioglu (Yomralioglu 1993). This land value map generated using ARCGIS was the actual reflection of the land value pattern observed in the primary study confirming the validity of the parameters. Therefore, the same parameters and weightages are used for the land valuation post settlement redevelopment. The increment in the land value due to improvement in the quality of living standards in the area post LPR is not accounted for in the calculation of the land value (Table 27.1).

The difference in the total value of the land before and after land pooling is calculated using the ArcGIS. The total value of the residential and commercial plots before LPR is estimated to be Rs 1037 Crore and the total value of the sellable

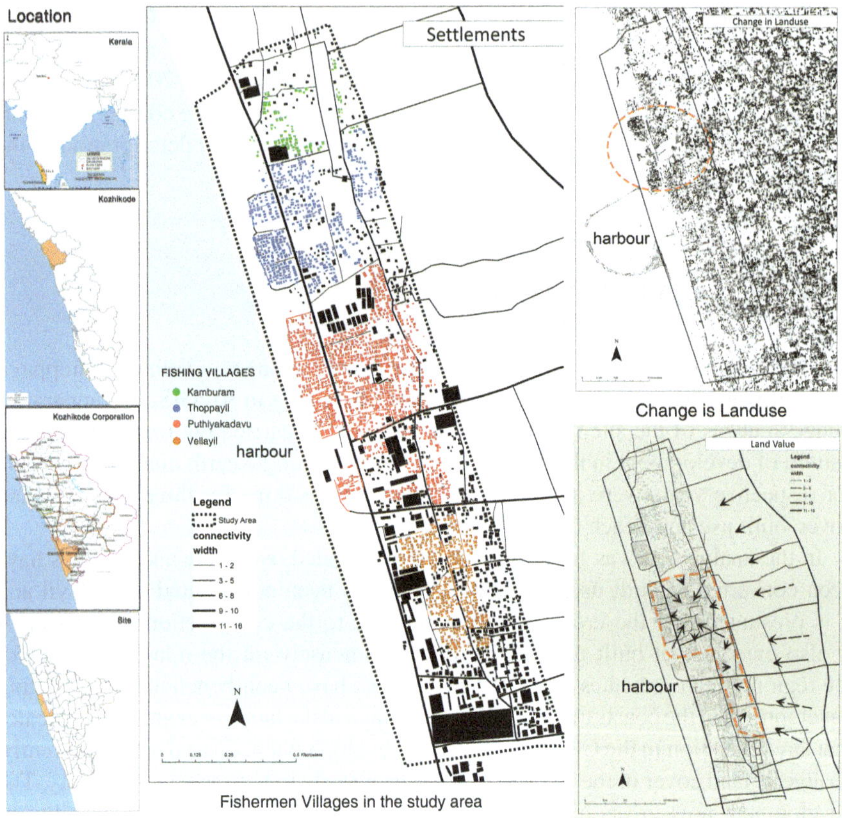

Fig. 27.1 Change in land cover is the last two decade

Table 27.1 Weightage of parameters used for weighted sum analysis

Parameters that determine land value	Weightages
Basic Services	87.1
Number of Floors that can be built	85.2
View	81.2
Access to streets	80.6
Nuisance	66 (negative)
Access to education	65.1
Access to shops	63.9
Access to health care	59.5
Access to religious institutions	55.4
Access to recreational parks	53.4
Availability of car parking	52.7

land (including residential land, commercial land and built value) after land pooling is estimated to be Rs 2298.2 Crores. Thus, the total difference in value before and after LPR is estimated to be Rs 1261.2 Crore. This LPR exercise is executed through the Fishermen Cooperative Society. The total savings of the FCS after paying compensation, investing in housing and infrastructure development is 847 Crores. This amount can be invested by FCS to diversify their traditional livelihood to secondary and tertiary sectors (Fig. 27.2).

Land value before land pooling				
Total area (m^2)	Total value of land (Lakh Rs)	Total area of road (m^2)	Value of Road (m^2)	Value of residential plot (Lakh Rs)
639646	116558	51417	12854	103704

Fig. 27.2 Land value of redevelopment area before and after LRR

Land value after land pooling								
Total area (m^2)	Total value (Lakh Rs)	Total area of roads (m^2)	Total value of Roads (Lakh Rs)	Total value of developable land (Lakh Rs)	Total value of built area for commercial and mixed residential (65% Coverage, FSI 2, Rs 1500/ Sqft)	Value of built area for high density residential plots (50% Coverage, FSI 4, Rs 4000/sqft, 50% of total available land is built) (Lakh Rs)	Total value of built area (Lakh Rs)	Total value after land pooling (Lakh Rs)
630130	148513	241826	60456	88056	10110	131660	141770	229826

Difference in value	Expenses for settlement development			Balance
Difference is value after land pooling (Lakh Rs)	Cost of construction of infrastructure (Lakh Rs)	Cost of construction of houses for 1450 families at the rate of Rs 1000/Sqft for 750 Sqft houses (Lakh Rs)	Compensation for houses and commercial units at the rate of Rs 500/Sqft for 1600 houses with an average of 700 Sqft each. (Lakh Rs)	Remaining value (Lakh Rs)
126122	25000	10800	5600	84722

27.6 Proposed Settlement Development Plan

A settlement development plan is proposed in an area of 0.63 km^2 within the study limits. The CRZ amendments of 2019 allow FSI up to 4 and other built uses like hospitals, residences, schools, etc. on the landward side of the existing road. It also allows certain tourism activities along the coast. In the proposed settlement development plan, the fishermen residential neighbourhood is zoned opposite to the Vellayil fishing harbour on the landward side of the beach road, with row houses and flatted units, and the harbour area is developed with transport, storage, handling and marketing facilities.

The land is pooled and registered under a fishermen cooperative society where all households who have given up land are members. The scattered fishermen houses are rebuilt to a higher density with better infrastructure through row houses. Row houses with intermediate hierarchical open spaces and pedestrian-friendly networks of roads to major work and leisure centres are considered in the planning for the neighbourhood. To ensure social mix, low-rise rental housing and plotted housing are planned within the residential neighbourhood. The surplus land generated in the LPR process is leased out or sold to the private developers for higher FSI constructions. The money generated by lease or sale of high-value serviced land along with external financial sources can be used by the FCS for starting new independent ventures in the fisheries sector.

Open spaces along water bodies, food streets, commercial shops, hotels and restaurants are planned along main roads to increase the interaction between the fishing community and the society. Several activity zones are created along the beach which are integrated with the proposed walkway along the beach, with spaces like fishing market, beach side walk, beach side shopping, eateries, sea food stalls, etc. that can diversify the employment horizon of the youngster in the fishermen community by providing jobs in secondary and tertiary sectors. Development of the harbour is proposed along with the development of fish processing units under the FCS in the West hill industrial zone.

The commercial and mixed commercial units are planned along the main roads. Heritage tourism is proposed in the traditional fishermen settlements by integrating with heritage route, cultural beach and proposed walkway along the beach (Fig. 27.3).

27.7 Fishermen Cooperative Society

Fishermen Cooperative society (FCS) is formed to organize the fishermen community for efficient and productive use of available resources. FCSs are generally inactive due to the poor management and insufficiency of funds. The land pooling exercise in the study area can unlock the value of the land, which can be monetized for the benefit of the community. Channelling of the educated people in the community towards the secondary and tertiary sectors of the fishing industry along with skill development can diversify the livelihood opportunities of the fishing community. FCS can invest in fishing-related activities like production, processing and marketing which can generate both revenue and employment. It can help in the effective utilization of the surplus money for becoming a strong competitor against corporate and private investors for recapturing the profit generated in fisheries. The independency of traditional fishermen will also play a significant role in the restoration of the coastal and marine ecosystem by applying sustainable traditional fishing practices (Fig. 27.4).

Fig. 27.3 Proposed settlement development plan and activity zone map

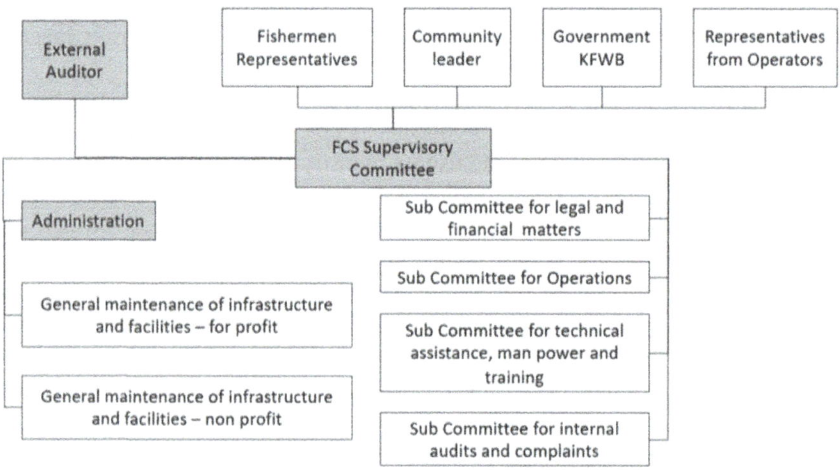

Fig. 27.4 Structure of Fishermen Cooperative Society

27.8 Conclusion

Land management exercise through techniques like LPR is useful in improving their living condition and livelihood of the fishermen communities through improvement in housing and infrastructure by unlocking the value of the land that they posses in the city core. Though the new amendments are perceived to be negligent of the demands of the indigenous fishing community, there are opportunities to leverage the benefits of higher FSI and relaxation in construction activities in the coastal area for the betterment of the deprived fishermen communities and restoration of the coastal ecology through planned interventions. Organizing the fishermen into cooperative societies can help in the effective utilization of the surplus money for becoming a strong competitor against corporates and private investors for recapturing the profit from fisheries and re-establish sustainable fishing practices. FCS can make planned investments in fish processing, education, training, tourism, restaurants, transportation, microfinance, etc. which can generate both revenue and employment. This model can be replicated in settlements in other coastal areas of similar context.

References

Bajracharya A, Pradhan P, Amatya P, Bhakta BK, Shrestha S, Hasan A (2015) Planning for affordable housing during densification in Kathmandu. IIED, London

Ministry Of Environment FA (2019) Notification G.S.R. 37(E). The Gazette Of India: Extraordinary, New Delhi

RS (2018) Vulnerability of coastal areas and livelihood challenges of fisher communities: a case study in Kerala. Int J Pure Appl Math:4739–4745

Yomralioglu T (1993). The investigation of a value-based urban land readjustment model and its implementation using geographical information systems. PhD. Thesis. Department of Surveying University of Newcastle upon Tyne, UK

Chapter 28
Rural Governance in Water and Sanitation Management: Empirical Observations and Policy Approach

Sunkari Satyam

Abstract Water is an essential and indispensable element to maintain a clean and sanitary environment. Without any dispute, promoting an equitable healthy society, management of sanitation and progress of further sanitation activities have been closely connected with the availability of water; and the larger the quantity and the better the quality of the water, the more rapid and extensive has been the advance of public health. At the grassroots of Indian states, availability and process of supplying vary from region to region and village to village within sub-regions, though involving grassroots' institutions is common across the country in the operation and maintenance of water distribution. Some areas are evidently of resourceful and it is quite in contrast in some other areas. Similarly, it is cost-effective to ensure sustainable provision of water to rural communities in some areas; however, it is very different from such examples in other parts of the country. The purpose of this study is to examine the effectiveness of the strategy of the institutional management for the water distribution and governance in selected villages of Andhra Pradesh and Telangana to identify the best governing practices for replication and highlight effective management methods in the framework of policy studies. The data for the study was gathered from 50 g Panchayats of two selected districts—Nizamabad and Prakasam—through direct interviews with elected presidents and local administration. Observations of

The research outcomes are part of the research project sponsored by the Indian Council of Social Science Research (ICSSR), New Delhi, the research project titled—Electoral Politics and Elected Presidents of Dalit and Adivasis Leaders: A Study in Andhra Pradesh and Telangana. I am thankful to ICSSR for the research project.

I am thankful to Merian-Tagore International Center of Advanced Studies-Metamorphosis of Politics, New Delhi, as part of fellow, I have spent considerable amount of time to write this paper apart from main project for which the fellowship was awarded.

I thank Prof. Kalpana Kannabiran, Director, Council for Social Development (CSD-SRC) for her continuous support.

Fellow, Merian-Tagore International Center of Advanced Studies-Metamorphosis of Politics, New Delhi (July–September 2019).

S. Satyam (✉)
Rajendranagar, Hyderabad, India
e-mail: satyamvvm@gmail.com

R. K. Mishra et al. (eds.), *Sustainable Climate Action and Water Management*, Advances in Geographical and Environmental Sciences,
https://doi.org/10.1007/978-981-15-8237-0_28

the data revealed that the village governance that is expected to drive the local administration to be effective was deserting away. Furthermore, experiences show that local governance was unable to institutionalize the water governance effectively to meet the demands of households due to the lack of accountability mechanism.

Keywords Governance · Management · Water · Rural and community

28.1 Introduction

This paper attempts to understand the village level water and sanitation conditions and governance process at Gram Panchayat institutions. It is based on the existing empirical study on water and sanitation conditions and functional aspects of village governance locating rural water supply in selected villages of Andhra Pradesh and Telangana. The data for the study was collected from 50 g Panchayats of two selected districts—Prakasam (Andhra Pradesh) and Nizamabad (Telangana)—through direct interviews with elected presidents and local administration. Observations of the data revealed that the grassroots governance that is expected to drive the local administration to be effective was deserting away. Furthermore, the local governance was unable to mobilize adequate revenue to defray the cost of water governance due to households' reluctance to pay coupled with the general lack of accountability on the part of the officials.

What are the basic issues in water governance and sanitation management in Gram Panchayats? How is the governance of the water supply functioning? How do village politics function in providing drinking water? What are caste equations in drinking water and sanitation? It is an attempt to answer these questions by providing empirical data. The analysis is sectioned based on these research questions such as issues in water governance and sanitation management in Gram Panchayats, governance of water supply, village politics in providing water and caste issues in drinking water and sanitation. The paper is organized based on these issues which provide a detailed account on water governance and sanitation management.

28.2 Water Governance and Sanitation Management

Availability of water is an essential source to provide clean drinking water and maintain a sanitary environment. These two elements—providing clean drinking water and maintaining a sanitary environment—are considered as biggest challenges of rural areas. The progress of providing drinking water and sanitation across Indian states has been closely associated with the availability of water. It is very clear that the provision of drinking water is highly dependent on better position of water availability in respective locations (rural as well as urban). It is considered to be rapid and extensive advancement in water governance. The programmes on drinking water are largely evident with both tragic and glorious milestones in which availability of water was an important factor. "Since all biological life is dependent on water, it must be obtained at intervals by all plants and animals, so that life develops around this need" (Wagenr et al. 1959: 13).

There have been various incidents which brought out since the last four–five decades such as water-borne epidemics and subsequent epidemiological incidents. There are studies which focussed attention on water quality and on the role of management. Water management as a practice for providing drinking water was not evident until the nineteenth century. Water management was evident in ancient Greece and Rome as these were the first urbanized ancient civilizations (Vuorinen 2007). Supply of water and maintenance of sanitation for military needs was a primary concern of Roman imperial power. Further, the idea of progress based on industrialization and urbanization became evident in Western countries and people were fascinated with the idea of progress.

Thus, the idea of progress which developed the concept of water closet was seen as a "victory for public health without any consideration as to where the human excreta went through sewer pipes" (Wagenr et al. 1959). Modern water supply and sanitation systems were developed initially by Germany and other parts of Europe, USA and later other non-western countries in the nineteenth century. The scheme of water supply is straightaway connected with "poor living conditions, a lack of democracy, poor hygiene, illiteracy, corruption and a lack of proper water and sanitation services" (Wagenr et al. 1959). Specifically, women and children are the worst sufferers from these constraints as mentioned in popular readings (Wagenr et al. 1959). It is a well-known fact that "there is a global shortage of potable water" (Wagenr et al. 1959). These observations obviousely reveal that it is indispensable to make effective institutional mechanism when any government plans to take fundamental decisions concerning water supply and sewerage. The water supply programmes which have been initiated so far have not achieved the desired results so far. Providing potable drinking water is a massive scheme that needs a challenging mechanism in place. Programmes that are now at the operational level are not achieving the targeted goals. This is more important to keep in mind when the policy is in the planning stage which requires stringent strategies.

The degree of water supply and sanitation is not necessarily bound with time, place and space as much as the capability of that state and society to take responsibility for developing the living conditions of its citizens and appropriating proper policies. Strategies of drinking water and sanitation must have been made concerning the situation and systems of water and sanitation. There have also been circumstantial situations where the choice of technology has been regarded as problematic. For example, installing pipe system was considered to be hazardous for health; however, it continued to be the usual supply of drinking water even today. It is sine qua non that effective water supply schemes and sanitation systems have always required continuous maintenance, effective management and adequate mechanism to reestablish the system.

There are various experiences that led to establish a fact that "there is a widespread recognition that sustainable and effective WASH service delivery is not only determined by the state of infrastructure, but also by complex institutional, governance and financial management systems" (Global Status Report 2019).

The GLASS report (2019) further states that "only 12% of countries reported that urban drinking-water surveillance is conducted at 100% of the required frequency and insufficient human resources are often a major constraint in drinking-water and sanitation surveillance and oversight. Most of the countries have not followed plans on budgetary allocations; this led governments to take irrational decisions. National Systems to Support Drinking-Water, Sanitation and Hygiene: Global Status Report (2019) reports that, "as of 2017, 2.2 billion people still lacked safely managed water, 4.2 billion lacked safely managed sanitation and 3 billion lacked access to basic hand washing facilities". It further states that achieving SDG 6[1] will require dramatic changes by countries.

Throughout the countries, the major problems seem to be lack of proper approach on maintenance, management and resources allocation. The lack of institutional mechanism affects poor management of water and sanitation system, this leads to lack of safe drinking water, health problems and poor education. There are other factors which also affect such as good governance and basic grassroots' democracy. The age-old tradition of fetching water from distant sources of water has been hindering women's empowerment and also impacting children, especially girls who are the victims of fetching water from distant sources of water. The daily task of fetching water is preventing girls especially from marginal social groups from going to school and higher studies.

The basic issues in water governance and sanitation management in Gram Panchayats are the availability of water resources and supply management at respective Gram Panchayats. There are three types of water resources available in rural areas, supply from river and other tanks are considered by both traditional and modern sources as well, though supply from the river is a recent phenomenon. The second category includes public tanks and individual taps while individual or private sources include borewells and the establishment of purified water machines and water supply from small-scale private enterprises. At the Gram Panchayat level, maintaining public resources such as management of water pipes of the river or linked with local tanks, community water tanks, individual taps and pipelines. The data based on a primary source reveals that about 70 percent of Gram Panchayats have established water

[1]. "SDG6: Ensure access to water and sanitation for all, Industry accounts for over 19% of global water withdrawal, and agricultural supply chains for 70% more. What gets measure gets managed, and at CDP, we've seen an increasing number of companies tracking, managing and implementing water solutions. SDG 6 cannot be achieved without corporate action, but significant progress will rely on robust data on the current landscape and what needs to change to ensure environmental sustainability on water. With the largest corporate water dataset in the world, CDP can offer insights to measure and track progress and accelerate action towards meeting Sustainable Development Goal 6" (https://www.cdp.net/en/policy-and-public-affairs/sustainable-development-goals#e7a7c86fc38fcdf21efa14482fc685c3, viewed on 17/9/2019).

tanks and individual taps apart from private sources which are held by economically affluent families.

28.2.1 Governance Matters in Water Supply and Sanitation

In many rural areas, community water tanks and individual taps are used as sources of domestic water supply which directly fall under the governance matters of Gram Panchayats. Previously, at least two decades back, rural areas were well-equipped with hand pumps which were used as sources of domestic supply for individual households. In spite of recent changes, no changes were observed by obtaining water being carried in cans from faraway places. In summer, it was pointed out by the Ward Member from Pajarla of Gudulur mandal that "though we have very good source of water, still some households do not have individual connections. We have to do a lot of things in this regard but funds are insufficient".

In terms of sanitation, "we provided facilities to construct individual toilets under the Swachh Bharat Abhiyan,[2] many families did not construct as the funds available under the scheme are insufficient. We have also provided sand and cement with our own money (Gram Panchayat) under scheme of construction of the rehabilitation center. But a majority of families did not use because they thought that it is not useful. The reason for such a thought is that, after completion of it (toilet room), what they will do without water supply. So many toilet rooms are being used as store rooms in our villages" as accounted by the President of Pajarla of Gudulur mandal of Prakasam district of Andhra Pradesh.

It was obvious that the amount of water supply is very limited and too small for effective encouragement of health and personal hygiene at the household level. The use of individual taps and community tanks in villages is inevitable and often many families depend on privately supplied water. Regarding public drainage and its maintenance, there is insufficient staff to cover all the areas of villages even within a week. About 90% of villages as revealed by respective Presidents of Gram Panchayats are covered entirely once in a month. Each ward gets its turn once in a month. This is the condition of sanitation in a village. "Special drives on sanitation

[2]. Objectives of the Mission: "To achieve 'Swachh Bharat' by 2 October, 2019, the main objectives of the SBM(G) are as follows: a) Bring about an improvement in the general quality of life in the rural areas, by promoting cleanliness, hygiene and eliminating open defecation b) Accelerate sanitation coverage in rural areas to achieve the vision of Swachh Bharat by 2 October, 2019 c) Motivate communities and Panchayati Raj Institutions to adopt sustainable sanitation practices and facilities through awareness creation and health education d) Encourage cost effective and appropriate technologies for ecologically safe and sustainable sanitation e) Develop, wherever required, community managed sanitation systems focussing on scientific Solid and Liquid Waste Management systems for overall cleanliness in the rural areas f) Create significant positive impact on gender and promote social inclusion by improving sanitation especially in marginalized communities" (Ministry of Drinking Water and Sanitation, Government of India (2018 Revised): Guidelines for Swachh Bharat Mission (Gramin), https://jalshakti-ddws.gov.in/sites/default/files/ SBM(G)_ Guidelines.pdf, viewed on 15/9/2019.

give better results but we can't do that regularly because people can't participate in it whenever we organized special drive but we have to organize them (drives) when people are at leisure which means when they don't engage with agriculture and other activities. People's participation is more important" added the President of Padgal of Velpoor mandal of Nizamabad district.

"Water from a river source by means of pipes to each village house is the goal of Telangana government. Moreover, many villages are connected with the programme but supply is not adequate. There should be more attention on supply and maintenance" responded a household member of Ankapur Gram Panchayat of Nizamabad district. The most important benefits of Mission Bhagiratha water are safe drinking water, convenience and an increased supply but it also needs greater attention in all villages of Telangana as the data confirmed that there are some problems, villages are struggling to get an adequate amount of water as it is observed in the study of villages of Telangana. "The government has been trying to supply regularly and adequately but there some problems like break up of pipelines and maintenance problem at village level and there is proper data how to supply to households and how much to supply and what amount of water each village requires" added a Sarpanch of the Gram Panchayat of Savel of Mupkal Mandal of Nizamabad district. There is a difference in water distribution systems between rural and urban areas in terms of protection of health and sanitation.

Water and sanitation and supply and management are largely dependent on the effectiveness of governance. What are issues of management concerning rural water and sanitation? Generally, community water tanks, individual pipes and public drainage system consist of essential issues of rural water and sanitation in the village.

Sufficient water sources, regular supply of water and setting-up of proper institutional arrangement to look after water supply related issues were initially collected from the Presidents of respective Gram Panchayats. As it is evident in Table 28.1, the majority of villages are reported to have insufficient water resources though community water tanks and individual water taps were largely evident. The situation is also not different in Nizamabad as it is evident that no trained staff is placed into look after water supply related matters (Table 28.2). As it was responded by a resident of Ramgopalpuram of Venigandla, "we have good water resources, water tanks and individual taps but don't have proper person to look after water supply. During rainy season, we always get inadequate water supply, sometimes they don't work to provide water". It is absolute dereliction what the GLASS report (2019) stated, "insufficient human resources are often a major constraint in drinking-water and sanitation surveillance and oversight".

There are some technological issues that may have been erroneous or less successful. For example, old pipes were continued to be used in connections from tanks to households, ultimately water supply and sanitation systems get disturbed. This needs continuous maintenance and adequate concentration on management and coordination. This was already evident in the GLASS report (2019) and the same is revealed in selected villages. Water is not just a commodity; it is an economic and social good. This places responsibility for its management and oversight in the

Table 28.1 Water supply management—Prakasam District

S. No	Mandal	Gram Panchayat	Sufficient water source	Regular	Staff
1	Venigandla	Nagireddypally	Sufficient	Yes	Adhoc
2	Guduluru	Potluru	Sufficient	No	Adhoc
3	Kandukuru	Machavaram	Insufficient	No	Adhoc
4	Venigundla	Hussainipuram	Insufficient	No	Adhoc
5	Valeti Varipalem	Ankabhupalapuram	Sufficient	Yes	Adhoc
6	Valeti Varipalem	Ayyavaripalli	Sufficient	No	Adhoc
7	Valeti Varipalem	Koddareddypalem	Sufficient	No	Adhoc
8	Guduluru	Pajarla	Insufficient	No	Adhoc
9	Guduluru	Swarnajipuram	Sufficient	No	Adhoc
10	Pamur	Laxminarsapuram	Sufficient	Yes	Adhoc
11	Kandukuru	Kondamudusupalem	Insufficient	No	Adhoc
12	Valeti Varipalem	Chundi	Sufficient	No	Adhoc
13	Valeti Varipalem	Zuppalapadu	Sufficient	Yes	Adhoc
14	Chandrasekarapuram	Chennapunayunipalli	Insufficient	No	Adhoc
15	Chandrasekarapuram	PeddaRaju Palem	Insufficient	No	Adhoc
16	Chandrasekarapuram	Boyamadugula	Sufficient	Yes	Adhoc
17	Chandrasekarapuram	Nallamadugula	Sufficient	No	Adhoc
18	Pamur	Vaggampalli	Insufficient	No	Adhoc
19	Pamur	Raviguntapalli	Insufficient	No	Adhoc
20	Pamur	Chilmkur	Insufficient	No	Adhoc
21	Guduluru	Guduluru	Sufficient	Yes	Adhoc
22	Venigandla	Ramgopalpuram	Sufficient	No	Adhoc
23	Kandukuru	Mopadu	Insufficient	No	Adhoc
24	Guduluru	Saalıpet	Insufficient	No	Adhoc
25	Kandukuru	Kovuru	Sufficient	Yes	Adhoc

Source Field Survey 2018–19

public sphere. Balancing water use priorities, water quantity and water quality is of high importance for the future.

28.2.2 Village Politics in Providing Drinking Water

"Why have public expenditures in India failed to deliver basic services to the majority of poor citizens, when political representatives taking decisions over public resource allocation depend upon the support of the poor to win office?" (Keefer, Philip Sruti Khemani 2004:935). This is the central question of why delivery governance has

Table 28.2 Water supply management—Nizamabad District

S. No	Mandal	Gram Panchayat	Sufficient water source	Regular	Staff
1	Vailpur	Vailpur	Sufficient	Yes	Adhoc
2	Armur	Manhani	Sufficient	No	Adhoc
3	Vailpur	Padgal	Sufficient	Yes	Adhoc
4	Jakranpally	Chinthaloor	Insufficient	No	Adhoc
5	Morthod	Thimmapur	Sufficient	Yes	Adhoc
6	Dichpally	Mallapur	Insufficient	No	Adhoc
7	Morthod	ThallaRampur	Sufficient	Yes	Adhoc
8	Balkonda	Savel	Sufficient	No	Adhoc
9	Balkonda	Renjarla	Insufficient	No	Adhoc
10	Dichpally	Nadpally	Insufficient	No	Adhoc
11	Balkonda	Chittapur	Insufficient	No	Adhoc
12	Dichpally	Indalwai	Insufficient	No	Adhoc
13	Morthad	Domchanda	Sufficient	Yes	Adhoc
14	Bheemgal	Pipri J	Insufficient	No	Adhoc
15	Armur	Govindpet	Insufficient	No	Adhoc
16	Nandipet	Siddapur	Sufficient	Yes	Adhoc
17	Sirikonda	Ravutla	Insufficient	No	Adhoc
18	Armur	Ankapur	Insufficient	No	Adhoc
19	Armur	Pipri	Sufficient	No	Adhoc
20	Balkonda	Bussapur	Sufficient	Yes	Adhoc
21	Nandipet	Ch Kondur	Sufficient	Yes	Adhoc
22	Nandipet	Badguna	Sufficient	No	Adhoc
23	Armur	Aloor	Insufficient	No	Adhoc
24	Velpur	Mothe	Insufficient	No	Adhoc
25	Jakranpally	Kaligote	Sufficient	Yes	Adhoc

Source Field Survey 2018–19

failed to deliver targeted goals by the elected government. Majority of democratically elected governments in India were formed by extensive participation of poor and disadvantaged citizens (Keefer, Philip Sruti Khemani 2004). Social fragmentation and marginalization provide some imperatives to weak concentration by governments on targeted goals. In Indian politics, it is more difficult to make credible pre-electoral promises. Inadequate information largely constitutes a major problem for the execution of credibility of electoral promises. Pre-election promises make marginal people believe in parties and post-formation of governments. There is a credibility problem in Indian politics apart from the concentration of marginal social groups. The marginal social group is in a weak position in terms of deliberation with the democratic process. This is another problem with regard to the provision of basic services in villages.

In selected villages, it was observed that the Presidents of Gram Panchayats were under the influence of rural elite and political force which constitutes a group of dominant social group. "I had a sufficient amount of funds to allocate providing better sanitation facilities in terms of rejuvenation of drainage, reconstruction of old drainage system and laying under pipe system in some wards where the population of Scheduled Castes concentrated. But, local party leader who has strong connection with local MLA and who is also same caste member what local MLA belongs to. He did not allow me to do what I want with regard to marginal areas. I just listened his advice. When I opposed him, he succeeded in removing my cheque power. I had to struggle a lot to regain my cheque power. They unnecessarily made corruption allegations on me and files of funds sanctioned were taken away from me. Local authorities also followed him though he is just a politician, but more important thing is he belongs to Kamma community as local MLA represents the same caste" narrated a woman Sarpanch who belongs to Scheduled Castes of Mopadu Gram Panchayat of Kandukur village of Prakasam district.

Another problem in village level politics is misguided political decisions. An observation from Nizamabad district locating an experience of a woman Sarpanch of Kudawanpur (or Kudanpur), one of total Gram Panchayats of the study, reveals that the Sarpanches were repeatedly changed. The discussion had taken place with the husband of the Sarpanch. The usual location of visit in the study village is the Office of Gram Panchayat. When the team of research visited and enquired about Sarpanch, people around the office informed about the location of Sarpanch's residence. Then, the team visited the place where people informed about the address of the Sarpanch. The house we found as we expected it, the location is not the present Sarpanch's address. The informant with whom the team interacted replied "I am not the present Sarpanch". Then, the study team again asked the question, "who is present Sarpanch?" The informant gave another address. Then again the team found the address. "Are you the Sarpanch of the village?" the question posed to the woman who stood at the open place of the house. She laughed at the team. "Madam, are you a Sarpanch?" "Yes, but I don't know anything. You have to wait till my husband comes. I don't know anything; I don't come to the office. He (my husband) only gives answers to your questions, gives you whatever the information you people want". Interrupting her, the question was asked. "Are you a Sarpanch, madam?" "Yes, I am" was the reply by the Sarpanch.

Her husband came after two hours of waiting. By this time, we visited the village and interacted with different persons asking about different issues such as local politics, MGNREGA, farming, education, youth employment and skill development programmes, and about Prime Minister Jan Dan Yojana and so on. People replied that they were aware of largely about both the programmes—MGNREGA and PMJDY—apart from the welfare schemes of the state government of Telangana. By that time, her husband came. We were standing at a small hotel having Chai (Tea). The tea shop was owned by a person who belonged to the Muslim community. We had very good discussions with him, it was about 11.30 AM. He gave very good information about village politics and how the village Sarpanch was elected. What are the invisible reasons for frequent change of Sarpanch?

The discussion which took place with the Chai vendor will be discussed after the presentation of the discussion with the Sarpanch's husband. The Sarpanch's husband came and took us off to his residence saying 'nobody will be available at the office of GP." We initiated the discussion asking, "Why the person of Sarpanch was changed?" The response was that "in this village, generally elections will not take place for the position of the Sarpanch because we don't want to see unnecessary political situation. So, during election time, 3 persons came forward to contest in the election. But our local leader Dattadri Goud, a very senior leader in the politics, advised us not to contest by saying why you people should contest in the election and why should you people waste money by making debts from others. It would not serve the purpose. So, he advised to share the position by all persons who were in the race of contest. Since then, the three persons shared the position. Now, this is our term. We will be in the position till the tenure end. This is another political drama which hinders the local governance and its effective use of governing tools."

28.2.3 Caste Issues in Drinking Water and Sanitation

In India, a struggle for water also equally amounts to a struggle for power. "Water being a limited and imperative resource has a close relationship with the three categories such as class, caste and gender, especially when it comes to that of distribution of water. At the grassroot level, this striving for power is much more than a process of increasing control over the water resource and control over the behaviour of the actors involved. It has more to do with the feeling of hatred towards the lower castes; in traditional Hindu setting, Dalits were denied all these facilities and were expected to work as slaves for the dominant castes" (Johns, Hannah, Nd). It is even today obvious, predominantly in locations where land is concentrated highly in the hands of other castes who have land holdings traditionally. Though the Indian Constitution guarantees under Article 14 the right to equality to every person irrespective of caste, creed, gender, status and religion, Indian society is still a brand ambassador of caste discrimination. Positioning caste in rural water supply context, it has its own layered experiences in the governance spheres of rural areas.

Thus, looking from the point of view of caste and social inclusion, water is also most important in India. Water was the source of discrimination and exclusion. It was the symbol of dominancy because till recently major community water bodies were under the control of feudal families or landholding classes who have typical dominant mindsets. Moreover, affluent water channels were closer to the lands and houses of main water resources. Even today, mini reserve water tanks are located nearby affluent families in villages. Apart from wellbeing, it is essential that water has caste layers regardless of whether it is from piped systems or point sources like wells. The same is the case with sanitation. The toilets were constructed but the use of them was severely affected by the lack of supply of water from the tanks which were largely mismanaged by local authorities. It was evident in many villages of Nizamabad districts and Prakasam, pipeline was established but an equal amount of water supply is a question. There is the question of caste in distribution. It is an important task to operate and maintain the systems properly with adequate skill. Though skill is adequately evident, politics cannot allow performing the skill because the position of skilled water management is largely held by marginal communities.

The distribution of water and allocation of funds for sanitation have always been a major factor influencing the delivery system. The strategies and ways how the system of water supply has been adequately maintained by local institutions are directly connected with local caste politics and caste identities. As it was already mentioned in the paper that the caste influence is highly visible in decision-making spheres, the distribution of water to marginal locality has also caste connections. In the areas where Scheduled Castes' people are concentrated, the supply of water is also marginal. The construction of tanks, pipelines and supply has also caste connections. The supply of water looks like equally distributed to all areas but when you keenly observe, the difference of caste matter could be understood. The location of community tanks is more affluent to get more water whereas the tanks which are concentrated in marginal localities are located at different places where it is difficult to receive water supply from these tanks. The importance of a better supply of drinking water for marginal areas is realized through better and inclusive governance.

Water is an essential and indispensable element to maintain a clean and sanitary environment for all classes and castes. Without any dispute, promoting healthy society, management of sanitation and progress of further sanitation activities have been closely connected with the availability of water, and the larger the quantity and the better the quality of the water, the more rapid and extensive has been the advance of public health.

At the grassroots of Indian states, availability and process of supplying of drinking water is vary from region to region and village to village within sub-regions, though involving grassroots' institutions is common across the country in the operation and maintenance of water facilities. Some areas are evidently resourceful and in some other areas, it is quite in contrast. Similarly, it is cost-effective to ensure sustainable provision of water to rural communities in some areas; however, it is very different from such examples.

World Water Development Report 2003, produced practically by all the UN family and thus almost all the sectors of human life and society, pointed out how the water crisis is largely a crisis of governance. The report pointed further "many of the leading obstacles to sound and sustainable water management: sector fragmentation, poverty, corruption, stagnated budgets, declining levels of development assistance and investment in the water sector, inadequate institutions and limited stakeholder participation" (World Water Development Report 2003). People should be allowed and encouraged to use their own experiences and abilities to solve their problems. Such empowerment is most probable and gets more sustainable results than any mere top-down approach. In overall effective water governance, it obviously needs to balance the supply requirements such as management of tanks and decentralizing water requirements.

Interestingly enough, some of the basic principles of sustainable water governance were written evidently in the pre-modern era. Understanding present problems, availability, storage capacity and distribution including inclusive issues are some of the important points.

28.3 Conclusion

It has been rightly said that the degree to which a water supply system fulfils its public health function varies almost directly with the efficiency and effectiveness of its management. This statement is substantiated by field evidence and experience shared by elected presidents of Gram Panchayats. In the planning of Panchayat level water systems, it sometimes takes place that enthusiastic local leaders and officials give too little consideration to the management of water supply and to the development of effective governing tools for smart and simple operations. To a great extent, people's opinion with regard to provision and management, and functions of grassroots water supply staff as a regular job which consists of handling daily supply issues is neglected.

References

Frioux S (2014) Environmental history of water resources. In: Agnoletti M, Neri Serneri S (eds) The basic environmental history, vol 4. Springer, Cham

Johns H (Nd) Stigmatization of dalits in access to water and sanitation in India. National Campaign on Dalit Human Rights (NCDHR), New Delhi. https://idsn.org/wp-content/uploads/user_folder/pdf/New_files/UN/HRC/Stigmatization_of_dalits_in_access_to_water_sanitation.pdf. Accessed 9 Sept 2019

Juuti PS, Katko TS, Vuorinen HS (eds) (2006) environmental history of water-global view of community water supply and sanitation. IWA Publishing, London

Keefer P, Khemani S (2004):Why do the poor receive poor services? Econ Polit Weekly:935–943. http://www1.worldbank.org/publicsector/anticorrupt/PoliticalEconomy/PREMCourse07/PoorServicesforPoor.pdf. Accessed 11 Sept 2019

Vuorinen HS (2007) Water and health in antiquity: Europe's legacy. In: Juuti PS, Katko TS, Vuorinen HS (eds) Environmental history of water -global view of community water supply and sanitation. IWA Publishing, London, pp 49–71

Wagner EG, Lanoix JN, World Health Organization (1959) Water supply for rural areas and small communities, Geneva: World Health Organization (Edmund G, Lanoix JN). http://www.unesco.org/new/en/natural-sciences/environment/water/wwap/wwdr/wwdr1-2003/Wagenr, https://apps.who.int/iris/handle/10665/41697

World Health Organization (2019) The UN-Water Global Analysis and Assessment of Sanitationand Drinking-Water (GLAAS): National Systems to support drinking-water, sanitation and hygiene: global status report (2019). World Health Organization, Geneva

World Water Development Report (2003) Water for People, Water for LifeWorld Water Assessment Programme (UNESCO WWAP), http://www.unesco.org/new/en/natural-sciences/environment/water/wwap/wwdr/wwdr1-2003/. Accessed 10 Sept 2019

The manufacturer's authorised representative in the EU is Springer Nature Customer Service Centre GmbH, Europaplatz 3, 69115 Heidelberg, Germany. If you have any concerns regarding our products, please contact ProductSafety@springernature.com

Printed and bound by CPI Group (UK) Ltd, Croydon, CR0 4YY
30/09/2026
02225293-0010